Photonics Elements for Sensing and Optical Conversions

This book covers a number of a rapidly growing areas of knowledge that may be termed as diffractive nanophotonics. It also discusses in detail photonic components that may find uses in sensorics and optical transformations.

Photonics Elements for Sensing and Optical Conversions covers a number of rapidly growing areas of knowledge that may be termed as diffractive nanophotonics. The book examines the advances in computational electrodynamics and nanoelectronics that have made it possible to design and manufacture novel types of photonic components and devices boasting unique properties unattainable in the realm of classical optics. The authors discuss plasmonic sensors, and new types of wavefront sensors and nanolasers that are widely used in telecommunications, quantum informatics, and optical transformations. The book also deals with the recent advances in the plasmonic sensors based on metal-insulator-metal waveguides for biochemical sensing applications. Additionally, nanolasers are examined in detail, with a focus on contemporary issues; the book also deals with the fundamentals and highly attractive applications of metamaterials and metasurfaces. The authors provide an insight into sensors based on Zernike optical decomposition using a multi-order diffractive optical element, and explore the performance advances that can be achieved with optical computing.

The book is written for opticians, scientists, and researchers who are interested in an interesting section of plasmonic sensors, new types of wavefront sensors and nanolasers, and optical transformations. The book will be bought by upper graduate and graduate level students looking to specialize in photonics and optics.

Photonics Elements for Sensing and Optical Conversions

Edited by
Nikolay L. Kazanskiy

CRC Press
Taylor & Francis Group
Boca Raton London New York

CRC Press is an imprint of the
Taylor & Francis Group, an **informa** business

Designed cover image: Nikolay L. Kazanskiy

First edition published 2024
by CRC Press
2385 NW Executive Center Drive, Suite 320, Boca Raton FL 33431

and by CRC Press
4 Park Square, Milton Park, Abingdon, Oxon, OX14 4RN

CRC Press is an imprint of Taylor & Francis Group, LLC

© 2024 selection and editorial matter, Nikolay L. Kazanskiy; individual chapters, the contributors

ISBN: 978-1-032-57294-9 (hbk)
ISBN: 978-1-032-57391-5 (pbk)
ISBN: 978-1-003-43916-5 (ebk)

DOI: 10.1201/9781003439165

Typeset in Times
by KnowledgeWorks Global Ltd.

Contents

Preface

Topics of this book cover a number of a rapidly growing areas of knowledge that may be termed as diffractive nanophotonics. Advances made in computational electrodynamics and nanoelectronics have made it possible to design and manufacture novel types of photonic components and devices boasting unique properties unattainable in the realm of classical optics. Significantly, procedures for designing and numerical simulation of such elements and devices do not involve the consideration of quantum effects and can do with solving exact and approximate Maxwell's equations. The monograph authors discuss in detail photonic components that may find uses in sensorics and optical transformations.

ABSTRACT

This book deals with nanophotonic structures and components as well as applications thereof for sensorics and optical transformations. Special attention is given to various waveguides and possibilities of their design and practical uses. Plasmonic sensors and nanolasers are discussed, methods for design and prospective applications of metasurfaces and other nanostructures intended to perform optical transformations and computation are given. Capabilities of new types of wavefront sensors are analyzed. The monograph is intended for scientific researchers and specialists working in optics, nanophotonics, information technology, and optical instrumentation. It may also prove useful for postgraduate students specializing in these research areas.

Nikolay Lvovich Kazanskiy

About the Author

Nikolay L. Kazanskiy graduated with honors (1981) from Kuibyshev Aviation Institute (presently, Samara National Research University), majoring in Applied Mathematics. He received his Candidate in Physics & Mathematics (1988) and Doctor in Physics & Mathematics (1996) degrees from Samara National Research University. He is the director of Image Processing Systems Institute of the RAS – Branch of the Federal Scientific-Research Centre "Crystallography and Photonics" of the Russian Academy of Sciences, and also holds a part-time position of professor at Technical Cybernetics department of Samara National Research University. He is a member of OSA, SPIE, and IAPR. He has co-authored 290 scientific papers and 14 monographs, and has 57 inventions and patents. His current research interests include diffractive optics, computer vision, optical sensors, mathematical modeling, lighting devices design, and nanophotonics. Website: http://www.ipsiras.ru/staff/kazanskiy.htm. Email: kazanskiy@ssau.ru.

List of Abbreviations

ADC	analog-to-digital converter
AI	artificial intelligence
ANN	artificial neural network
AOC	analog optical computing
AOI	angle of incidence
AR	autorefraction
BC	boundary condition
BOX	buried oxide
BPSK	binary phase-shift key
BS	beam splitter
BS–PS	beam splitter–phase shifter
BWs	bandwidths
CCD	continuity of care document
CCT	central corneal thickness
CFG	command voltage generator
CG	chalcogenide glass
CGH	computer generated holograms
CMOS	complementary metal-oxide-semiconductor
CNN	convolutional neural network
CPP	channel plasmon polariton
CR	contrast ratio
CRf	cycloplegic refraction
CROW	coupled-resonator optical waveguide
CRR	circular ring resonator
DFB	distributed feedback
DLs	dynamic light scattering
DMD	digital micromirror device
DMSO	dimethyl sulfoxide
DNA	deoxyribonucleic acid
DOC	digital optical computing
DOE	diffractive optical element
DRS	dual rotating Scheimpflug
DUV	deep UV, deep ultraviolet
EASLM	electrically addressed spatial light modulator
EBL	electron beam litho
EFR	evanescent field ratio
EIT	electromagnetically induced transparency
EM	electromagnetic
EO	electro-optic
EO-SLM	electro-optic spatial light modulator
ER	extinction ratio
FBG	fiber Bragg gratings

FDTD	finite difference time domain
FEM	finite element method
FF	fill factor
FFT	fast Fourier transform
FIB	focused ion beam
FOBG	fiber-optic Bragg gratings
FOM	Figure of Merit
FP	Fabry–Perot
FPGA	field-programmable gate array
FR	Fano resonance
FSR	free spectral range
FSS	frequency selective surface
FT	Fourier transform
FWHM	full width at half maximum
GRIN	graded index
GST	$Ge_2Sb_2Te_5$
HCG	high-contrast gratings
HMDS	hexamethyldisilane
HOA	higher-order aberrations
HPWG	hybrid plasmonic waveguide
I/P	input
IFTA	iterative Fourier transform algorithm
IgG	immunoglobulin G
IPA	isopropyl alcohol
IR	infrared
ITO	indium tin oxide
JTC	joint transform correlator
KC	keratoconus
KR	Kretschmann
LASE	laser particle stimulated emission
LASIK	laser-assisted in situ keratomileusis
LCoS	liquid crystal on silicon
LCVWP	liquid crystal variable wave plate
LESPR	lasing-enhanced surface plasmon resonance
LGs	logic gates
LOD	limit of detection
LP	linearly polarized
LSP	localized surface plasmon
LSPR	localized surface plasmon resonator
LW	leaky wave
MA	meta-atom
MAC	multiply-accumulate
MBBs	meta building blocks
MDM	mode division multiplexing
MIBK	methyl isobutyl ketone
ML	machine learning

MM	metamaterial
MMI	multimode interference
MODAN	DOE for laser mode generation (or laser mode generator)
MS	metasurface
MS-OASLM	metasurface-based optically addressed spatial light modulator
MTF	modulation transfer function
MTS	metasurface
MWv	microwave
MZI	Mach–Zehnder interferometer
NA	numerical aperture
NIR	near-infrared
NN	neural networks
NPs	nanoparticles
NV	nonvolatile
NW	nanowall
O/P	output
OASLM	optically addressed spatial light modulator
OC	optical computing
OCT	optical coherence tomography
ODS	optical differentiation sensor
OF	optical fiber
OFBG	optic fiber Bragg grating
OLG	optical logic gates
OLO	optical logic operation
ONC	optical neural chip
OSI	objective scattering index
OTDR	optical time-domain reflectometry
PA	perfect absorber
PhA	photoacoustic
PB	Pancharatnam–Berry
PBG	photonic bandgap
PC	photonic crystal
PCC	photonic crystal cavities
PCM	phase change material
PDA	personal digital assistant
PDMS	polydimethylsiloxane
PH	pinhole
PIC	photonic-integrated circuit
PL	photolithography
PML	perfectly matched layers
PMMA	polymethyl methacrylate
POL	polarization
PR	photoresist
PS	phase shifter
PScF	point scattering function
PSF	point spread function

PWFS	pyramid wave front sensors
R&D	research and development
RFID	radio frequency identification
RI	refractive index
RIE	reactive ion etching
RIU	refractive index unit
RMS	root mean square
RR	ring resonator
SCB	self-collimated beam
SEM	scanning electron microscope
SERS	surface-enhanced Raman scattering
SERsp	surface-enhanced Raman spectroscopy
SLED	superluminescent light-emitting diode
SLM	spatial light modulator
SMF	single mode fiber
SOA	semiconductor optical amplifier
SOI	silicon on insulator
SP	surface plasmon
SPP	surface plasmon polariton
SPR	surface plasmon resonance
SR	square ring
SRR	square ring resonator
SSRR	split square ring resonator
SW	surface waves
SWG	subwavelength gratings
SWM	surface wave medium
TBP	time-bandwidth product
TE	transverse electric
TEM	transverse electromagnetic mode
TF	transfer function
TIA	transimpedance amplifier
TIR	total internal reflection
TM	transverse magnetic
TMAH	tetramethylammonium hydroxide
UC	upper cladding
UV	ultraviolet
VIS	visible
WDM	wavelength division multiplexing
WFS	wave front sensors
WG	waveguide
μ-RR	micro-ring resonator
Γ	field con

1 Silicon Photonic Waveguides
Comparison and Utilization in Sensing Applications

Muhammad Ali Butt[1,2], Svetlana Nikolaevna Khonina[1,3], and Nikolay Lvovich Kazanskiy[1,3]
[1]Samara National Research University, Samara, Russia
[2]Warsaw University of Technology, Institute of Microelectronics and Optoelectronics, Warszawa, Poland
[3]Image Processing Systems Institute – Branch of the Federal Scientific Research Centre "Crystallography and Photonics" of Russian Academy of Sciences, Samara, Russia

1.1 INTRODUCTION

Electric wires are still the source to transmit data in roughly all modern devices, such as the world's fastest supercomputers, laptops, or phones. The salient features of copper wire include reliability and low price. For decades, copper wire can provide massive quantities of transistors to interconnect and maintain circuitry on smaller chips. However, the imperfections of copper wire, such as limited bandwidth, current leakage, and cross talk between the contiguous wires come into the spotlight as we have swiftly approached the physical limit of chip miniaturization. Additionally, the power intake, the heat generated, and the space occupied by copper are also practical concerns that should be addressed. One of the possible solutions to the aforementioned problems is based on light-based technologies such as optical fiber cables, which utilize photons to transport the information at a much faster speed and bandwidth. Nevertheless, the excessive cost of replacing the electric wires in integrated circuits with cutting-edge photonics is a barrier to commercialization.

It is interesting to realize integrated optics in silicon (Si) due to well-developed technology and economy. It is a well-studied material, and the processing has been advanced by the electronics industry to a point that is highly adequate for most integrated optical applications. At present, the smallest feature size for most applications is ~1–2 μm, which in terms of microelectronics is very old technology, while there is a tendency toward miniaturization. Besides, novel and enhanced processing becomes available as Si microelectronics continue to progress. Photonic circuits can be fabricated by direct patterning Si to create optical interconnects to broadcast data-carrying laser signals. The biggest advantage is that it can transmit massive data while consuming low power with no heating up or causing any deterioration in the signal.

Si is transparent to low energy in the infrared (IR) region of the wavelength spectrum due to its electronic bandgap; however, it is opaque to photons in the visible section of the spectrum [1]. It is an indirect bandgap, which makes Si a bad emitter. Even though Si photonics is an ideal solution on paper, scientists have struggled for more than 30 years to overcome its different shortcomings [2].

For easy understanding of the attributes of Si, we have listed some major advantages of Si photonics, along with a few disadvantages associated with both material and device characteristics, in Figure 1.1. Mainly all the points are easy to understand. However, we will give a little description of the "Pockels effect," which is an important phenomenon that allows changing the optical properties of a certain material by applying an external electric field. Mostly high-speed modulators and switches are available in lithium niobate, which is used in optical communication systems. However, these devices are massive; that's why their application is limited to complex optical networks such as data centers and telecom networks, though this effect is inherently zero in single-crystalline Si because of the inversion symmetry of the crystal structure. Recently, a record-high electro-optic (EO) response in Si photonic devices was demonstrated by applying nanometer-thick, crystalline layers of a material exhibiting the Pockels effect [3]. This can be foreseen as the next step toward an application area mostly beyond communication technologies.

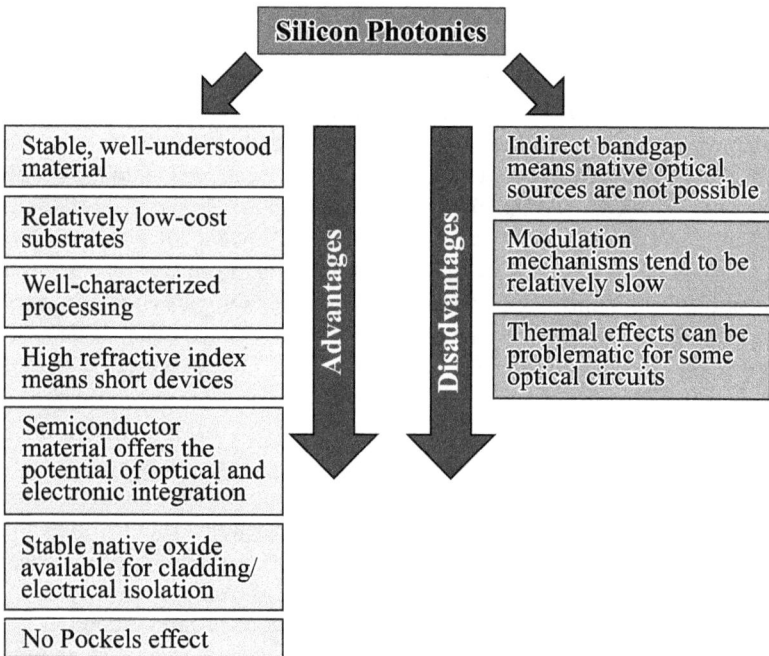

Silicon Photonics

Advantages

- Stable, well-understood material
- Relatively low-cost substrates
- Well-characterized processing
- High refractive index means short devices
- Semiconductor material offers the potential of optical and electronic integration
- Stable native oxide available for cladding/ electrical isolation
- No Pockels effect

Disadvantages

- Indirect bandgap means native optical sources are not possible
- Modulation mechanisms tend to be relatively slow
- Thermal effects can be problematic for some optical circuits

FIGURE 1.1 A few major advantages and disadvantages of Si photonics.

1.2 OPTICAL WAVEGUIDE

Integrated photonics contribute to the integration of numerous photonic components, such as beam splitters, polarizers, couplers, sensors, interferometers, and detectors on a single platform. The basic theory of light confinement in the optical waveguides (WGs) is the same as that in optical fibers where a material having a high refractive index (core) is surrounded by a material with a lower refractive index (cladding) that can act as a light trap. The light injected into the core undergoes a phenomenon of total internal reflection between the interfaces; thus, light cannot escape the core. As a result, light can be transported from point to point in an integrated photonic chip. In this section, we focus on several configurations of optical WGs built on a Si on insulator (SOI) platform such as ridge, rib, slot, Si-based hybrid plasmonic WG (hereafter represented as HPWG), and suspended WGs.

1.2.1 RIDGE WAVEGUIDE

Ridge WG is one of the primitive components of Si photonics that has been comprehensively studied over the years [4]. They are generally used in different applications such as lasers, modulators, polarizers, switches, and amplifiers. It comprises a Si core embedded on a silica (SiO_2) substrate as presented in Figure 1.2(a), where W_{Si} and H_{Si} are the width and height of the core, respectively.

The single-mode condition is vital in the realization of functional photonic devices; therefore the size of the WG core should be determined to obey the single-mode condition. The principal requirement is the single-mode propagation of the

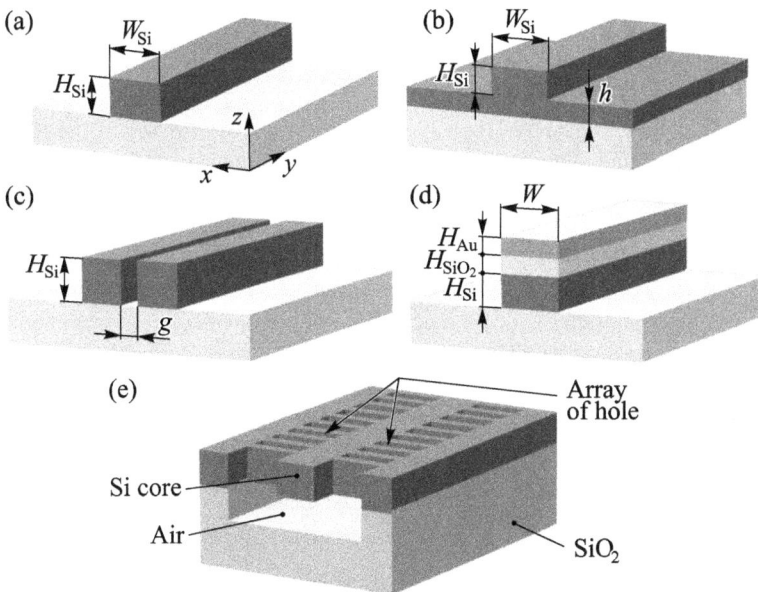

FIGURE 1.2 Graphic representation of (a) ridge WG, (b) rib WG, (c) slot WG, (d) Si-based HPWG, and (e) suspended Si WG.

Effective refractive index (n_{eff})   Effective refractive index (n_{eff})

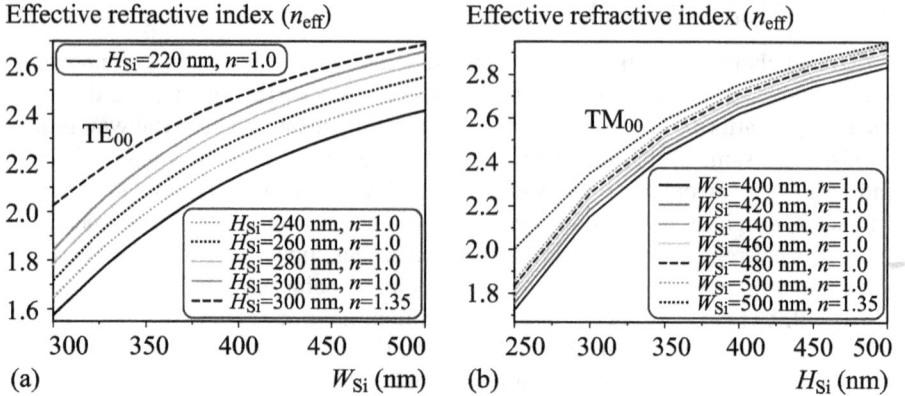

(a) W_{Si} (nm)

(b) H_{Si} (nm)

FIGURE 1.3   The effective refractive index of a ridge WG: (a) TE_{00} mode, (b) TM_{00} mode.

TE_{00} and TM_{00} mode. In a WG, each mode propagates with a phase velocity of c/n_{eff}, where c represents the speed of light in a vacuum and n_{eff} is the effective refractive index of that mode. It implies the optical power of the mode in the WG core. The mode is guided when n_{eff} is higher than the cladding and lesser than the core; otherwise it will be radiated into the substrate. Moreover, the higher the n_{eff}, the more the mode will be energetically guided. The dependence of the mode n_{eff} on the W_{Si} and H_{Si} is calculated for TE_{00} and TM_{00} modes via the finite element method (FEM), which is executed in a COMSOL Multiphysics software, as displayed in Figure 1.3.

1.2.2   RIB WAVEGUIDE

The rib WG can be considered as a special case of ridge WG where the Si layer is not completely etched. The guiding layer principally consists of the slab with a strip placed onto it. As shown in Figure 1.2(b), W_{Si} is the rib width, H_{Si} is the total height of the rib, where h is the slab height. Single-mode silica WGs are comparatively easy to design. On the other hand, SOI WGs with dimensions larger than a few nanometers support multiple modes [5]. These kinds of WGs are typically unattractive in photonic circuits as their working mechanism can be acutely compromised by the existence of multiple modes. However, large rib WGs in SOI can be designed to be monomodal. In 1991, Soref et al. [6] proposed a straightforward equation to obtain a single-mode condition of such WGs:

$$\frac{W_{Si}}{H_{Si}} \leq 0.3 + \frac{h}{\sqrt{1-h^2}},$$

where h is the ratio of the slab height to overall rib height, and W_{Si}/H_{Si} is the ratio of the WG width to total rib height. The dependence of mode effective refractive index on h is studied in Figure 1.3 at a fixed value of W_{Si} and H_{Si}. The E-field mapping of quasi-TE and quasi-TM mode is displayed in the inset of Figure 1.4.

Effective refractive index (n_{eff})

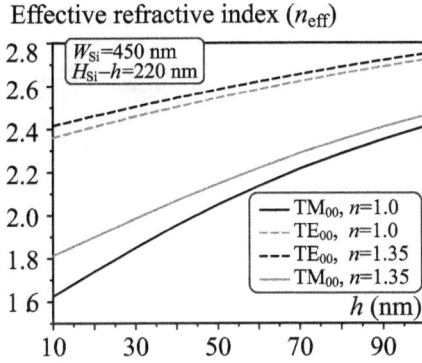

FIGURE 1.4　The effective refractive index of a rib WG at TE_{00} and TM_{00} mode.

1.2.3　SLOT WAVEGUIDE

Slot WG was anticipated by Vilson Rosa de Almeida and Carlos Angulo Barrios [7] in 2003. It is proficient in confining light in a low-index material between two high-index rail WGs by improving the gap (g), width (W), and height (H_{Si}) of the strip WGs. The design of the slot WG is shown in Figure 1.2(c). The normal E-field component of quasi-TE undergoes high interruption at the interface between a high and a low refractive index material, which oversees providing an eminent amplitude in the low-index slot region that is relative to the ratio of the refractive indices of the cladding material to that of the Si:

$$E_{x,g} = \frac{n_{\text{Si}}^2}{n_{\text{clad}}^2} E_{x,\text{Si}}.$$

Here, $E_{x,g}$ and $E_{x,\text{Si}}$ stand for the E-field confined in the slot (which is air) and residual field in the Si, respectively. The dependence of mode effective refractive index on g is presented in Figure 1.5(a) at a fixed value of W and H_{Si}. The image of the slot WG fabricated via focused ion beam (FIB) milled is shown in Figure 1.5(b). Conversely, this WG scheme is not able to confine the quasi-TM mode as the E-field remains continuous at the boundary.

As compared to other conventional WG schemes, slot WG can provide high field confinement in the slot, which is in general not possible to attain. This attribute makes slot WG an eye-catching replacement for applications that encompass sturdy light–matter interaction, for instance sensing and nonlinear photonics. By utilizing EO polymer material in slot WG it is possible to obtain the Pockels effect. As a result, it gathered high potential in the field of optical switching and high-speed modulation even at frequencies of 100 GHz. Several devices like Mach–Zehnder interferometers (MZIs) and ring resonators (RRs) have been realized utilizing slot WG phase shifters.

Light is typically injected into a slot WG by matching the propagation constants of the ridge WG and the slot WG. Nonetheless, due to the mode mismatch and

Effective refractive index (n_{eff})

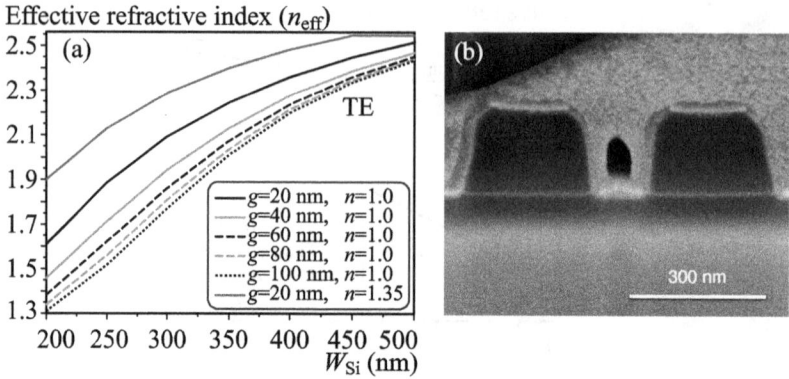

FIGURE 1.5   (a) The effective refractive index of slot WG; (b) scanning electron microscope (SEM) image of the FIB milled cross section of a fabricated device (the small hole in the center was due to the incomplete coating of platinum deposited to protect the WGs during the FIB milling) [8].

scattering loss, effective coupling between the ridge WG and slot WG is a difficult challenge. As a result, extensive study is being conducted to leverage the use and development of slot WGs for integrated optical sensors.

1.2.4   HYBRID PLASMONIC WAVEGUIDE

A new type of plasmonic WG recognized as HPWG has gathered significant interest due to its ability to offer both subwavelength confinement and long propagation lengths. It appears like the insulator-metal (IM) WG formation, apart from that a thin layer of low-index material (SiO_2) is sandwiched between the metal layer (Au or Ag) and a higher-index dielectric layer (Si), as displayed in Figure 1.2(d), where H_{Si}, W, H_{SiO_2}, H_{Au} denote the layer thickness of the Si core, the core width, a layer thickness of the SiO_2 layer, and a layer thickness of the Au layer, respectively. Theoretical studies have shown that a low-index layer can hold a low-loss compact mode whose propagation length robustly relies on its thickness [9, 10].

The mode-coupling theory may be used to explain how the HPWG works. In general, the Si core confines the dielectric WG mode, but the metal surface confines surface plasmons (SPs), which are restricted to the metal surface. When these two structures are kept in close vicinity, the dielectric WG mode supported by Si ridge couples to the SP mode supported by the metal surface. Due to the mode coupling, the light is transferred to the region between the metal and the low index medium. The real part of n_{eff} is calculated by maintaining H_{Au} and W at fixed values, whereas H_{SiO_2} and H_{Si} are varied, as demonstrated in Figure 1.6.

HPWG integrates the characteristics of the dielectric and plasmonic WGs. The excitation of surface electron fluctuation in the hybrid plasmonic WG focuses the photons to reside near the metal surface. This helps in the reduction of radiation loss. The high optical confinement offered by this WG scheme is beyond the diffraction limit, while the propagation loss is comparatively low as compared to the Si slot WG.

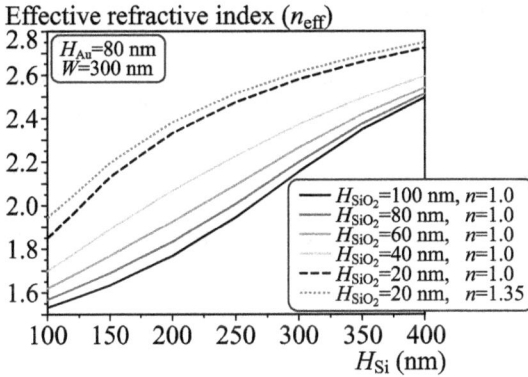

FIGURE 1.6 The effective refractive index of HPWG.

Si photonics has turned out to be very interesting due to its manufacturing compatibility with the regular complementary metal oxide semiconductor (CMOS) microelectronics technology. For that reason, it is exciting to realize Si-based HPWG with basic fabrication processes.

1.2.5 SUSPENDED SI WAVEGUIDES

The mid-infrared wavelength (2–20 μm) is interesting as it contains several so-called water windows: wavelengths of light that can propagate through the Earth's atmosphere without being absorbed by water molecules. It is vastly applicable in spectroscopy and gas-sensing applications. SOI is an attractive platform for the realization of integrated Si photonic devices at near-infrared wavelengths; however, it is not suitable for the devices operating in the mid-infrared region for the wavelengths between 2.75 and 3 μm and beyond 4 μm due to the strong absorption of the buried oxide (BOX) layer at longer wavelengths. For that reason, other WG platforms such as Si-on-sapphire, Si-on-nitride or germanium-on-Si are being studied for mid-infrared applications. Recently, a new technique is proposed that involves the selective removal of the BOX layer in the SOI platform, resulting in Si membrane WGs as displayed in Figure 1.7(a). As a result, the operational wavelength range of the SOI platform is stretched to cover the full Si transparency window up to 8 μm. This allows the ease of manufacturing of such devices via standard SOI fabrication processes.

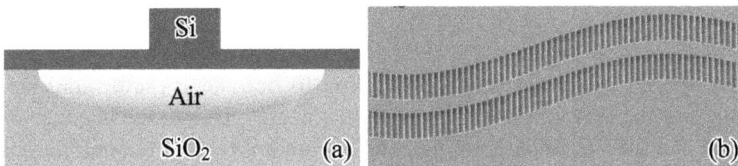

FIGURE 1.7 (a) Cross section of suspended Si WG, (b) SEM image of S-bend WG [11].

The manufacturing of such WGs involves two etch steps: In the first step, the WG core and an array of holes along the sides of the core are patterned and etched. In the second step, the BOX layer is removed using a hydrofluoric acid passing through the venting holes. The propagation loss of 3 dB/cm at 2.75 μm of such WGs is reported. In 2016, J. Soler Penades [11] demonstrated several fundamental photonic functional blocks such as low loss 90° bends, S-bends, 2×2 multimode interference couplers, and MZIs based on suspended SOI WGs. An extraordinary propagation loss of as low as 0.82 dB/cm at 3.8 μm is achieved. The SEM image of the S-bend WG is presented in Figure 1.7(b).

1.3 WAVEGUIDE COUPLING TECHNIQUES

Coupling light to an integrated optical circuit is conceptually straightforward, but, it is a non-trivial problem. This is attributed to the small size of the optical WGs, usually a few microns at most in any cross-sectional dimension. There are a variety of techniques available to perform the coupling task, the most common being butt coupling, end-fire coupling, prism coupling, and grating coupling. The end-fire and the butt coupling are very similar, with light just shining at the end of the WG. The difference normally made between these two approaches is that the butt coupling entails merely butting the two devices or WGs close to each other, such that the mode field of the transmitting device lands on the end face of the second device; whereas the end-fire coupling uses a lens to concentrate the input beam on the end face of the receiving device. Consequently, light is transmitted into the end of the WG and can theoretically excite all modes of the WG. Nevertheless, prism coupling and grating coupling are distinctly different methods, since they pass an input beam at a specific angle through the surface of the WG. This allows for phase matching within the WG to a particular propagation constant, thus enabling a particular mode to be excited.

Prism coupling is not particularly useful for semiconductor WG assessment. Coupling conditions are such that the material from which the prism is produced should have a higher refractive index than the WG; this severely limits the possibilities, particularly for Si with a high refractive index of 3.5. Nevertheless, there are materials available that could be used, such as germanium, although other drawbacks also indicate that the technique is inferior to other techniques. Such drawbacks are that the prism coupling will affect the WG's formation, and it is not suitable if a surface cladding is to be used. It is best suited to planar WGs, and it is not ideal for material systems that utilize rib WGs such as Si technology. The remaining three techniques can be beneficial and will be addressed in effect with Si-based technology. Figure 1.8 illustrates the concepts of these four coupling methods.

1.3.1 GRATING COUPLING

Grating couplings provide a way for coupling to individual modes and are useful for coupling to the WG of a wide range of thicknesses. Due to the need to insert the input beam at a specific angle, grating couplers are not reliable enough for consumer applications, but they are a powerful development tool. To couple light into a WG, as displayed in Figure 1.8(a, b), the components of the phase velocities must be the

Prism coupling

Incident light

— Prism

Waveguide

(a)

Grating coupling

Incident light

Waveguide

(b)

Butt coupling

Waveguide

(c)

End-fire coupling

Incident light → Waveguide

Lens   (d)

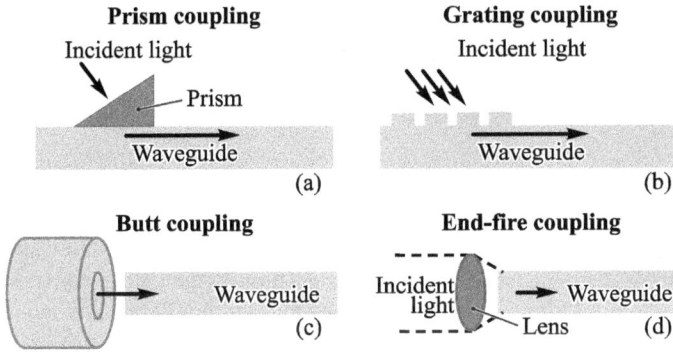

FIGURE 1.8 Four light coupling techniques to optical WGs: (a) prism coupling, (b) grating coupling, (c) butt coupling, and (d) end-fire coupling.

same in the propagation direction (z-direction). This is referred to as the state of phase-match, and, in this case, the propagation constant must be the same in the z-direction. Consider an incident beam on the WG's surface at an angle θ, as displayed in Figure 1.9.

The beam propagates in the medium n_{clad} with a propagation constant $k_o n_{\text{clad}}$, in the direction of propagation. The z-directed propagation constant in the n_{clad} medium will therefore be as follows:

$$k_z = k_o n_{\text{clad}} \sin \theta. \tag{1.1}$$

Therefore, the phase-match condition will be:

$$\beta = k_z = k_o n_{\text{clad}} \cdot \sin \theta, \tag{1.2}$$

where β is the WG propagation constant and is greater than $k_o n_{\text{clad}}$. Hence, the condition of Eq. (1.2) can never be satisfied since $\sin\theta$ will be less than unity. Therefore, a prism or grating is needed to couple light into the WG, since both can satisfy the condition of phase-match if properly designed.

A grating is a periodic structure, and it is usual to fabricate it on the WG surface if it is to be used as an input or output coupler. The periodic design of the grating induces a period modulation of the effective index of the WG. For optical mode

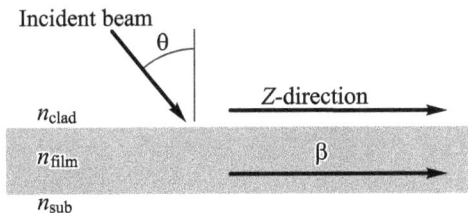

Incident beam

θ

Z-direction

n_{clad}

n_{film} β

n_{sub}

FIGURE 1.9 Light beam incident on the surface of the WG.

with propagation constant β_{WG} when the grating is not present, modulation results in a series of possible propagation constants β_p, given by:

$$\beta_p = \beta_{WG} + \frac{2p\pi}{\Lambda}, \tag{1.3}$$

where Λ is the grating period and $p = \pm1$, ±2, ±3, etc. Such modes are analogous to diffraction grating orders. Obviously, in the WG, the propagation constants referring to the positive values of p cannot occur because the propagation constant β_p is still less than $k_o n_{clad}$. Therefore, only the negative values of p will result in a phase-match. It is normal to render the grating so that only the value $p = -1$ results in a phase-match with a WG mode. Thus, the propagation constant of the WG becomes:

$$\beta_p = \beta_{WG} - 2\pi/\Lambda. \tag{1.4}$$

Therefore, the phase-match condition becomes:

$$\beta_{WG} - 2\pi/\Lambda = k_o n_{clad} \cdot \sin\theta. \tag{1.5}$$

Expressing β_{WG} in terms of the effective index N, Eq. (1.5) becomes:

$$k_o N - 2\pi/\Lambda = k_o n_{clad} \cdot \sin\theta, \tag{1.6}$$

and on substituting for k_o we obtain:

$$\Lambda = \frac{\lambda}{\left(N - n_{clad} \cdot \sin\theta\right)}. \tag{1.7}$$

Since the medium with refractive index n_{clad} is usually air ($n_{clad} = 1$), therefore Eq (1.7) becomes:

$$\Lambda = \frac{\lambda}{\left(N - \sin\theta\right)}. \tag{1.8}$$

Accordingly, Eq. (1.8) can be used to calculate the grating period for the desired input angle in the air for mode coupling with β_{WG} propagation constant.

Surprisingly little work has been done on Si grating couplers, mainly due to production difficulties. Since the refractive index of Si is high, the necessary grating period for the input/output coupling is 400 nm. Ang et al. [12] obtained an output coupling efficiency of approximately 70% for rectangular grating and 84% for non-symmetrical profile grating [13].

1.3.2   BUTT COUPLING AND END-FIRE COUPLING

The variables that affect the performance of the butt coupling and the end-fire coupling are identical, so these methods will be discussed together. When the incident beam is shined at the end-face of the optical WG, the strength with which the light is coupled to the WG depends on the following aspects. There may also be a numerical aperture discrepancy where the optical WG's input angles are not well matched to the range of angles of excitation, but the latter concept is ignored here.

1.3.2.1   The Excitation and Waveguide Modes Match

Matching the excitation fields is generally determined by conducting the overlap integral between the excitation field and the WG field. The most straightforward situation is considered when both the fields are single moded. The overlap integral Γ between the exciting field E_{ex} and the guided field E_g is indicated as:

$$\Gamma = \frac{\int_{-\infty}^{\infty} dy \int_{-\infty}^{\infty} E_{ex} E_g \, dx}{\left[\int_{-\infty}^{\infty} \int_{-\infty}^{\infty} E^2 \, dx \, dy \int_{-\infty}^{\infty} \int_{-\infty}^{\infty} E_g^2 \, dx \, dy \right]^{1/2}}, \tag{1.9}$$

where the denominator is a standardizing value that guarantees the value of Γ lies between 0 and 1. A straightforward approach to gain a reasonable idea of the overlap between excitation and WG fields is to use a Gaussian approximation. Eq. (1.9) simplifies this considerably, and the result gives a good understanding of the overlap values if the fields for both the excitation beam and the WG mode are presumed to have Gaussian shapes, as described in the following equations:

$$E_g = \exp\left[-\left(\frac{x^2}{w_x^2} + \frac{y^2}{w_y^2} \right) \right]. \tag{1.10}$$

This represents a WG field with 1/e widths in the x and y directions of $2w_x$ and $2w_y$ respectively. If we suppose a circularly symmetrical input beam, then E_{ex} is expressed as:

$$E_{ex} = \exp\left[-\frac{\left(x^2 + y^2 \right)}{w_o^2} \right]. \tag{1.11}$$

Utilizing the mathematical identity for a definite integral:

$$\int_o^{\infty} \exp\left[-r^2 x^2 \right] dx = \frac{\sqrt{\pi}}{2r}. \tag{1.12}$$

Eq. (1.9) reduces to express the coupling efficiency:

$$\Gamma = \frac{\dfrac{2}{w_o} \left[\dfrac{1}{w_x w_y} \right]^{1/2}}{\left[\dfrac{1}{w_x^2} + \dfrac{1}{w_o^2} \right]^{1/2} \left[\dfrac{1}{w_y^2} + \dfrac{1}{w_o^2} \right]^{1/2}}. \tag{1.13}$$

The power coupling efficiency is given by:

$$\Gamma^2 = \frac{\dfrac{4}{w_o^2}\left[\dfrac{1}{w_x w_y}\right]}{\left[\dfrac{1}{w_x^2}+\dfrac{1}{w_o^2}\right]\left[\dfrac{1}{w_y^2}+\dfrac{1}{w_o^2}\right]}. \tag{1.14}$$

1.3.2.2 The Degree of Reflection from the Waveguide Facet

The incoming confined light wave is partially transmitted and partially reflected at the end facet of a dielectric WG, where the guided backwards traveling fraction of the optical power strongly depends on the geometry of the WG and the refractive index contrast. The reflection coefficient r for TE and TM polarization is given as:

$$r_{TE} = \frac{n_1 \cos\theta_1 - n_2 \cos\theta_2}{n_1 \cos\theta_1 + n_2 \cos\theta_2}, \tag{1.15}$$

$$r_{TM} = \frac{n_2 \cos\theta_1 - n_1 \cos\theta_2}{n_2 \cos\theta_1 + n_1 \cos\theta_2}. \tag{1.16}$$

Using Snell's law, Eq. (1.15) reduces to:

$$r_{TE} = \frac{-\sin(\theta_1 - \theta_2)}{\sin(\theta_1 + \theta_2)}. \tag{1.17}$$

Consequently, the power reflectivity is determined by $R = r_2$. Hence the reflectivity for TE and TM polarization is given by:

$$R_{TE} = r_{TE}^2 = \frac{\sin^2(\theta_1 - \theta_2)}{\sin^2(\theta_1 + \theta_2)} \quad \text{and} \quad R_{TM} = r_{TM}^2 = \frac{\tan^2(\theta_1 - \theta_2)}{\tan^2(\theta_1 + \theta_2)}.$$

Both the reflectivities are the same for normal incidence, $\theta_1 = 0$. Additionally, end-fire coupling allows light to be incident almost normal to the face. In that case:

$$R_{TE} = R_{TM} = R = \left|\frac{n_1 - n_2}{n_1 + n_2}\right|^2. \tag{1.18}$$

The reflectivity is about 31% for a Si/air interface, which causes an extra loss of 1.6 dB/facet, which is significant and is minimized using the anti-reflecting coating in commercial devices. The coating thickness should be equal to $\lambda/4$ so that the waves reflected from the front of the coating and the facet is in anti-phase. In this way, the reflections are cancelled to some extent, reduced, or eliminated. It may be shown that for normal incidence, the net reflectivity is given by:

$$R = \left|\frac{n_1 n_2 - n_{ar}^2}{n_1 n_2 + n_{ar}^2}\right|^2, \tag{1.19}$$

where n_{ar} is the refractive index of the anti-reflection coating. The net reflectivity can be zero, if $n_1 n_2 = n_{ar}^2$.

1.3.2.3 The Quality of the Waveguide End-Face

The WG end-face quality is highly dependent on the preparation technique. The purpose of any surface preparation technique is to create sufficiently smooth facets to reduce optical scattering. There are three main options used for the end-face processing of semiconductor WGs: cleaving, polishing, and etching. Cleaving is achieved by manually placing a small crack at the edge of the sample and then applying pressure to the sample to crack along the crystallographic plane. This procedure is difficult to master and is limited to use in the research laboratory, as it is unsuitable for commercial applications due to the nature of the procedure. The results may be promising but appear to be volatile in SOI because the addition of a buried oxide layer implies that the surface and the substrate layers of Si are not directly connected. The most common method of preparing a WG facet is probably polishing. Lapping with abrasive materials with sequentially diminishing grit sizes polishes the sample end-face. A specified recipe is usually followed, which may result in an excellent surface finish, although the rounding of the end-face is a common failure. End-faces may also be prepared using chemical or dry etching. It goes beyond the scope of this chapter to describe how this is achieved, but it is sufficient to say that it is a technique that can be developed to the highest level for commercial application.

1.3.2.4 The Spatial Misalignment of the Excitation and Waveguide Fields

When the two fields are slightly misaligned, the overlap between the exciting and receiving fields can change. The effect of spatial misalignment is calculated by measuring the overlap integral described by Eq. (1.9), when the exciting and WG fields are displaced by an amount, relative to each other. Let us assume that the WG field is still represented by Eq. (1.10), but the exciting field has an offset of X, so:

$$E_{ex} = \exp\left[\frac{x^2 + (y-X)^2}{w_o^2}\right]. \tag{1.20}$$

The offset is assumed to be along the y-direction, to simplify the calculation. In the presence of misalignment, the overlap integral may now be written as:

$$\Gamma' = \exp\left[-\frac{X^2}{w_y^2 + w_o^2}\right] \cdot \frac{\int_{-\infty}^{\infty} dy \int_{-\infty}^{\infty} E_{ex} E_g \, dx}{\left[\int_{-\infty}^{\infty}\int_{-\infty}^{\infty} E^2 \, dx \, dy \int_{-\infty}^{\infty}\int_{-\infty}^{\infty} E_g^2 \, dx \, dy\right]^{1/2}}, \tag{1.21}$$

$$\Gamma' = \exp\left[-\frac{X^2}{w_y^2 + w_o^2}\right]. \tag{1.22}$$

1.4 CONFINEMENT FACTOR

It is a clear fact that not all the power propagating in a WG mode stays inside the core of the WG. It is practical to calculate the fraction of power that stays inside the core, which enables comparisons between WG structure and modes. The field confinement factor (Γ) is used as a Figure of Merit (FOM) to characterize the optical WGs to describe the interaction of light with the ambient medium, and their ability to confine an optical mode within a certain region. Γ is defined as the ratio of the time-averaged energy flow through the area of interest ($A_{interest}$) to the time-averaged energy flow through the total area, A_{total}.

$$\Gamma = \frac{\iint_{A_{interest}} \text{Re}\left\{\left[E \times H^*\right].e\right\} dx\,dy}{\iint_{A_{total}} \text{Re}\left\{\left[E \times H^*\right].e\right\} dx\,dy}, \tag{1.23}$$

where E and H^* are the electric and magnetic field vectors, respectively, and e is the unit vector in the propagation direction of light. Depending on the application, the area of interest can differ. For example, in the case of ridge and rib WG, $A_{interest}$ is Si core, whereas, in the case of slot WG, $A_{interest}$ is a gap (air). Finally, in an HPWG, $A_{interest}$ is a thin layer of SiO_2 sandwiched between Si and Au, respectively, whereas A_{total} is the integral of energy flow in the core and cladding. The Γ of ridge, rib, slot, and HPWG is calculated depending on geometric variables of the WGs as displayed in Figure 1.10.

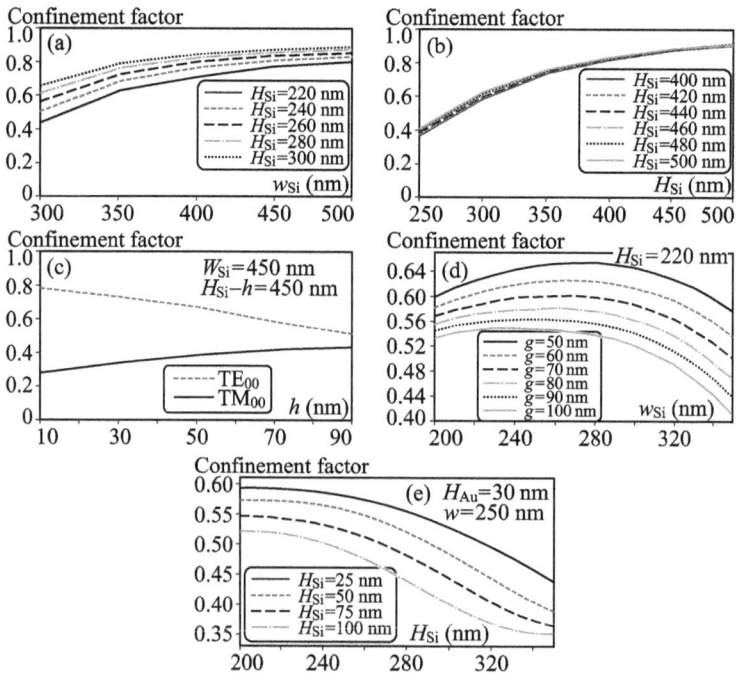

FIGURE 1.10 Γ of (a) ridge WG at TE polarization, (b) ridge WG at TM polarization, (c) rib WG at TE and TM polarization, (d) slot WG, (e) HPWG.

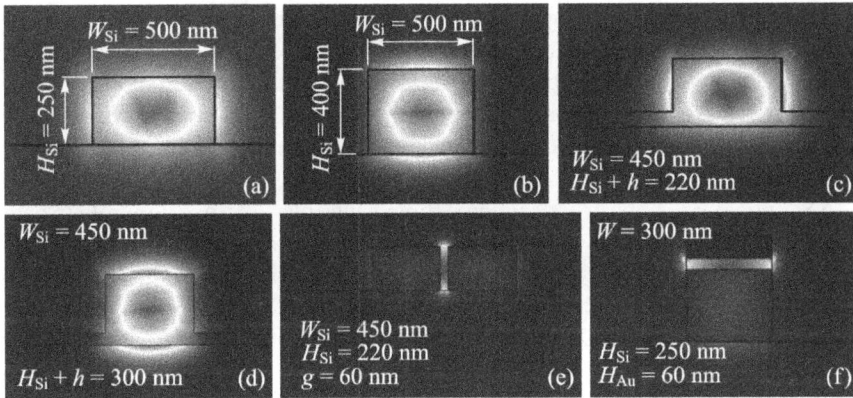

FIGURE 1.11 E-field mapping of mode in (a) ridge WG at TE_{00} mode, (b) TM_{00} mode, (c) rib WG at TE_{00} mode, (d) TM_{00} mode, (e) slot WG, and (f) hybrid plasmonic WG.

The maximum Γ of 0.91, 0.78, 0.65, and 0.59 is obtained for ridge, rib, slot, and HPWG, respectively. This shows that ridge and rib WGs offer a better Γ than slot and HPWG. However, in the slot and HPWG, ~40% of the mode power is residing in the cladding; this fact can be visualized as an opportunity to utilize these WGs in sensor applications.

The E-field mapping of propagating mode in the WG is displayed in Figure 1.11. The geometric variables of the WGs are labeled in the figure.

1.5 LOSSES OF SI WAVEGUIDES

Optical WGs are vulnerable to losses that occur from three sources: scattering, absorption, and radiation. The relative contribution of each of these effects is dependent upon the WG design and the quality of the material in which the WG is manufactured. Integrated photonics based on a heterogeneous Si platform is very attractive as it combines the best attributes of Si and III-V platforms in the realization of numerous high-performing active and passive optical devices on a single chip [14, 15].

1.5.1 SCATTERING LOSS

There are two types of scattering losses associated with optical WGs: volume scattering and interface scattering. The former deals with the imperfections in the bulk WG material, such as voids, inclusions, or crystalline defects, and the latter is connected to the roughness at the interface between the core and the cladding layer. Generally, volume scattering is irrelevant in a well-developed WG technology because the material has been upgraded to an adequate point before the fabrication process of the WGs. On the other hand, interface scattering may not be neglected, even for relatively smooth interfaces. The SOI high index contrast platform permits small WG dimensions and tight bends to realize compact photonic devices with a small footprint. But regrettably, the high index contrast also leads to high propagation loss primarily subjugated by scattering, which limits the performance of the devices requiring on-chip delays larger than a few centimeters. The Si WGs are extremely sensitive to fabrication imperfections such as sidewall roughness, which induces high loss and phase error. Additionally, these WGs are robustly birefringent, have considerable temperature dependence, and the chip to fiber coupling loss is significantly high.

In Si optical WGs, the cross-sectional dimensions play an important role in determining the propagation loss of the mode. The major sources of propagation loss in an optical WG are related to the absorption within the materials and scattering loss from the sidewalls, etc. [16]. As demonstrated in [17–19], typical propagation loss of single-mode ridge WGs is in the range between 0.5 dB/cm and 2 dB/cm. It is probable to accomplish low propagation loss utilizing advanced immersion 193 nm lithography (litho), but presently it might not be as commonly available in Si photonics.

The manufacturing of slot WGs is a challenging task, as sidewall roughness can induce high scattering loss. The lowest demonstrated losses of the conventional [20] and segmented slot WGs [21], both fabricated with E-beam litho, are 10 dB/cm and 4 dB/cm, respectively. Recently, A. Spott et al. [20] demonstrated the photolithographically fabricated low-loss asymmetric Si slot WGs having losses as low as 2 dB/cm. Moreover, light injection into such a nanoscale structure is a hefty job.

1.5.2 Absorption

In semiconductor WGs, absorption loss appears due to two probable sources: interband absorption and free carrier absorption. Interband absorption comes into play when photons having energy superior to the bandgap are used to stimulate the electrons from the valence band to the conduction band. In Si, the band edge wavelength is roughly 1.1 μm. Below this wavelength, Si is a perfect absorber. That is why it is one of the most widespread materials used for photodetectors in visible and near-infrared wavelengths. For the waveguiding purpose, the wise selection of wavelength range is mandatory. For instance, the attenuation loss of 2.83 dB/cm and 0.004 dB/cm is demonstrated at the wavelengths of 1.15 μm and 1.52 μm, respectively. This shows that interband absorption can be neglected by selecting a suitable wavelength of operation.

Free carrier absorption is a phenomenon that deals with the transfer of photon energy to an electron or hole in the conduction or valence band of a semiconductor, respectively. Due to the absorption of photon energy, excited electrons and holes transition to a state in the same or another conduction and valence band, respectively. It is considered a parasitic process, as it hurts the performance of the devices such as Si solar cells and thermophotovoltaic cells. Unsurprisingly, free carrier absorption is much stronger in heavily doped semiconductors. In the case of pure Si, this effect appears at approximately 1 μm.

1.5.3 Radiation Loss

The leakage of optical mode from the guiding layer (core) into the surrounding region of the WG comes under the category of radiation loss. If the WG geometry is well optimized, this loss can be minimized. Radiation loss from a straight channel WG should ideally be ignored. However, when an EM wave (electromagnetic wave) propagates along with a WG with a uniform bending radius, some portion of its energy is lost. A part of this lost energy is redistributed among the guided modes of the WG; the residue is swiftly radiated away as tunneling leaky modes.

For WG structures as depicted in Figure 1.1, the probability of mode leakage exists, if the thickness of the lower cladding layer is finite. In SOI-based WGs, the BOX layer must be adequately thick, roughly 400 nm for operation in the wavelength range of 1.3–1.6 μm to avoid optical modes from penetrating the oxide layer and coupling to the Si substrate. As the dimension of the WG core reduces, resulting in a less-confined mode, the requirement for a thick BOX layer increases. However, the required BOX thickness can differ from mode to mode, as each mode penetrates the cladding to a different depth.

1.6 MANUFACTURING OF SI WAVEGUIDES ON SILICON ON INSULATOR PLATFORM

In the early 1960s, the semiconductor manufacturing process was originated in Texas, and CMOS was patented by Frank Wanlass in 1963. Usually, Si WGs are manufactured from SOI wafers, which consist of a buried oxide (BOX) layer between the Si wafer and a thin Si layer, using standard CMOS processes. This includes the patterning of Si via optical litho and etching techniques to form WG structures.

There are many attractive attributes connected to the manufacturing of optical devices on an SOI platform, such as low cost of the material, mature and well-developed processing technique, development and manufacturing in the microelectronics industry, and the prospect of easy integration with electrical components in the same substrate [22, 23]. The BOX layer has a refractive index of 1.46, considerably lower than that for the crystalline Si layer of 3.5. Therefore, this type of platform setup is a conventional WG structure. In this section, we will briefly explain the steps involved in the manufacturing of Si WGs based on the SOI platform.

1.6.1 PHOTOLITHOGRAPHY AND LASER LITHOGRAPHY

Photolithography (PL) is a method for transferring patterns from a photomask to a thin layer of photosensitive material called a photoresist (PR) that covers the sample's surface. These patterns are only transitory features on the PR that must be transmitted to the device's underlying layers (Si) to gain the WG characteristics. When dealing with PL, the photomask is an essential element of transferring patterns onto PR coated samples. To provide high-quality, application-specific PL, a variety of recipes and process modifications have been devised, but the final device structure is dependent on the accuracy of all the fabrication processes involved [24]. PL should be done in a clean processing room because dust particles in the air might stick to the sample or mask, causing flaws in the circuit characteristics. When a dust particle clings to the photomask, it behaves like an opaque pattern and is transmitted

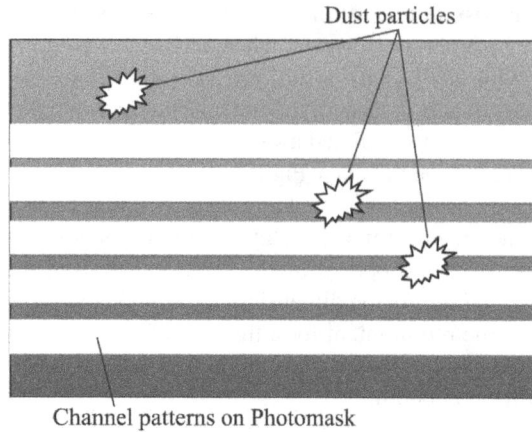

FIGURE 1.12   Dust particles contaminating the photomask patterns.

to the PR, resulting in a fault on the circuit features, as displayed in Figure 1.12. A class 100 cleanroom is required for the manufacturing of the most integrated optical structures, which means that the dust count must be four orders of magnitude lower than that of regular room air. For PL, however, a class 1 or class 10 cleanroom is necessary, as the dust count is even lower. UV (ultraviolet) litho, extreme UV litho, X-ray litho, electron beam litho, and nanoimprint litho are only a few of the lithographic methods available. The resolution limit is the key distinction between these methods.

The mask is in direct touch with the wafer in contact litho, which is also identified as contact printing. The PR is subjected to a collimated beam of UV light for a set amount of time. This results in a resolution of ~1 micron, which is quite high. However, there is one significant downside to this method: when the mask is brought close to the wafer, a dust particle may become embedded in it. This dust particle has the potential to irreversibly harm the mask. The second kind of litho is proximity litho, often known as proximity printing, which includes a tiny space between the photomask and the wafer during exposure, typically 10–50 microns. Due to the optical diffraction at the feature edges, this approach is utilized to limit the risks of photomask damage, but it reduces the resolution to a 2- to 5-micron range. Figure 1.13 depicts a contact and proximity litho scheme.

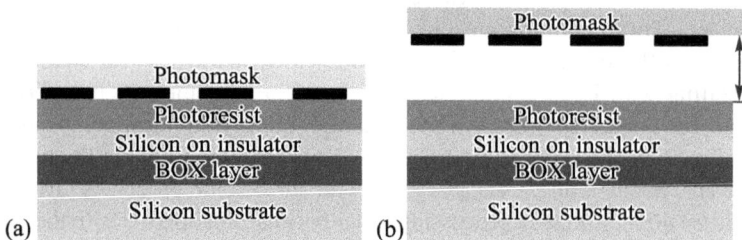

FIGURE 1.13   (a) Contact litho, (b) proximity litho.

Laser litho is a maskless method that uses a focused laser beam to write complicated patterns with tiny components. It offers significantly improved quality and throughput. On nearly any material, this method can write arbitrary microstructures with minimum feature sizes as small as 100 nm. It is frequently utilized in the manufacture and research and development (R&D) of masks and direct writing of microstructures.

1.6.2   WAFER PREPARATION

Before PR spin coating, we should make sure that the wafer is free from dust particles and has been desorbed of any moisture. Removing any traces of moisture is particularly vital because wafer cleaning is performed with a wet process ending in a deionized water wash and dry. A completely dry surface can be obtained by baking the wafer at a temperature above 432 K for several minutes before the spin coating of the PR. To improve the PR adhesion on the Si wafer, hexamethyldisilane (HMDS) is used.

1.6.3   SPIN COATING AND EXPOSURE

To produce micrometer and submicrometer structures, PR is primarily employed in microelectronics and microsystems technologies. A thin layer of PR is coated before litho to transfer the specified patterns to the Si layer. A PR is a radiation-sensitive chemical that may be used to create a patterned coating on the sample. A photosensitive chemical, a base resin, and an organic solvent are the three components. There are two sorts of PRs: negative and positive.

The exposed portions of a negative PR become less soluble in the developer, resulting in the reverse patterns on the PR as shown on the mask. The underlying phenomenon behind this type of feature is that following exposure, the negative PR initiates a chemical reaction that induces polymer crosslinking. The increased molecular weight of the crosslinked polymer causes it to become insoluble in the developer. Consequently, as illustrated in Figure 1.14, the unexposed portions are eliminated during the development process.

The UV-exposed areas of the positive PR, on the other hand, become soluble in the developer and are removed. The photosensitive compound is insoluble in the developer solution before exposure. The photosensitive compound, on the other hand, may absorb radiations in the exposed patterns during exposure. As illustrated in Figure 1.14, this helps to modify its chemical structure, which becomes soluble in the developing solution and is eliminated throughout the development process.

After cleaning, the wafer is straightaway coated with PR. A few drops of resist, approximately 1–10 mL, are applied to the center of the wafer, which is placed in a spin coater held via a vacuum seal on a polymer chuck. The wafer is spun at a usual speed of ~2,000–4,000 rpm for thin resists and between 800 and 2,000 rpm for thick resists. Once the spin coating is done, a soft bake is used to get rid of most of the solvents in the resist. Additionally, it can improve resist uniformity and bonding to the wafer. Soft baking is carried out at 373 K for a couple of minutes. Once the sample is ready, it is transferred to the mask-aligner where it is exposed with a UV light passing through a permanent patterned metallic mask.

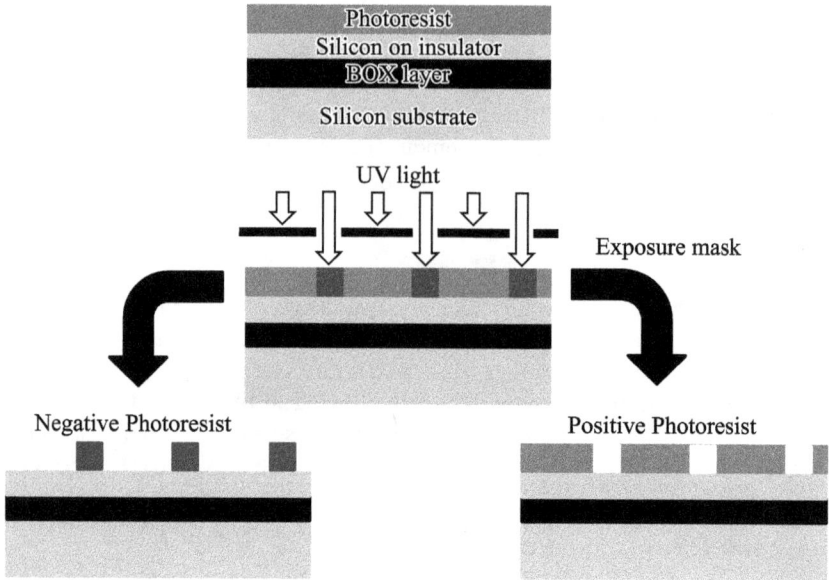

FIGURE 1.14 Demonstration of negative and positive PRs. The UV light-exposed region of the PR is displayed in a dark gray color, whereas the unexposed part is presented in a medium gray color.

1.6.4 Photoresist Developing and Hard Bake

The PR patterns are formed on the wafer during the developing stage. Depending on the type of PR used, the solution dissolves the exposed part of the resist (or unexposed part of resist), leaving behind the resist pattern. The final hard bake process evaporates the residual solvents in the resist and promotes the resist adhesion to the wafer surface. Usually, it is carried out at a temperature of 363–413 K for up to several minutes.

1.6.5 Etching Process

Etching involves the selected and controlled removal of material from the Si wafer using a chemically reactive or physical process. Two common methods are used to etch Si: wet and dry etching. Each approach has its limitations, but dry etching is preferred for the sub microns device features. Crystalline Si can be wet-etched by using a strong aqueous alkaline media such as NaOH, KOH, or tetramethylammonium hydroxide (TMAH) solutions via:

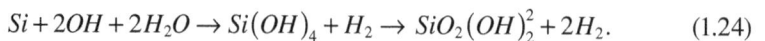

$$Si + 2OH + 2H_2O \rightarrow Si(OH)_4 + H_2 \rightarrow SiO_2(OH)_2^2 + 2H_2. \qquad (1.24)$$

Reactive ion etching (RIE) is a dry etching technique that uses a mix of chemical and physical interactions between the etching gas and the substrate [25]. The physical mechanism involves bombarding the material with high-energy ions that break it down, while the chemical mechanism involves induced interactions on the material

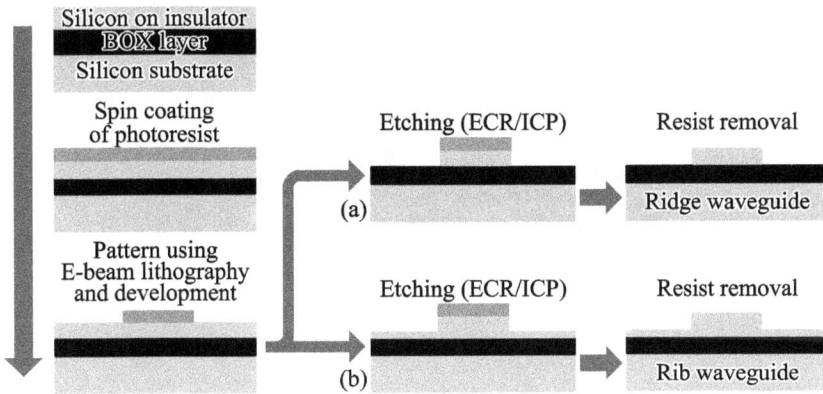

FIGURE 1.15 Manufacturing steps of SOI WGs: (a) ridge WG, (b) rib WG.

surface with plasma species to produce volatile species. By adjusting the chemical and physical variables of the etching process, the etch rate and sidewall angle may be adjusted. Because RIE-produced integrated optical devices have minimal propagation losses and smooth sidewalls, the sidewall slope is an important element in determining device performance. Some pollutants produced by the RIE method are deemed inappropriate for microelectronic processing. For RIE of Si WGs, the plasma is usually derived from CF_4 gas, which is a stable gas but dissociates into CF_3 and F atoms in a plasma with the single fluorine atom being used as the active etch element for both Si and Si dioxide materials. If the plasma is based on CF_4, then the etch rate is slow due to the swift recombination of CF_3 and F. By adding O_2 gas in the chamber, the Si etch rate can be considerably improved due to the reaction of oxygen with CF_3. It helps in repressing F recombination and consequently increasing the free F concentration. Fluorine-based chemical reactions are considered to be a foundation of many different Si etching processes. The whole manufacturing process of Si WGs is presented in Figure 1.15. Ridge WG is formed when the Si layer is completely etched till the BOX layer (Figure 1.15a), whereas rib WG is produced when the Si layer is partially etched (Figure 1.15b).

1.7   ADVANCES IN SI PHOTONICS SENSORS

There are several platforms available for the realization of photonic sensors, however Si technology is one of the powerful and promising tools. Most of the research is conducted on Si photonic devices for the telecommunication field and little attention is given toward the sensing applications. Not long ago, Si photonics attracted a lot of attention toward photonic sensing with applications in environmental, chemical, and biomedical sensing. An important advantage of integrated photonic sensors is their miniaturized size, mass producibility, and low cost. In Si technology, it is possible to integrate several building blocks on the same chip, such as light generation, WGs, light encoding, grating, detection, packaging the devices, and smart electronics to control all the building blocks, as presented in Figure 1.16.

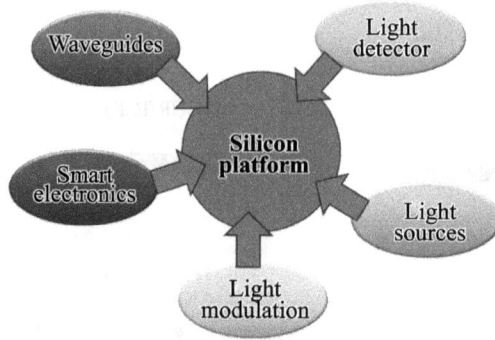

FIGURE 1.16   Integration of different building blocks on a Si platform.

In principle, the light-guiding for sensing applications can be based on WGs as presented in Section 1.2 of this chapter. However, for the realization of a complete photonic lab-on-a-chip microsystem device, several units ought to be integrated on the same platform, i.e. the micro/nanodevices, the flow cells and the flow delivery system, for interferometric sensors, a phase modulation system to convert the periodic output signal in direct phase measurements, integration of the light sources and the photodetectors, and CMOS processing electronics. Here we are presenting SOI-based sensors relying on two types of interrogation methods: wavelength interrogation and intensity interrogation.

1.7.1   WAVELENGTH INTERROGATION METHOD

For label-free detection of analytes, RRs and MZIs are highly sensitive sensors based on the interaction between the guiding light and the analytes through its evanescent field. The light–matter interaction can be enhanced by employing the slot WGs, as the analytes can propagate in the core and can directly interact with the strongly confined light in the low-index region. The sensor signal of these devices results from a refractive index change of the ambient medium. For RR sensors, the signal is a shift of the well-separated and sharp resonance peaks, while for the MZI, it is a shift of the continuous sinusoidal transmission spectrum. SOI slot WG-based RRs have been demonstrated to achieve more than three times larger RR sensitivities compared to conventional ridge WG-based RRs. The history of MZI goes back to 1891 with the pioneering work of Zehnder [26] and Mach [27] on free-space optics interferometers, which can be used for sensing macroscopic samples. Integrated photonics versions of MZIs have also been realized in photonic crystals, using self-collimation of light, where the ratio of the two MZI outputs is the sensing signal.

Refractive index sensors have several applications in the field of biochemical domain and have been intensively studied in recent years, such as solution concentration and pH value, which can be measured on the basis of refractive index change [28, 29]. A semiconductor nanowire sensor for biosensing offers a sensitivity as high as 235 nm/RIU (refractive index unit) [30]. It has been established that the highest experimental sensitivity for quasi-TM RRs was 135 nm/RIU [31]. However, by optimizing the WG thickness, the bulk sensitivity of 270 nm/RIU was demonstrated [32]. This value is within 90% of the maximum achievable sensitivity for a resonator sensor using a ridge WG.

Here, the sensing capabilities of ridge, slot, and a Si-based HPWG that relies on wavelength interrogation method are presented. For this purpose, we simulated RR designs based on the aforementioned WG structures, as presented in Figure 1.17 (first row). The geometric variables of the three designs are kept constant to perform a reasonable comparison for their spectral behavior. The device height (H_{Si}) and the gap (g) between the bus WG and the ring are fixed at 220 nm and 10 nm, respectively. The ring radius is $R = 1500$ nm (from the outer edge), which is similar for all three designs. The width of WGs used in RRs is varied between 300 and 350 nm. For HPWG RR, the Si circular ridge encircles a gold disk in the center. The slot width (s) of 40 nm is used for RRs based on slot and HPWG. And the material filled in the slot is air ($n = 1.0$). Figure 1.17 (second and third rows) presents the E-field mapping and transmission spectrum of a ridge, slot, and HPWG RR at their respective λ_{res}. The measurement of changes in λ_{res} is the most common interrogation method used in RRs. The optical resonances are obtained by loading the ambient medium with a material of $n = 1.03$ for all three designs. The transmission spectrum and E-field mapping are simulated using the finite element method. From the analysis of slot and HPWGs presented in Section 1.2, we can expect that RRs based on these

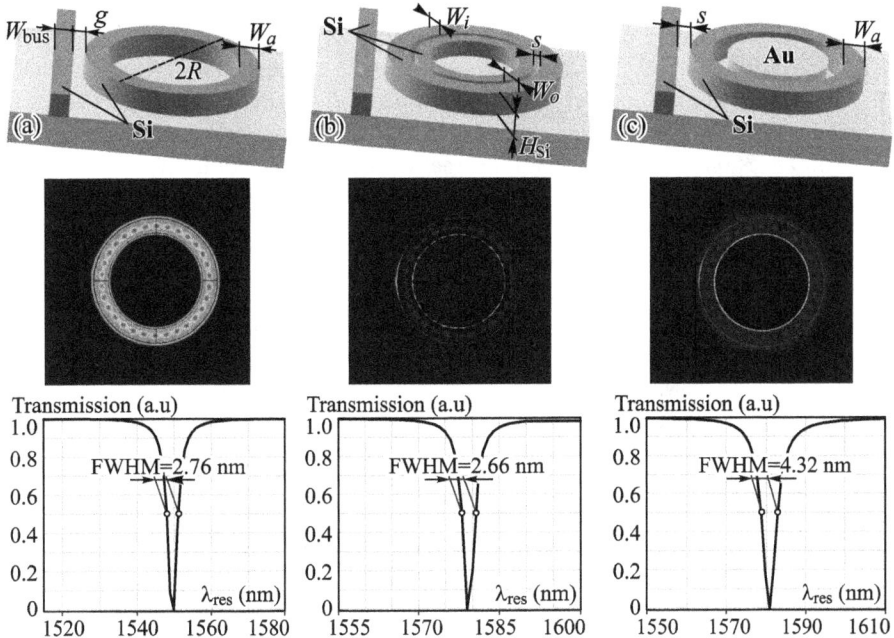

FIGURE 1.17 RR design based on (a) ridge WG, (b) slot WG, (c) HPWG. E-field mapping and transmission spectrum of respective RR designs are presented in the second and third row, respectively.

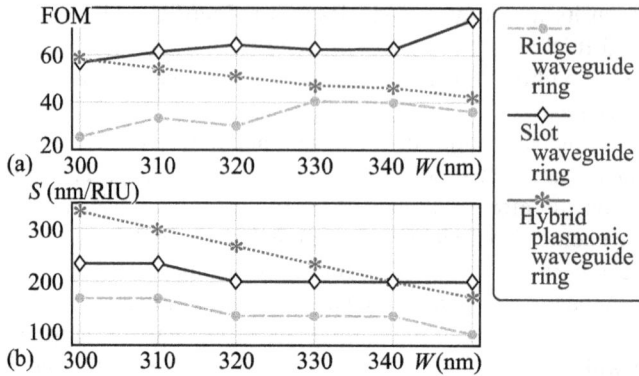

FIGURE 1.18 (a) The Figure of Merit, (b) sensitivity comparison.

WGs can enhance the evanescent field as compared to a ridge WG RR having the same design variables. High sensitivity is always desirable in these sensors, which strongly depends on light polarization, optical loss, and the light–matter interaction. Consequently, elevated sensitivity can be expected [33–35].

In this case, sensitivity is calculated as $S = \Delta\lambda/\Delta n$, where $\Delta\lambda$ represents the change of the resonance wavelength in nm and Δn is the difference of the refractive index in the medium. The FOM is another variable that should also be thought of while constructing the RR sensor. FOM is expressed as $S/FWHM$. In Figure 1.18, the FOM and S of RRs are plotted for the width of the WG for $\Delta n = 0.03$. The HPWG RR offers maximum S as compared to the other two designs at $W = 300$ nm (where $W = W_o$ and W_a for slot and ridge/HPWG, respectively), which deteriorate fast as W increases. The S_{max} drops from 333.33 nm/RIU to 170 nm/RIU when W changes from 300 to 350 nm. This fact signifies that the hybrid plasmonic ring WG RR has less fabrication tolerance as compared to the other two ring schemes. In terms of FWHM, the slot RR has the lowest value, 2.66 nm at $W = 350$ nm, whereas the ridge WG RR has broad, FWHM = 6.65 nm at $W = 300$ nm. Even though the HPWG RR has the highest S, its broad FWHM makes its FOM lower than the slot WG RR design. As we have mentioned earlier, FOM is the ratio between S and FWHM; therefore, a design with a narrow FWHM can have a larger FOM, as presented in Figure 1.18. The FOM_{max} is approximately 75.2, 58.5, and 40.4 for a slot, hybrid, and ridge WG RR, respectively.

1.7.2 INTENSITY INTERROGATION

The ambient refractive index can alter the transmission power of the propagating mode in the WG. Here, we have proposed a transmission decay (dB) of all the proposed WGs at 1,550 nm when the ambient refractive index is varied between 1 and 1.35, as presented in Figure 1.19. The propagation length of the WGs is fixed at 3 μm. The power decay in the transmission for all the WGs under consideration is presented in Table 1.1, where HPWG offers the highest power decay of 2.45 dB when the ambient refractive index is changed from 1 to 1.35.

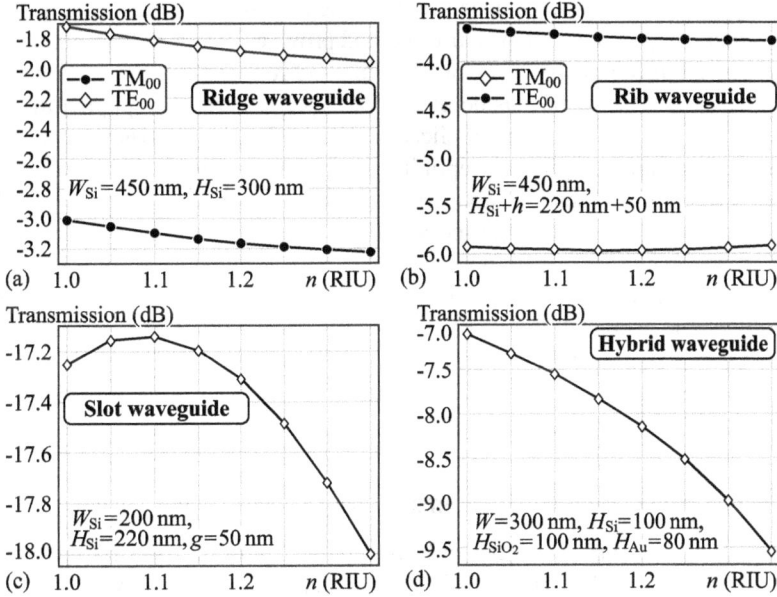

FIGURE 1.19 Transmission versus ambient refractive index in (a) ridge WG, (b) rib WG, (c) slot WG, (d) HPWG.

TABLE 1.1
Comparison of Transmission (dB) and S of Rib, Ridge, Slot, and HPWG

	Rib WG		Ridge WG		Slot WG	HPWG
	TE00	TM00	TE00	TM00	TE	TM hybrid
Type of WG	mode	mode	mode	mode	mode	mode
Transmission (dB)	0.127	0.016	0.234	0.216	0.75	2.45
Sensitivity (dB/RIU)	0.36	0.045	0.67	0.617	2.14	7

The sensitivity is calculated using $\Delta T(dB)/\Delta n$, where ΔT and Δn are the change in transmission power in dB and the change in ambient refractive index, respectively. The maximum sensitivity of rib, ridge, slot, and hybrid WG is obtained at 0.36 dB/RIU, 0.67 dB/RIU, 2.14 dB/RIU, and 7 dB/RIU, respectively. This undoubtedly displays the domination of slot and Si-based-HPWG for sensing applications due to their high light–matter interaction.

1.8 NUMERICAL SIMULATION OF SI RIDGE WAVEGUIDE BY COMBINING A THIN METAL FILM

In Section 1.2, we discussed a few highly recognized Si-based optical WGs that can be used as optical interconnects as well as in sensor applications. We present a new metal-supported Si ridge WG structure that may be utilized in gas-sensing applications. When compared to slot WG, this method provides a high evanescent field ratio (EFR), as well as simplicity of manufacturing and reduced complexity. The ability of slot WG to restrict light outside of the WG core material has piqued interest in chemical and biological sensing applications. Though, in comparison to ridge WG, these structures have a higher susceptibility to sidewall roughness-generated scattering loss, which raises questions about their efficacy. A thin metal layer is introduced as an under cladding in the suggested WG design to restrict light confinement in the substrate area, while the energy is raised in the upper cladding (hereafter represented as UC) region, enhancing the light–matter interaction. Using the 3D-FEM, Γ, power distribution, and propagation loss of the suggested structure and the conventional Si ridge WG are calculated. In comparison to the conventional ridge WG, numerical studies show that the suggested WG formation can achieve significant light confinement in the UC (upper cladding) area.

1.8.1   DESIGN OF A METAL-SUPPORTED SI RIDGE WAVEGUIDE

Figure 1.20 depicts the schematics of a conventional Si ridge WG (Subsection 1.2.1) and a metal-supported Si ridge WG. When compared to silver (Ag), gold (Au) is utilized in this design because of its biocompatibility and resistance to oxidation. A thin SiO_2 layer is inserted between Au and Si ridge WG in a metal-supported WG arrangement. Furthermore, this formation is compatible with current CMOS manufacturing methods and may be built in the following steps. On a Si wafer, an Au layer is deposited, followed by a thin coating of SiO_2. Finally, the Si layer is placed on the surface and etched to create the ridge WG. The layer thickness of Au and SiO_2 is represented by H_{Au} and H_{SiO_2}, respectively. In our example, $H_{Au} = 100$ nm was constant, while H_{SiO_2} was varied to improve the evanescent field in the UC area. H_{Si} and W_{Si} are the height and width of the ridge WG, respectively.

Figure 1.21 reveals the E-field mapping of a TE mode in the conventional Si ridge WG and metal-supported Si ridge WG at $\lambda = 1550$ nm, where $H_{Si} = 220$ nm, $W_{Si} = 400$ nm, $H_{SiO_2} = 50$ nm, and $H_{Au} = 100$ nm. The conventional ridge WG has a symmetric evanescent field, which is why a significant portion of light also dwells in the substrate, as shown in Figure 1.21(a), but the metal-supported ridge WG lifts the evanescent field toward the top cladding region, as seen in Figure 1.21(b). This property makes it easier to improve light–matter interaction, which is useful in gas and biochemical applications.

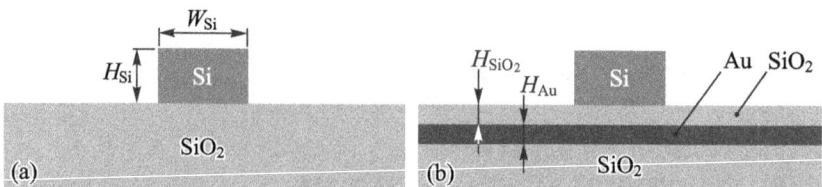

FIGURE 1.20   Schematic of Si ridge WG: (a) conventional design. (b) metal-supported design.

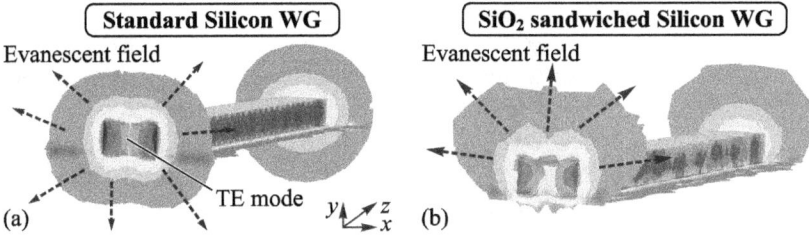

FIGURE 1.21 E-field mapping: (a) conventional WG design, (b) metal-supported WG design [40].

The sensitivity of the device is proportional to Γ, and a significant amount of light confinement in the UC improves the sensitivity. When the optical confinement is substantial, the concentration variation of a cover medium, for example an aqueous solution, has a significant impact on the effective refractive index. As a result, EFR is a crucial characteristic in the development of gas sensors established on evanescent field absorption. Sensors having a high EFR can intermingle with the absorbing medium more, improving the S of the sensor [36–39].

Figure 1.22 shows the power distribution in the conventional and metal-supported ridge core, cladding, and substrate for TE mode at 1,550 nm. The height and width of

FIGURE 1.22 The power distribution of the conventional (a) and metal-supported Si ridge WG (b–d) at 1,550 nm, where $W_{Si} = 400$ nm and $H_{Si} = 220$ nm [40].

the ridge WG are set at 200 nm and 400 nm, respectively, for both the WG configurations. Figure 1.22(a) shows that the maximum power in the UC for the conventional WG is 34.27%, whereas 18.31% power leaks to the substrate. The power distribution in a metal-supported WG is determined by the thickness of the SiO_2 layer inserted between the Au and Si ridge WG. For instance, when H_{SiO2} = 50 nm, the maximum power of 44% is obtained in the UC and only 9.77% power goes to the substrate, as shown in Figure 1.22(b). With an expansion in the SiO_2 cladding layer, the power in the UC drops (Figure 1.22c, d).

As a result, using a metal layer lowers the light leakage to the substrate. The light confinement in the suggested structure's UC region is greater than that of the conventional arrangement. The metal's high reflection allows for strong confinement to be achieved. By changing the WG's size, as discussed in the following section, one may build the WG for several combinations of power distribution.

1.8.2 THE CONNECTION BETWEEN THE WAVEGUIDE DIMENSION AND Γ

To address the impact of light confinement in the cladding area, the WG dimensions and distance between the WG core and the Au layer are critical factors. As a result, the impact of both these variables is explored. The Γ in WG cladding, core, and substrate is shown versus H_{SiO_2} in Figure 1.23(a–c) for thickness range of 50 nm to 100 nm, respectively.

At H_{SiO_2} = 50 nm, the cladding area achieves the highest confinement. Because these WGs are intended for TE polarization, the Γ is heavily influenced by the WG width (W_{Si}). To investigate the power distribution in the WG, we set H_{Si} at 220 nm

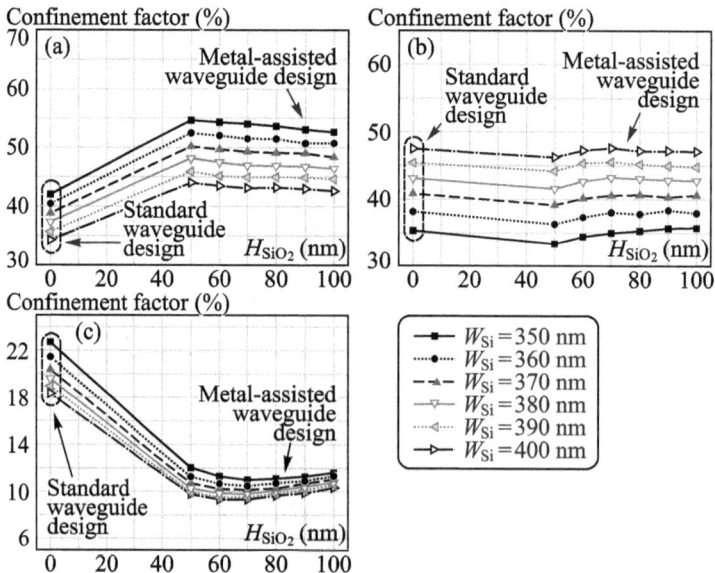

FIGURE 1.23 Influence of SiO_2 layer on Γ of (a) cladding region, (b) core, and (c) substrate [40].

and changed W_{Si} to between 350 nm and 400 nm. For two reasons, the optical Γ in the cladding area decreases when H_{SiO_2} and W_{Si} increase: (1) due to the large WG size, the TE mode is more restricted in the core region, and (2) the influence of metal reflection diminishes, as illustrated in Figure 1.23(a). Furthermore, it is worth noting that metal does not affect the light Γ in the core (Figure 1.23b). When W_{Si} is small, it even amplifies a few percentage points of light in the core. In metal-supported WGs, the majority of the light from the substrate moved toward UC, as illustrated in Figure 1.23(c). As a result, we can claim that for conventional and metal-supported WG, 10% power is improved in the UC at any given dimensions.

1.8.3 INFLUENCE OF METAL LAYER ON THE PROPAGATION LOSS

As the metal is incorporated into the suggested arrangement, it is necessary to evaluate the losses caused by the metal's optical absorption. When estimating the propagation loss, the WG dimensions, SiO_2 layer thickness, and Au thickness must all be considered. The following equation is used to determine the propagation loss: $10 \times \log(P_{out}/P_{in})$/propagation length in microns, where P_{in} and P_{out} are the input and output power of the WG, respectively. The propagation loss of the conventional and metal-supported ridge WG are computed for various values of H_{Au} and H_{SiO_2}, as shown in Figure 1.24.

As previously stated, $H_{Au} = 100$ nm is constant, but H_{SiO_2} ranged from 50–100 nm with a step size of 25 nm. The propagation loss tends to decrease when H_{SiO_2} grows from 50 nm to 100 nm, as seen in Figure 1.24. This is because the metal layer is well separated from the WG core, and as a result, the metal's absorption is reduced. It can, however, impact the WG's EFR. As a result, there is always a trade-off between propagation loss and EFR. We also discovered that H_{Au} had the smallest impact on losses. As a result, a thin layer of Au, such as 25 nm, may be employed in the sensor design from an inexpensive standpoint.

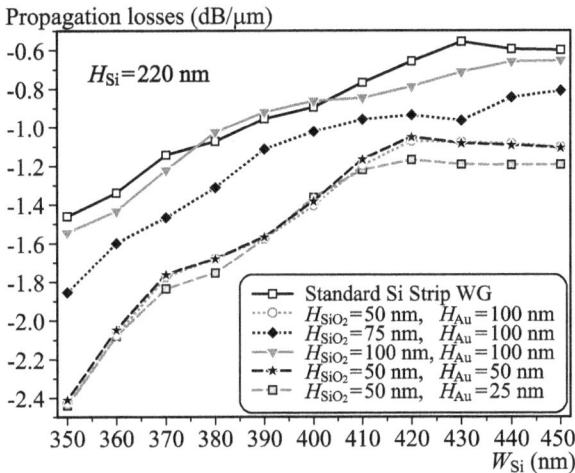

FIGURE 1.24 The propagation loss of conventional and metal-supported ridge WG [40].

1.9   CONCLUDING REMARKS

This chapter has discussed several attributes of Si photonics that can be fabricated by direct patterning of Si to create optical interconnects to broadcast data-carrying laser signals. The biggest advantage is that it can transmit gigantic data while consuming low power with no heating up or causing any deterioration in the signal. The modal characteristics are studied via the finite element method (FEM), which give a general guideline of different types of WGs. The effective refractive index, Γ, and losses are thoroughly studied. Based on their analysis, one can choose an appropriate WG arrangement for on-chip communication and sensing applications. The manufacturing of optical elements is also discussed in detail, which provides a general idea of the technology used for this platform. There are many attractive attributes connected to the manufacturing of optical devices on an SOI platform, such as low cost of the material; mature and well-developed processing technique, development, and manufacturing in the microelectronics industry; and the prospect for easy integration with electrical components in the same substrate. The sensing performance of the aforementioned WGs is presented, which shows the supremacy of slot and Si-based HPWG over conventional WG formations. At the end of the chapter, a novel WG scheme is proposed that utilizes a metal layer buried under an oxide layer that separates metal from the WG core. This scheme suppresses the light confinement in the substrate region, whereas the mode is elevated in the UC region, which enhances the light–matter interaction. This kind of WG scheme can be used as a highly sensitive evanescent field gas absorption sensor.

REFERENCES

1. Zhang, Y. et al. 2018. High transparent mid-infrared silicon "window" decorated with amorphous photonic structures fabricated by facile phase separation. *Opt. Express.* 26(14):18734–48. https://doi.org/10.1364/OE.26.018734.
2. Reed, G. T., and Knights, A. P. 2008. *Silicon photonics: The state of the art.* Wiley. ISBN: 978-0-470-02579-6.
3. Abel, S. et al. 2019. Large Pockels effect in micro-and nanostructured barium titanate integrated on silicon. *Nat. Mater.* 18:42–7. https://doi.org/10.1038/s41563-018-0208-0.
4. Dong, P. et al. 2010. Low loss shallow-ridge silicon waveguides. *Opt. Express.* 18(14):14474–9. https://doi.org/10.1364/OE.18.014474.
5. Butt, M. A., Kozlova, E. S., and Khonina, S. N. 2017. Conditions of a single-mode rib channel waveguide based on dielectric TiO_2/SiO_2. *Computer Optics.* 41(4):494–8. https://doi.org/10.18287/2412-6179-2016-40-4-494-498.
6. Soref, R. A., Schmidtchen, J., and Petermann, K. 1991. Large single-mode rib waveguides in GeSi-Si and Si-on-SiO_2. *IEEE J. Quantum Electron.* 27(8):1971–4. https://doi.org/10.1109/3.83406.
7. Almeida, V. R., Xu, Q., Barrios, C. A., and Lipson, M. 2004. Guiding and confining light in void nanostructure. *Opt. Lett.* 29(11):1209–11. https://doi.org/10.1364/OL.29.001209.
8. Wang, X. et al. 2013. Silicon photonic slot waveguide Bragg gratings and resonators. *Opt. Express.* 21(16):19029–39. https://doi.org/10.1364/OE.21.019029.
9. He, X. et al. 2018. Ultralow loss graphene-based hybrid plasmonic waveguide with deep-subwavelength confinement. *Opt. Express.* 26(8):10109–18. https://doi.org/10.1364/OE.26.010109.

10. Zenin, V. A. et al. 2017. Hybrid plasmonic waveguides formed by metal coating of dielectric ridges. *Opt. Express.* 25(11):12295–302. https://doi.org/10.1364/OE.25.012295.
11. Penades, J. S. et al. 2016. Suspended silicon mid-infrared waveguide devices with subwavelength grating metamaterial cladding. *Opt. Express.* 24(20):22908–16. https://doi.org/10.1364/OE.24.022908.
12. Ang, T. W., Reed, G. T., and Vonsovici, A. et al. 1999. Grating couplers using silicon on insulator. *Proc. SPIE.* 3620:79–86.
13. Ang, T. W., Reed, G. T., and Vonsovici, A. et al. 1999. Blazed-grating coupler in unibond SOI. *Proc. SPIE.* 3896:360–8. https://doi.org/10.1117/12.370333.
14. Pollock, C. R., and Lipson, M. 2003. *Integrated photonics.* New York: Springer Science+Business Media. ISBN: 978-1-4419-5398-8.
15. Liu, J.-M. 2005. *Photonic devices.* Cambridge: Cambridge University Press. ISBN: 978-0-521-55195-3.
16. Rickman, A. G., Reed, G. T., and Namavar, F. 1994. Silicon-on-insulator optical rib waveguide loss and mode characteristics. *J. Lightw. Technol.* 12(10):1771–6. https://doi.org/10.1109/50.337489.
17. Selvaraja, S. K. et al. 2014. Highly uniform and low-loss passive silicon photonics devices using a 300 mm CMOS platform. *Optical Fiber Communication Conference:* Th2A.33. https://doi.org/10.1364/OFC.2014.Th2A.33.
18. Kobayashi, N., Sato, K., Namiwaka, M., Yamamoto, K., Watanabe, S., Kita, T., Yamada, H., and Yamazaki, H. 2015. Silicon photonic hybrid ring-filter external cavity wavelength tunable lasers. *J. Lightw. Technol.* 33(6):1241–6. https://doi.org/10.1109/JLT.2014.2385106.
19. Gao, F., Wang, Y., Cao, G., Jia, X., and Zhang, F. 2005. Improvement of sidewall surface roughness in silicon-on-insulator rib waveguides. *Appl. Phys. B.* 81:691–4. https://doi.org/10.1007/s00340-005-1951-x.
20. Baehr-Jones, T., Hochberg, M., Walker, C., and Scherer, A. 2005. High-q optical resonators in silicon-on-insulator based slot waveguides. *Appl. Phys. Lett.* 86(8):081101. https://doi.org/10.1063/1.1871360.
21. Wang, G., Baehr-Jones, T., Hochberg, M., and Scherer, A. 2007. Design and fabrication of segmented, slotted waveguides for electro-optic modulation. *Appl. Phys. Lett.* 91(14):143109. https://doi.org/10.1063/1.2793618.
22. Spott, A. et al. 2011. Photolithographically fabricated low-loss asymmetric silicon slot waveguides. *Opt. Express.* 19(11):10950–8. https://doi.org/10.1364/OE.19.010950.
23. Zhang, L. 2014. Silicon process and manufacturing technology evolution: An overview of advancements in chip making. *IEEE Consum. Electron. Mag.* 3(3):44–8. https://doi.org/10.1109/MCE.2014.2317896.
24. Tiginyanu, I., Ursaki, V., and Popa, V. 2011. *Nanocoatings and ultra-thin films: Technologies and applications.* Sawston, Cambridge: Woodhead Publishing Limited. ISBN: 978-1-84569-812-6.
25. Jansen, H. et al. 1996. A survey on the reactive ion etching of silicon in microtechnology. *J. Micromech. Microeng.* 6:14–28. https://doi.org/10.1088/0960-1317/6/1/002.
26. Zehnder, L. 1891. Ein neuer Interferenzrefraktur. *Z. Instumentenkd.* 11:275–85.
27. Mach, L. 1892. Uber einen Interferenzfractur. *Z. Instrumentenkd.* 12:89–93.
28. Vaiano, P. et al. 2016. Lab on fiber technology for biological sensing applications. *Laser Photonics Rev.* 10(6):922–61. https://doi.org/10.1002/lpor.201600111.
29. Guo, T. 2017. Fiber grating-assisted surface plasmon resonance for biochemical and electrochemical sensing. *J. Lightw. Technol.* 35(16):3323–33. https://doi.org/10.1109/JLT.2016.2590879.
30. Wang, Y. L. et al. 2017. Refractive index sensor based on leaky resonant scattering of single semiconductor nanowire. *ACS Photonics.* 4(3):688–94. https://doi.org/10.1021/acsphotonics.7b00064.

31. Xu, D. X. et al. 2010. Real-time cancellation of temperature induced resonance shifts in SOI wire waveguide ring resonator label-free biosensor arrays. *Opt. Express.* 18(22):22867–79. https://doi.org/10.1364/OE.18.022867.

32. Fard, S. T., Schmidt, S., Shi, W., and Wu, W. X. 2017. Sensitivity of silicon-on-insulator (SOI) strip waveguide resonator sensor. *Biomed. Opt. Express.* 8(2):500–11. https://doi.org/10.1364/BOE.8.000500.

33. Butt, M. A., Khonina, S. N., and Kazanskiy, N. L. 2019. A Serially cascaded micro-ring resonator for simultaneous detection of multiple analytes. *Laser Phys.* 29(4):046208. https://doi.org/10.1088/1555-6611/ab0371.

34. Butt, M. A., Khonina, S. N., and Kazanskiy, N. L. 2018. Hybrid plasmonic waveguide-assisted metal-insulator-metal ring resonator for refractive index sensing. *J. Mod. Opt.* 65(9):1135–40. https://doi.org/10.1080/09500340.2018.1427290.

35. Butt, M. A., Khonina, S. N., and Kazanskiy, N. L. 2018. Highly sensitive refractive index sensor based on hybrid plasmonic waveguide microring resonator. *Waves Random Complex Media.* 30(2):292–9. https://doi.org/10.1080/17455030.2018.1506191.

36. Butt, M. A., Khonina, S. N., and Kazanskiy, N. L. 2018. Silicon on silicon dioxide slot waveguide evanescent field gas absorption sensor. *J. Mod. Opt.* 65(2):174–8. https://doi.org/10.1080/09500340.2017.1382596.

37. Butt, M. A., Khonina, S. N., and Kazanskiy, N. L. 2018. Modelling of Rib channel waveguides based on silicon-on-sapphire at 4.67 μm wavelength for evanescent field gas absorption sensor. *Optik.* 168:692–7. https://doi.org/10.1016/j.ijleo.2018.04.134.

38. Butt, M. A., Degtyarev, S. A., Khonina, S. N., and Kazanskiy, N. L. 2017. An evanescent field absorption gas sensor at mid-IR 3.39 μm wavelength. *J. Mod. Opt.* 64(18):1892–7. https://doi.org/10.1080/09500340.2017.1325947.

39. Butt, M. A., Khonina, S. N., and Kazanskiy, N. L. 2019. Enhancement of evanescent field ratio in a silicon strip waveguide by incorporating a thin film. *Laser Phys.* 29(7):076202. https://doi.org/10.1088/1555-6611/ab1414.

40. Butt, M. A., Khonina, S. N., and Kazanskiy, N. L. 2019. Sensitivity enhancement of silicon strip waveguide ring resonator by incorporating a thin metal film. *IEEE Sens. J.* 20(3):1355–62. https://doi.org/10.1109/JSEN.2019.2944391.

2 Photonic Crystal Cavities in Integrated On-Chip Optical Signal Processing Components

Pavel Grigoryevich Serafimovich[1] and
Nikolay Lvovich Kazanskiy[1,2]
[1]Image Processing Systems Institute – Branch of the
Federal Scientific Research Centre "Crystallography and
Photonics" of Russian Academy of Sciences, Samara, Russia
[2]Samara National Research University, Samara, Russia

2.1 INTRODUCTION

Optical signal processing components, such as optical modulators, sensors, all-optical integrators, and differentiators implemented on a chip, are important for developing computer technology. All-optical, fully integrated, on-chip computing components will increase the speed of information processing by several orders of magnitude. Moreover, such components enable the processing of not only real but also complex values. In this regard, it is important to implement the basic computing operations optically. In recent years, all-optical integrators and differentiators based on Bragg gratings [1] and ring resonators [2] have been proposed. Such elements can be used in both digital and analog signal processing. Among the digital signal-processing applications are the use of optical integrators and differentiators as pulse counters and ultrafast memory elements [3]. Analog signal-processing applications include all-optical solution of differential equations of various orders [4].

Integrators and differentiators based on Bragg gratings are a few millimeters in size. Integrators and differentiators based on ring resonators are more compact. Their size is on the order of tens of micrometers on the chip plane. Photonic crystal structures are currently the subject of intensive research [5–7]. In this chapter, we numerically describe and study the most compact optical integrators and differentiators based on photonic crystal (hereafter represented as PC) cavities [8].

The model of a two-component nanocavity is also described. In this model, the minimum details of the structure are found only in the periodic component of the resonator. The advantages of such a structure include a promising way to construct an electrically pumped photonic cavity, the ease of introducing nonlinear optical materials in the nanocavity, the possibility of formation of the desired energy distribution in the far zone, and the possibility to develop dynamic systems based on nanocavities.

DOI: 10.1201/9781003439165-2

The optical modulator is one the most important optical signal processing components. In recent years, these devices have been significantly improved [9]. Nevertheless, some problems remain. In particular, the size of the device and the structural complexity must be reduced to provide functional flexibility and enable effective electrical modulation. Photonic crystal structures are widely used as platforms for implementing various nanophotonic elements [10]. Several optical modulators based on a single PC cavity structure have been proposed, including high-frequency modulating PC lasers [11], thermal tuning PC modulators [12], and optical [13, 14] and electrical [15] pumping PC modulators. An electro-optic modulator constructed with two-dimensional (2D) PC slab cavity has been described in [16]. A two-component structure with several 2D PC slab cavities has been used in [17] to increase the frequency bandwidth of the device by employing wavelength division multiplexing. Electro-optic modulators built with a PC nanobeam cavity have been demonstrated in [18, 19]. The switching energy is an important characteristic of any electro-optic modulator. The value of this energy depends, in particular, on the resistance and the capacitance of the device. Reducing the footprint of the modulator decreases the resistance and capacitance values and therefore diminishes the total energy consumption. The footprint of the PC nanobeam cavity is, as a rule, several times smaller than the one of the 2D PC slab cavity. Thus, a PC nanobeam cavity has the potential to exhibit reduced switching energy compared to its 2D PC counterpart [19].

Today, much attention attracts the developing of the refractive index sensors based on optical microcavities [20–22]. Such sensors are used, for example, in biological [23] and chemical investigations [24], when measuring the temperature and mechanical stresses in acoustics. The values of the Q-factor of the cavities at the level of 10^6 and higher are reached. The application of the active cavities with light and electron pumping is the way to increase the values of the Q-factor. The high values of the Q-factors of the optical cavities ensure the high sensitivity of the sensors on their basis.

Moreover, the special features of the resonant sensors' functioning offer the possibility to obtain the dynamic response in real time, as well as the wide range of the examined samples. The samples, for example, can be different liquid or gaseous chemical compounds or biomolecules. The optical resonant sensors demonstrate good compatibility with the requirements of the microfluidics. All these make them suitable for using as a part of integrated solutions in the forms of biochips or "lab-on-a-chip." Another advantage of the optical resonant sensors is the eliminating of the need to use the fluorescent tags.

2.2   TEMPORAL INTEGRATION OF OPTICAL SIGNALS

Figure 2.1 shows a scheme of the coupled-resonator optical waveguide (CROW). The variable a_i, $i = [1, N]$ is the complex amplitude of the resonant mode in the i-th resonator; κ_{i-1} and κ_i, $i = [1, N]$, are the left and right coupling coefficients of the i-th resonator, respectively; r_i, $i = [1, N]$ is the energy loss of the i-th resonator to the exterior space; and p_{in}, p_{rf}, and p_{tr} are the amplitude of the input, reflected, and transmitted fields, respectively.

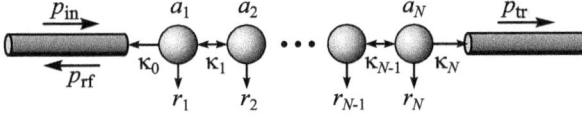

FIGURE 2.1 Scheme of coupled-resonator optical waveguide.

Consider an array in which the resonators have identical resonant frequencies. Then, according to the temporal coupled-mode theory, the transmission function of the system can be written as:

$$T_N(s) \equiv \frac{p_{tr}}{p_{in}} = -\frac{2(-i)^{N-1}\sqrt{\kappa_0\kappa_N}\,\kappa_1\kappa_2\cdots\kappa_{N-1}}{\det(\mathbf{M})},$$ (2.1)

where $\mathbf{M}$ is the corresponding tridiagonal matrix [9],

$$p_{tr} = -i\sqrt{2\kappa_N}\,a_N = -2\sqrt{\kappa_0\kappa_N}\left[\mathbf{M}^{-1}\right]_{N,1} p_{in},$$

and $\det(\mathbf{M})$ is the determinant of the matrix $\mathbf{M}$.

For $N = 1$, Eq. (2.1) is reduced to the form

$$T_1(s) = -\frac{2\kappa_0}{s+2\kappa_0}.$$ (2.2)

Hereafter, for simplicity, we neglect the losses to the exterior space.

Let us consider how accurately Eq. (2.2) approximates the integrator of the first order. The polarized electric field with envelope $P_{in}(t)$ can be written as:

$$E(x,t) = P_{in}\left(t - x/v_g\right)\exp(im_0 x - i\omega_0 t) =$$
$$= \int_{-\infty}^{\infty} R(\omega - \omega_0)\exp(im(\omega)x - i\omega t)d\omega,$$ (2.3)

where $R(\omega)$ is the envelope spectrum signal, $m(\omega)$ is the wave number [$m_0 = m(\omega)$], and v_g is the group velocity.

A linear system described by the complex transfer function (TF) $H(\omega)$ converts the envelope of the input pulse [Eq. (2.3)] to

$$P_{tr}(t) = \int_{-\infty}^{\infty} R(\omega)H(\omega)\exp(i\omega t)d\omega = P_{in}(t)*h(t),$$ (2.4)

where the symbol "*" denotes the convolution operation, and $h(t)$ is the spectrum of the TF $H(\omega)$.

The impulse response of a linear system with the TF $T_1(s)$ is:

$$h_1(t) = -\kappa_0 \exp(-\kappa_0 t) u(t), \tag{2.5}$$

where $u(t)$ is the Heaviside step function.

Substituting Eq. (2.5) into Eq. (2.4), we obtain an expression for the envelope of the output pulse:

$$P_{tr}(t) = -\kappa_0 \int_{-\infty}^{t} P_{in}(T) \exp(-i\kappa_0(t-T)) dT. \tag{2.6}$$

The right side of this equation expresses the integral of the input pulse envelope with exponential weight.

Figure 2.2 shows the result of integration of the envelope of an optical pulse with a duration of 100 ps by resonators with Q-factors of 3×10^4 and 5×10^4. The Q-factor is related to κ_0 by the ratio $Q = \omega_0/(4\kappa_0)$. Figure 2.2 shows that the higher the Q-factor of the resonator is, the more slowly the integrated signal envelope decays. For resonators with Q-factors of 3×10^4 and 5×10^4 we estimate an integration time window (defined as the decay time required to reach 80% of the maximum intensity) of 12.5 ps and 19.5 ps, respectively.

For $N = 2$, Eq. (2.1) can be written as

$$T_2(s) = \frac{i2\kappa_0\kappa_1}{(s+\kappa_0)^2 + \kappa_1^2}. \tag{2.7}$$

Let us calculate the parameters of the PC nanobeam cavity that integrates the optical signal. Compared with the resonators in the 2D PC layer [25], PC nanobeam cavities [26] have a smaller area and are naturally integrated into the waveguide geometry of the chip.

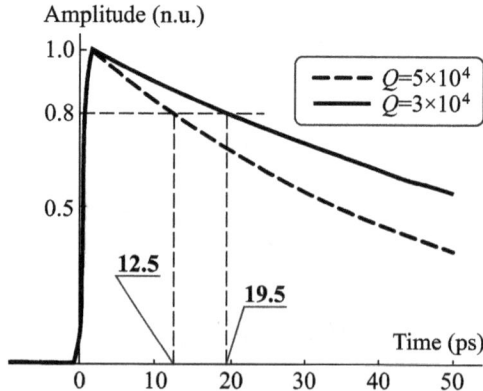

FIGURE 2.2 Result of first-order integration of optical pulse with duration of 1 ps by resonators with Q-factors of 3×10^4 and 5×10^4.

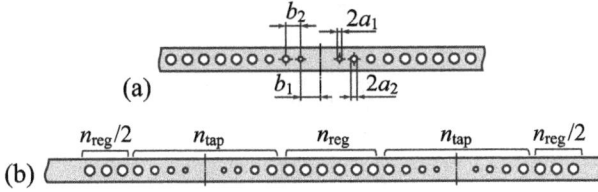

FIGURE 2.3 Schemes of (a) PC nanobeam cavity and (b) array of two such cavities.

Figure 2.3(a) illustrates a possible embodiment of a resonator based on a PC nanobeam. The decreasing radius of holes in the tapering region forms a defect in which the resonant mode is excited. An array of two PC resonators is shown in Figure 2.3(b), where n_{tap} is the number of holes in the tapering region. The coupling value between resonators in the array is determined by n_{reg}, the number of holes with the maximum radius between defects.

The resonance cavity characteristics were computed using the parallel three-dimensional (3D) finite-difference time-domain method. The waveguide in our simulations has a width of 490 nm and a height of 220 nm. It is composed of silicon (Si) and deposited on silica substrate. Air-filled holes in the regular part of the waveguide have a radius of 100 nm and are spaced 330 nm apart. The lattice parameters and the radii of the holes near the defect ($n_{tap} = 12$) are (in nm): $a_1 = 40$, $b_1 = 255$, $a_2 = 55$, $b_2 = 350$, $a_3 = 65$, $b_3 = 365$, $a_4 = 75$, $b_4 = 375$, $a_5 = 85$, $b_5 = 385$, $a_6 = 95$, $b_6 = 395$. These parameters demonstrate the existence of an energy bandgap for transverse electric polarization in the waveguide. The left and right borders of the bandgap are 1.46 μm and 1.67 μm, respectively. The size of the bandgap is about 210 nm. The resonant mode wavelength (1.57 μm) is placed in the center of this region. The free spectral range (FSR) of the cavity can be estimated to 100 nm (~12.5 THz). The Q-factor of the resonator is 3.6×10^4, and $n_{reg} = 6$. The integration time window for the first-order integrator based on such cavity is about 14 ps. Figure 2.7 shows the frequency response of the integrator.

The time-bandwidth product (TBP) of the integrator can be defined as the multiplication of the integration time window (in ns units) and the FSR of the cavity (in GHz units) [27]. Thus, the TBP of the calculated integrator is $0.014 \times 12,500 = 175$. That value is comparable to the TBP of the passive integrators based on Bragg grating [1] and ring resonator [2]. The suggested integrator based on PC cavity has a large potential of increasing of the achieved TBP value. At least two methods can be proposed. Both of them are related to increasing of Q-factor of the resonator while the size of the bandgap remains the same. The first one is based on the fine-tuning of the tapering region and increasing of n_{reg}. By this way, the Q-factor value achieved in this work can be enhanced by several orders of magnitude [28]. Another way is to use an active cavity [29]. The development of the active photonic crystal cavities with the optical and electrical pumping is discussed in Sections 2.2 and 2.3, respectively.

Figure 2.4(a) shows the result of integration of the first derivative of a Gaussian pulse with a duration of 150 fs. The Q-factor of the resonator is 3.6×10^4, and $n_{reg} = 6$. The result of integration of the second derivative of a pulse with the same duration by

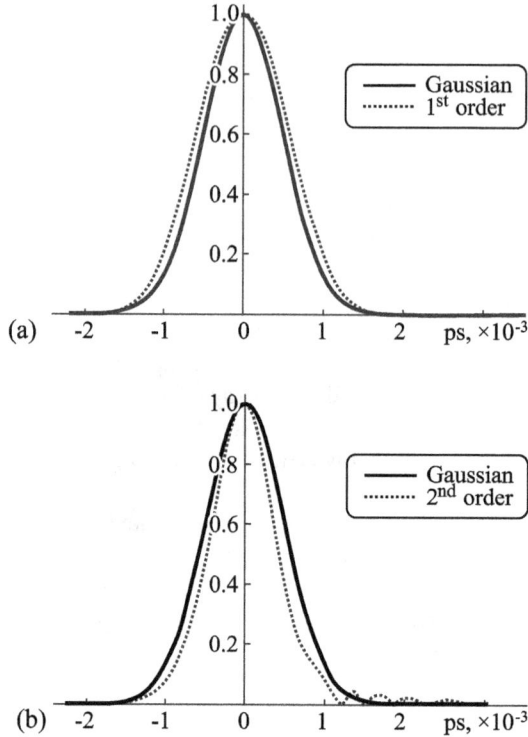

FIGURE 2.4 Results of integration of corresponding derivatives of Gaussian pulse with duration of 150 fs by (a) one PC resonator and (b) an array of two PC resonators.

second-order integrator is shown in Figure 2.4(b). The second-order integrator consists of two resonators, as shown in Figure 2.3(b). The coupling coefficients between the resonators in each integrator are given by Eq. (2.8).

PC cavities can also be used as optical differentiators. This application is described in [30].

2.3 IN-PLANE OPTICAL PUMPING OF PHOTONIC CRYSTAL CAVITY

The cavity's Q-factor can be defined as:

$$Q \equiv 2\pi \frac{E_0}{E_1}, \tag{2.8}$$

where E_0 is the quantity of energy in the cavity and E_1 is the quantity of energy lost per oscillation. The cavity's mode volume V_m is defined as [31]:

$$V_m \equiv \frac{\int dV \varepsilon |E|^2}{\varepsilon_{max} \max\left[|E|^2\right]} \left(\frac{n_{max}}{\lambda}\right)^3, \tag{2.9}$$

where the integration is performed over the cavity volume; ε_{max} and n_{max}, respectively, denote the permittivity and the refractive index at the point of the maximal value of the square of the electric field modulus $|E|^2$; and λ is the wavelength of light.

In an optical cavity, light is confined in the form of one or several resonance modes. Note that the nonlinear optical transformations require that the mode spatial overlap should be maximal. For two orthogonally polarized resonance modes, the coefficient of spatial overlap of orthogonal resonance modes can be derived from the formula:

$$\gamma \equiv \frac{\varepsilon_{NL} \int\limits_{NL} dV \sum\limits_{i,j,i\neq j} E_{TE,i} E_{TM,j}}{\sqrt{\int dV \varepsilon |E_{TE}|^2} \sqrt{\int dV \varepsilon |E_{TM}|^2}}, \tag{2.10}$$

where $\int_{NL}$ denotes the integration over the volume of a nonlinear material in the resonance cavity. $E_{TE,i}$, $E_{TM,j}$ are the electric field components of the orthogonal resonance modes.

Two orthogonally polarized modes can be excited in such a nanocavity, with the modes overlap coefficient γ found in the range 0.76–0.78 [32]. The disadvantages of the nanocavity include a large waveguide thickness (about three operating wavelengths) and a complicated process of independently tuning the frequencies of two resonance modes.

In such a structure, there are two separate light input channels for two orthogonally polarized resonance modes, thereby simplifying the implementation of optical switches [16]. Note that the waveguide thickness is not larger than a quarter of the operating wavelength. Besides, the structure under discussion provides for the flexible tuning of frequencies corresponding to two orthogonal resonance modes. However, the overlap coefficient γ, defined according to Eq. (2.3), is not larger than 0.07.

Two general approaches can be used to construct nonlinear optical elements. For structures studied in [32, 33] the PC waveguide itself should be fabricated from an optically nonlinear material to employ nonlinear effects. Under another approach, nonlinear materials are introduced into the cavity [34, 35] and its shape is optimized. In this chapter, we use the latter approach and propose a resonance cavity that has the coefficient γ [33].

In the ridge PC waveguides, the total internal reflection (TIR) impedes the propagation of light in the transverse directions. As far as the longitudinal direction is concerned, it is the PC that ensures the propagation of light in the nanocavity [36].

To design an optical nanocavity with high Q-factor, we employ a structure composed of three parts (Figure 2.5a). First, it includes PC mirrors containing identical equidistant holes in the waveguides. The operating wavelength is assumed to be 1.55 μm. The crossing waveguides each have width $w = 530$ nm and height $h = 227$ nm. The waveguide holes of radius 95 nm are air-filled and located 365 nm apart. Such geometric parameters provide the formation of a bandgap for the TE and TM waves in the horizontal and vertical waveguides, respectively. The region under simulation contains five holes in each of four cavity arms. The cavity is made from Si ($n = 3.48$) placed in the ambient air.

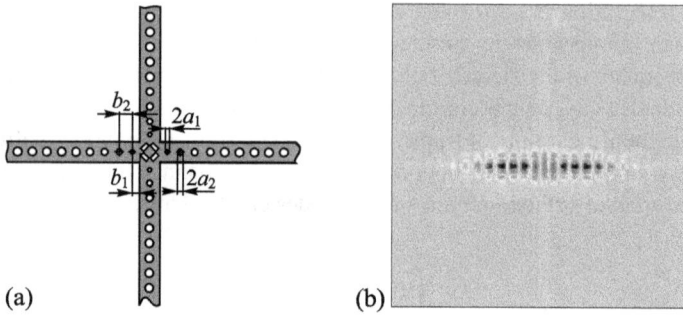

(a)   (b)

FIGURE 2.5 Schematic structure of the optical nanocavity under study (a) and the distribution of the E-field modulus for one of the orthogonal modes (b).

TABLE 2.1

Positions and Radii of the Holes in the Transition Region for a Basic Resonance Cavity (Nm)

a_1	b_1	a_2	b_2	a_3	b_3	a_4	b_4	a_5	b_5
50	285	55	400	65	400	75	400	85	390

The second system's component is a transition region between the PC waveguide and the resonance cavity region, which serves to reduce the energy loss in the cavity. The holes of designed radii are arranged in it in such a manner as to provide the maximal Q-factor. The transition region contains five air-filled holes. The holes' radii and spacing between them are given in Table 2.1. In Table 2.1, the hole nearest to the cavity of radius a_1 is offset from the cavity by the distance b_1 (Figure 2.5a).

The third system's component is the proper resonance cavity. An optical resonance cavity like that shown in Figures 2.5(a) and 2.6(a) [33]. What we have changed in the cavity under study is the waveguide's geometric parameters and the holes' shape (which were assumed to be rectangular in [33]).

The resonance cavity characteristics were calculated using the parallel FDTD (finite difference time domain) method [37]. Absorbing layers were put at the boundaries of the 3D region under calculation. The computational mesh spacing was chosen in such a way as to obtain a converging solution.

Figure 2.5(b) shows the distribution of the E-field modulus for one of the orthogonal modes. The cavity's Q-factor is ~9,000.

Figure 2.6(a) depicts the configuration of the basic resonance cavity. A detailed pattern of the E-field modulus of the resonance mode is depicted in Figure 2.6(b). With the geometry of the horizontal and vertical waveguides being identical, it is only possible to generate the degenerate resonance modes in the form of TE and TM waves. However, as reported [33], the use of waveguides with different geometry enables the excitation of nondegenerate resonance modes. For the degenerate resonance mode shown in Figures 2.5(b) and 2.6(b), the value of γ is 0.03, which is equal to that reported in [33].

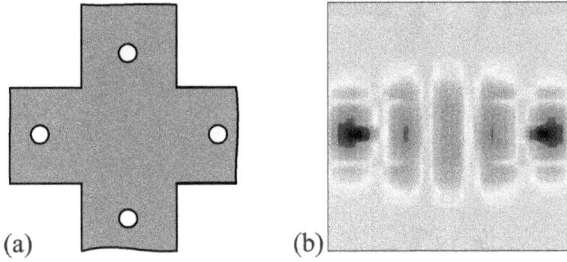

FIGURE 2.6 (a) A detailed scheme of the basic resonance cavity and (b) the pattern of the E-field modulus for the resonance mode.

The resonance cavity regions that contain the nonlinear material are characterized by a smaller refractive index than that of the waveguide. With this approach, the electromagnetic waves can efficiently be concentrated in low refractive-index regions of the resonance cavity. This results in the enhanced interaction of light with the filling material of the said regions. In addition, the waveguide internal energy loss, e.g. due to two-photon light absorption in Si, is also reduced.

Next, the resonator shown in Figure 2.7(a) is investigated. There are four holes of radius 100 nm in the resonance cavity's central region. The diagonal distance between the holes is 245 nm. The holes' positions and radii have been calculated from the condition of the maximal Q-factor. Table 2.2 gives the positions and radii of the transition region holes.

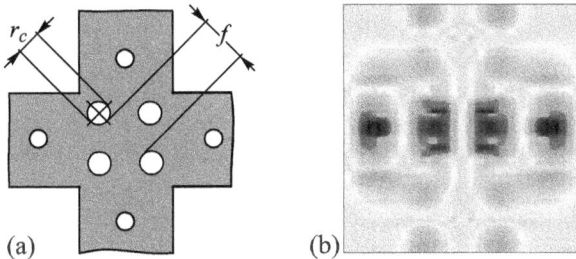

FIGURE 2.7 (a) Schematic of a resonance cavity with circular holes and (b) the pattern of the resonance mode E-field modulus.

TABLE 2.2

Positions and Radii of the Transition Region Holes for the Cavity in Figure 2.3(a) (Nm)

a_1	b_1	a_2	b_2	a_3	b_3	a_4	b_4	a_5	b_5
50	245	55	395	65	395	75	395	85	390

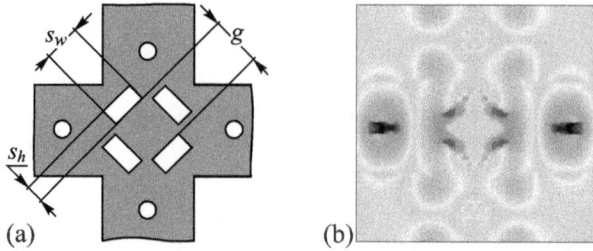

FIGURE 2.8 (a) Schematic of the slit resonance cavity and (b) the pattern of the E-field modulus of a resonance mode.

Shown in Figure 2.7(b) is the pattern of the E-field modulus for one of the orthogonal modes. The mode energy is seen to be concentrated in the cavity's central region, which is where the holes are located. The Q-factor is ~5,000. The value of γ is 0.05.

Let us consider the resonance cavity with a different shape of holes in the form of slits, as shown in Figure 2.8(a). Using a slit resonance cavity [38], the E-field within the slit can be enhanced by the value of n_{wg}^2/n_{sl}^2, where n_{wg} is the waveguide refractive index and n_{sl} is the slit-filling material refractive index.

To make use of the enhanced E-field in the slit resonance cavity, four rectangular holes are made in it, as shown in Figure 2.8(a). For each rectangular slit, the length is given by s_w and the width is s_h. The diagonal distance between the slits is $g = 245$ nm. Table 2.2 gives the positions and radii of the slits of the resonance cavity.

Shown in Figure 2.8(b) is the pattern of the E-field modulus for one of the orthogonal modes for the rectangular slits of size 160 nm × 115 nm, $g = 210$ nm. The slits' size and positions have been calculated from the condition of the maximal Q-factor. The cavity Q-factor is ~8,000. The value of γ is 0.15.

Changes in the slit width affect the cavity's Q-factor, the mode overlap coefficient, and the mode volume. Shown in Figure 2.9(a) are the distributions of the E-field modulus $|E|$ along a line that passes through the cavity's center and is rotated by 45° relative to the waveguide axis for several slit widths. The peak of the field amplitude is achieved when the slit width is small. Figure 2.9(b) shows in which way the cavity's Q-factor and the mode volume depend on the slit width.

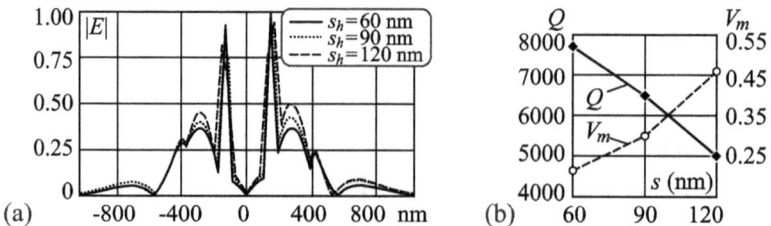

FIGURE 2.9 The distribution of the E-field modulus for (a) a resonance mode on the cavity's central diagonal for several slit widths and (b) the cavity's Q-factor and mode volume for several slit widths.

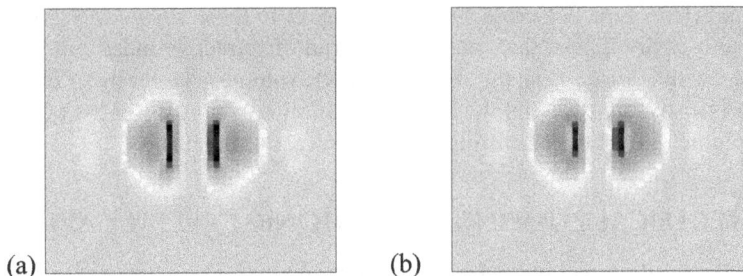

FIGURE 2.10 The distribution of the E-field modulus for a resonance mode along the central diagonal of the cavity plane for the slits of width 120 nm and height of (a) 225 nm and (b) 170 nm.

Figure 2.10 shows that it is possible to reduce the transverse losses in the cavity. Up to this point in the work, the slit height in the cavity has been assumed to be fixed and equal to the waveguide width. With decreasing slits' height, they are transformed into intra-waveguide hollows. Although these structures are more difficult to fabricate, such hollows allow the cavity's Q-factor to be increased and the mode volume to be made smaller.

Figure 2.10(a) shows the distribution of the E-field modulus $|E|$ in a plane that passes through the cavity's center and is rotated by 45° relative to the waveguide axis for the slits that go as far as the entire waveguide length of 225 nm. Figure 2.10(b) shows the distribution of $|E|$ for the slits of height 170 nm.

For the through-length slits in Figure 2.10(a), the coefficient γ is 0.15, the Q-factor is 7,800, and the mode volume is 0.32. For the closed slits in Figure 2.10(b), these values are, respectively, 0.14, 7,900, and 0.29.

Figure 2.11 illustrates the simulation results when the slits are filled with some material, e.g. chalcogenide glass (CG) [36]. The chalcogenide glasses have a refractive index ranging from 2.3 to 2.8 at wavelength 1.5 µm. Note that the CGs have the optical nonlinearity factor by three orders of magnitude higher than for Si, the low-level two-photon absorption, and fast response time (<100 fs).

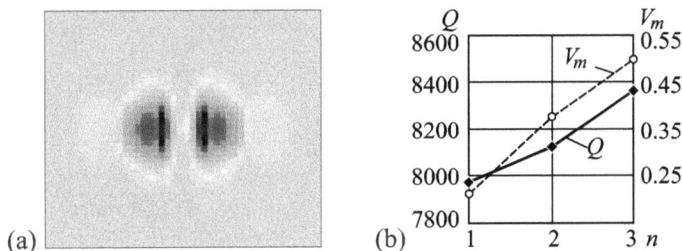

FIGURE 2.11 (a) The distribution of the E-field modulus for a resonance mode on the central diagonal of the cavity plane. The slit material refractive index is 2.5, slit width is 120 nm, slit height is 240 nm. (b) The plots for the cavity's Q-factor and mode volume at different values of the slit-filling material refractive index.

Figure 2.11(a) depicts a cross section analogous to those shown in Figure 2.6 for the through-cavity holes filled with the material of refractive index $n = 2.5$. When compared with Figure 2.6(a), the simulated mode volume is larger by a factor of two. Q-factor is also increased slightly. Figure 2.11(b) shows the plots for the Q-factor and mode volume against the slit-filling material refractive index.

2.4 ELECTRICAL PUMPING OF PHOTONIC CRYSTAL CAVITY

Most of the existing technologies in use to create high-Q PC nanocavities suggest fine-tuning of the resonance chamber geometry by changing the parameters of the PC. Such parameters may be, for example, the radius of the hole in the PC period and/or the hole periodic spacing. To simplify the solutions for the problems cited, the authors theoretically investigated a two-component PC cavity. The first component of such a cavity is a periodic structure based on a PC nanobeam. Compared with the 2D structure based on a PC slab, the area of PC nanobeam is smaller and is naturally integrated into the waveguide geometry of connections on a chip. The second component is a fragment of a complementary material that occupies an area of several lattice constant of the PC. The shape and size of the fragment were determined from the given parameters of PC cavity. While combining the two components, a defect forms in the resulting nanostructure. The resonant mode of the corresponding frequency can be excited in this defect.

The proposed approach to creating two-component PC cavities is illustrated through the structure shown in Figure 2.12. The first component of the resonator was a PC nanobeam. The nanobeam was made of Si and was placed on a silica substrate. The holes in the nanobeam were of the same radius; they were equidistant from each other and filled with air. PC nanobeam parameters are given in the caption to Figure 2.12. With these parameters, PC band gap was created for TE-dominant polarization radiation in the wavelength range of 1.4–1.7 μm. The second component of the nanocavity was an elliptical piece of Si, placed on a silica substrate.

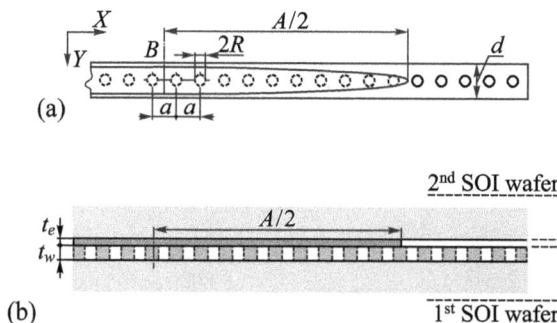

FIGURE 2.12   The geometry of the resonator calculated by (a) top view and (b) side view. PC nanobeam ($n = 3.46$) lies on the substrate ($n = 1.45$). PC nanobeam width is $d = 0.5$ μm, thickness $t_w = 0.26$ μm. Circular holes have a radius of $R = 75$ nm and are filled with air; distance between holes $a = 0.34$ μm. The elliptical shape (ellipse parameters A and B) ($n = 3.46$) lies on the substrate ($n = 1.45$). Thickness of ellipse $t_e = 100$ nm.

Therefore, both the cavity's components have a structure of silicon on insulator (SOI) wafers. The surface roughness of Si wafers can be as low as several hundred picometers at 1–300 μm length scales. This makes it possible to tightly combine two Si surfaces of wafers, as shown in Figure 2.12.

In subsequent calculations, the thicknesses of the PC nanobeam and elliptical defect were assumed to be 260 nm and 100 nm, respectively. These thicknesses produced an optimal FF (fill factor) change in the cavity. Increasing the thickness of the nanobeam necessitates increase in the thickness of the elliptical defect.

The resonance cavity characteristics were computed using the parallel FDTD method. In particular, the cavity ($Q = 3.05 \times 10^4$) was calculated with the parameters of the ellipse $A = 6.8$ μm (20 holes below the ellipse) and $B = 0.5$. To achieve a high-Q nanocavity, five additional holes were placed in the PC nanobeam at both ends of the ellipse. Thus, the total length of the cavity was $(20 + 5 \times 2) \times 0.34 = 10.2$ μm. Figure 2.13(a) shows the distribution of H_z in the vertical plane passing through the axis of the nanobeam. There is some vertical asymmetry of the resonance mode due to flow of energy in the elliptical defect. Figure 2.13(b) shows the distribution in the horizontal plane, just above the elliptical Si fragment (in silica). H_z values along the intersection line of these two planes are represented by the dotted line in the graph of Figure 2.13(c), and the values directly below the PC nanobeam (in silica) by the dashed line on the same chart. The solid line represents the function $\cos(\pi x/a)$ $\exp(-\sigma x^2)$ with $\sigma = 0.23$, $a = 0.34$ microns. Good agreement between the distributions of H_z and an analytic function demonstrates that the shape of the resonant mode's envelope is Gaussian. Assuming a linear dependence of γ on x, the relation $\gamma(x) = a/\pi \int \sigma dx \approx x/40$ can be obtained. In [39], quadratic tapering of PC nanobeam width was used to form the defect. In the paper [28], for a nanocavity with a length of 60 periods of the PC, the relation $\gamma(x) \approx x/120$ was implemented. Thus, it can be

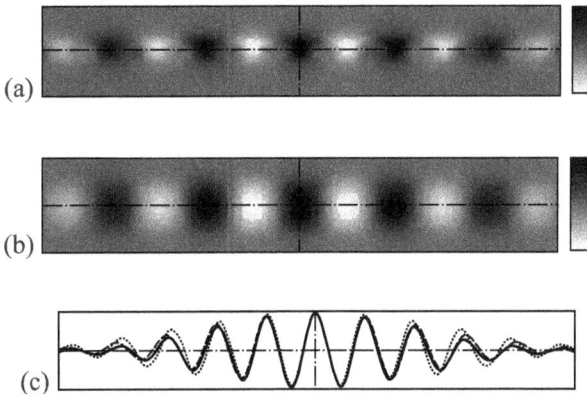

(a)

(b)

(c)

FIGURE 2.13   (a) The distribution of H_z in the vertical plane passing through the axis of the waveguide, (b) the distribution of H_z in the horizontal plane just above the elliptical fragment (in quartz), (c) the dotted line – H_z values along the line of intersection of the planes (a) and (b) the dashed line – H_z values just below PC nanobeam (in quartz), the solid line – function $\cos(\pi x/a)\exp(-\sigma x^2)$ for $\sigma = 0.23$, $a = 0.34$ μm.

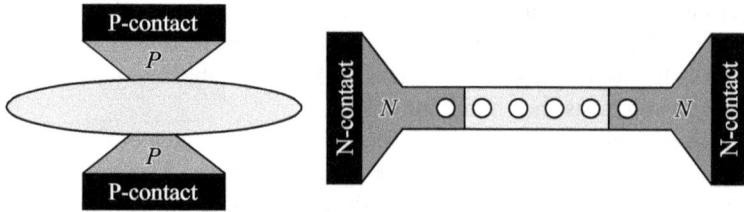

FIGURE 2.14 An example of geometry (not to scale) for P-type (left) and N-type (right) doping regions.

concluded that the two techniques used in creating a defect are almost equivalent. The nanocavity with an elliptical defect is three times shorter than the one with variable nanobeam width. Accordingly, the rate of change γ in the nanocavity with an elliptical defect is three times faster [40].

The two-component nanocavity proposed in this chapter has in our opinion two main advantages when compared to existing solutions. First, the proposed structure allows for the development of an integrated on-chip light source with vertical electrical pumping. Integrated on-chip light-emitting diodes with a laterally doped p-i-n structure, based on the nanobeam PC cavity, were demonstrated recently. Electron beam lithography steps can be used to implant N- and P-type dopants to the first and second components of the structure, respectively. Figure 2.14 shows an example of geometry for P-type and N-type doping regions. Such geometry makes it possible to focus current flow to the active region of the cavity, thereby, in comparison with lateral electrical pumping, improving efficiency and reducing threshold. The P-type parts that adjoin the elliptical defect have a small intersection with the resonance mode. Therefore, Q-factor suffered low degradation, especially in case of $B > d$ (Figure 2.12a).

Using hybrid metal/PC nanocavities is another possible approach to realize vertical electrical pumping. The two-component structure of the cavity assumes additional flexibility in choice of electrical current pathways. Although Q-factor in this case can be reduced to several hundreds, this could be enough for development of an optical amplifier integrated on-chip.

The second advantage is that the creation of nanocavities with nonlinear properties is simplified. The supplementary component of the structure can be used to bring a nonlinear or optically active material directly into the nanocavity region.

2.5 OPTICAL MODULATOR BASED ON COUPLED PC CAVITIES

We describe a compact optical modulator based on two PC nanobeam cavities coupled by a waveguide. Compared with previously proposed single-cavity modulators there are two potential advantages of the suggested modulator. First, the shift range of the modulator frequency is scalable and depends on switching energy level. Second, it is possible to modulate pumping in the low-intensity region of the resonant mode, allowing for effectively electrically controlled modulation.

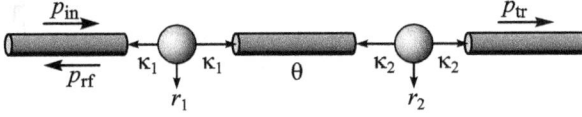

FIGURE 2.15 Scheme of two resonators coupled through a waveguide.

Figure 2.15 shows the general scheme of two resonators coupled by a waveguide. The variable κ_j, $j = \{1,2\}$ is the coupling coefficient of the j-th resonator of the waveguide; r_j, $j = \{1,2\}$ is the energy loss of the j-th resonator to the exterior space; and p_{in}, p_{rf}, and p_{tr} are the amplitudes of the input, reflected, and transmitted fields, respectively. To simplify the notation, we characterize the resonators with the inverse values of the resonator lifetime parameters.

The system depicted in Figure 2.15 may be interpreted as a Fabry–Perot (FP) etalon. The transmission function of the mirrors in the etalon is defined by the transmission function of the single resonator. According to temporal coupled-mode theory, the transmission function of the single resonator can be written as:

$$T_1(s_j) = \frac{2\kappa_j}{s_j + 2\kappa_j}, \quad s_j = i(\omega - \omega_j) + r_j, \quad j = \{1,2\}, \tag{2.11}$$

where ω_j, $j = \{1,2\}$ is the resonant frequency of the j-th resonator. The Q-factor of the j-th resonator Q_j is given by $1/Q_j = 1/Q_{j\kappa} + 1/Q_{jr}$, where $Q_{j\kappa} = \omega_j/(4\kappa_j)$ and $Q_{jr} = \omega_j/(2r_j)$.

The transmission function of an FP etalon, with mirror transmission function T_1 and reflection function $(1 - T_1)$, can be written as:

$$T_2(s_1, s_2) \equiv \frac{p_{tr}}{p_{in}} = \frac{T_1(s_1) T_1(s_2) \ell^{i\theta}}{1 - (1 - T_1(s_1))(1 - T_1(s_2)) \ell^{i2\theta}}, \tag{2.12}$$

where θ is the phase shift in the waveguide connecting the resonators. We neglect the waveguide dispersion. Equation (2.12) describes the FP etalon with two direct-coupling resonators. A similar structure but with two side-coupling resonators in the 2D PC slab was considered in [41] and the effect of electromagnetically induced transparency (EIT) was observed. This effect is missed in the case of the FP etalon with direct-coupling resonators.

The dashed line in Figure 2.16 shows the transmission function of a single resonator, T_1. The Q-factor of this resonator is $Q = 1,200$; the resonance wavelength is $\lambda_0 = 1.488$ μm. The energy loss of the resonator is determined by the value of $Q_r = 3.2 \times 10^5$. The solid line in Figure 2.16 shows the transmission function of two such resonators coupled by a waveguide with a phase shift of $\theta = \pi/2$. This phase shift corresponds to the situation in which the length of the connecting waveguide is equal to 0.

Dynamic adjustment of the variable θ can be implemented by changing the refractive index of the waveguide, e.g. by thermal tuning or free carrier generation. Free carrier generation provided dynamic control of the PC cavity Q-factor in [12].

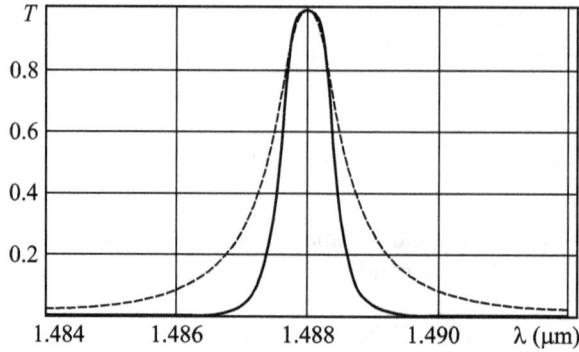

FIGURE 2.16 The transmission function of one cavity (dashed line) and two coupled resonators with $\theta = \pi/2$ (solid line).

Pulsed optical pumping was used to change the phase shift in a Si PC waveguide over the range of 0 to π. Free carrier generation can provide an ultrafast nonlinear optical tuning on the order of 10 s of GHz [14].

Figure 2.17(a) shows the transmission function of two resonators with $\theta_1 = \pi/25$ (solid line) and $\theta_2 = \pi/25 + \pi/125$ (dotted line). The width of the mode splitting equals 5 nm. Thus, the proposed optical modulator can scale the resonant frequency

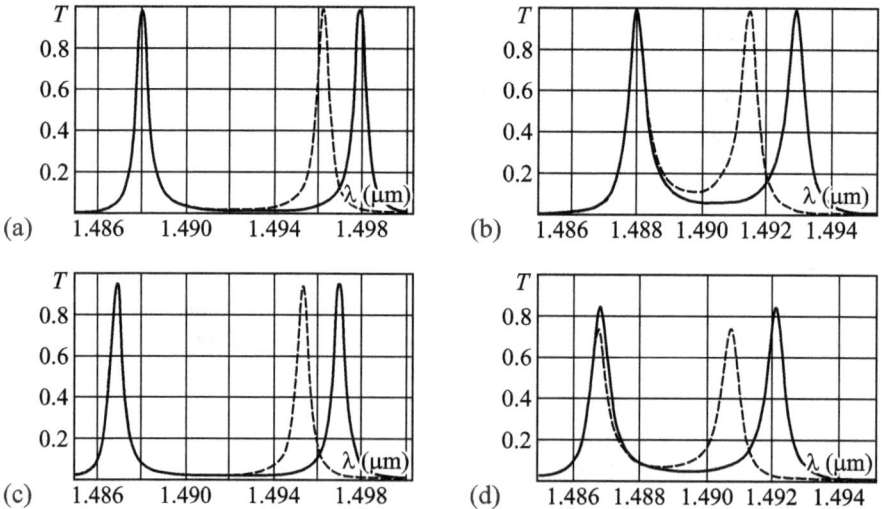

FIGURE 2.17 (a) The transmission functions of two resonators with phase shifts in the waveguide of $\pi/25$ (solid line) and $\pi/25 + \pi/125$ (dashed line). (b) The transmission functions of two resonators with phase shifts in the waveguide of $\pi/50$ (solid line) and $\pi/50 + \pi/250$ (dashed line). (c) The transmission function of two mismatched resonators (resonance wavelengths of 1.486 μm and 1.488 μm) with phase shifts in the waveguide of $\pi/50$ (solid line) and $\pi/50 + \pi/250$ (dashed line). (d) The transmission function of two mismatched resonators (resonance wavelengths of 1.486 μm and 1.488 μm) with phase shifts in the waveguide of $\pi/25$ (solid line) and $\pi/25 + \pi/125$ (dashed line).

shift range using the initial pumping energy, which corresponds to the initial phase shift θ_0 [14]. Thermal pumping can be used for fine-tuning θ_0. The timescale for this process is on the order of microseconds. The scaling feature of this optical modulator can be used to increase the frequency range of optical communication systems, optical sensors, and spectrometers integrated on a chip [42, 43].

The optical modulator proposed here does not require such a large range of phase shifts. Figure 2.17(b) shows the transmission function of the two resonators with $\theta_1 = \theta_0 = \pi/50$ (solid line) and $\theta_2 = \theta_0 + \theta_s = \pi/50 + \pi/250$ (dashed line). It can be seen that for values of $\theta \ \pi / 2$, the two degenerate modes of the resonators split. The width of the mode splitting equals 10 nm for $\theta_0 = \pi/50$. In addition, shifting the resonant mode by one line width requires an additional phase shift of only $\theta_s = \pi/250$.

In the next step we explore how the resonance mode wavelength shift affects device function. Such a shift can appear due to the errors in PC resonator manufacturing. Figure 2.18 demonstrates the case of $\theta = \pi/2$. A mismatch between resonant modes wavelengths leads to a significant change in the transmission function. The solid line in Figure 2.18 shows the transmission function of two resonators with degenerate resonant modes at a wavelength $\lambda_0 = 1.488$ μm, i.e. no mode wavelength mismatch. If the resonance mode wavelength of one of the resonators is shifted by 1 nm, such that $\lambda_0 = 1.487$ μm, then the transmission function undergoes significant changes (dashed line in Figure 2.4). These changes become even greater if the mismatch of the resonant modes is 2 nm ($\lambda_0 = 1.486$ μm), as shown by the dotted line in Figure 2.18.

Figures 2.17(c) and (d) demonstrate the stability of the transmission function to resonance wavelength shifts of mismatched resonators coupled by θ sufficiently less than $\pi/2$. The parameters of the coupled resonators in Figures 2.17(c) and (d) are the same as those in Figures 2.3(a) and (b), except for the resonator resonant frequencies. In Figures 2.17(c) and (d), the resonant frequencies are tuned to 1.488 μm and 1.486 μm in the first and second resonators, respectively. Errors in PC resonator manufacturing usually produce some shift of the resonant frequency. Thus, the proposed optical modulator allows for high tolerance in manufacturing. The differences

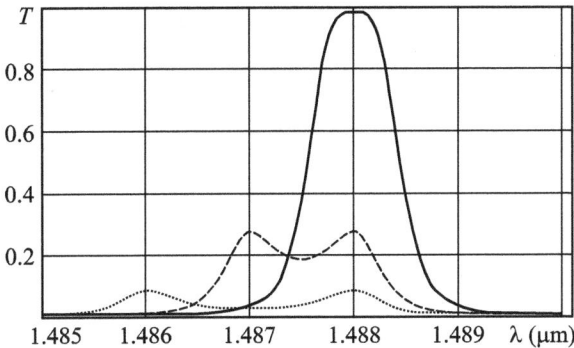

FIGURE 2.18 The transmission function of two resonators with resonance wavelengths 1.488 μm and 1.488 μm (solid line), 1.487 μm and 1.488 μm (dashed line), and 1.486 μm and 1.488 μm (dotted line).

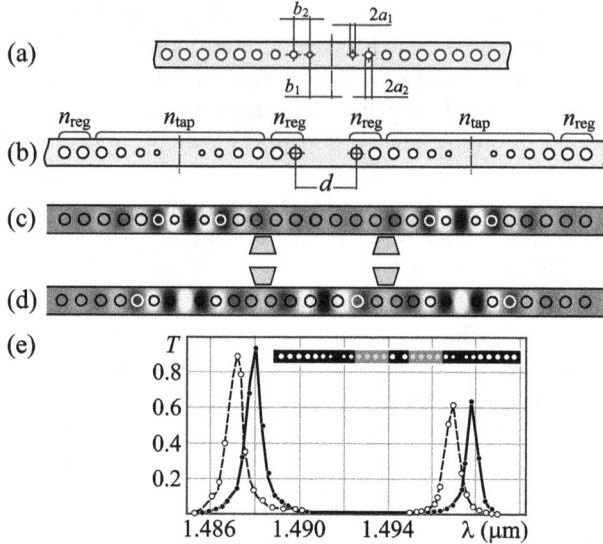

FIGURE 2.19 (a) Scheme of PC nanobeam cavity. (b) Scheme of array of two PC nanobeam cavities. (c) The electromagnetic field distribution of the "basic" resonant mode ($\lambda_0 = 1.488$ μm). (d) The electromagnetic field distribution of the "shifted" resonant mode ($\lambda_0 = 1.492$ μm). (e) The FDTD calculation of the transmission function of two resonators coupled by a waveguide of length $d = 490$ nm (solid line) and the transmission function of the tuned modulator; the figure inset shows the tuned parts of the waveguide (gray rectangles).

between the transmission functions in Figures 2.17(a) and (b) and Figures 2.17(c) and (d) are the smaller for smaller phase shift θ. This result can be explained by using theory of FP etalon transmission.

Figure 2.19(a) illustrates a possible embodiment of a resonator based on PC nanobeam cavities. Compared with resonators in a 2D PC slab [8], PC nanobeam cavities have a smaller area and are naturally integrated into the waveguide geometry of the chip. In the tapering region, the radii of the holes decrease, forming a defect in which the resonant mode is excited. An array of two PC resonators is shown in Figure 2.19(b), where n_{tap} is the number of holes in the tapering region. The coupling between resonators in the array is determined by n_{reg}, the number of holes in the regular part of the PC waveguide, and d, the length of the coupling waveguide. Let us calculate the parameters of the particular PC nanobeam cavities that would realize our suggested optical modulator.

The 2D FDTD method is used to calculate the characteristics of the resonator. The waveguide has a width $w = 500$ nm and an index of refraction $n = 2.97$. This value corresponds to the effective refractive index of a Si waveguide with thickness of 270 nm, which lies on a glass substrate and is surrounded by air. Air-filled holes in the non-tapering part of the waveguide have a radius of 93 nm and are spaced 350 nm apart. Table 2.3 lists the lattice parameters and the radii of the holes near the defect ($n_{tap} = 4$). These parameters demonstrate the existence of an energy bandgap for transverse electric polarization in the waveguide. For $n_{reg} = 4$,

TABLE 2.3

Geometric Parameters of PC Resonator Shown in Figure 2.19(a)

a_1	b_1	a_2	b_2	a_3	b_3	a_4	b_4
65	200	80	290	90	310	93	304

the resonance wavelength is 1.488 µm and $Q = 1{,}230$. The energy loss of the resonator is $Q_r = 3.19 \times 10^5$.

Several methods can be used to tune the phase shift θ, e.g. optical and electrical pumping used in [13] and [15], respectively, to generate free carriers. In both cases, the pumping is performed in the cavity region, where the amplitude of the resonant mode is maximal. Increasing the pumping energy sharply reduces the value of Q-factor [13, 15]. Thus, shifting the resonant wavelength by greater than one line width becomes problematic.

Next, we analyze the electromagnetic field distribution in the suggested modulator. Figure 2.19(c) shows the amplitude of the "basic" resonant mode ($\lambda_0 = 1.488$ µm). Figure 2.19(d) demonstrates the amplitude of the "shifted" resonant mode ($\lambda_0 = 1.492$ µm). The resonators share two regions of low amplitude of both resonant modes (denoted in Figures 2.19(c) and (d) by gray trapezoids). Increasing the value of n_{reg} can increase the size of those regions. The overlapping of the contacts and the doped areas with the field of the resonant mode increases the optical absorption and reduces the Q-factor. In previously suggested electro-optic modulators, the doped areas were placed near regions of the maximal amplitude of resonant modes. Increasing the optical absorption in those regions drastically deteriorates the Q-factor of the cavity. There are at least two approaches to resolve this problem. First, the side slabs were used in [18, 19] to separate the PC nanobeam cavity and the doped areas. The thickness and lateral size of those slabs are the trade-off between the high Q-factor and the low switching energy. Second, a two-component parabolic PC nanobeam cavity was suggested in [40] to separate the region of the maximal amplitude of resonant mode and the doped areas. Both approaches can be used with the modulator described in this work. The regions of low resonant modes amplitude represent the part of the device that can be electrically tuned without deteriorating the Q-factor of the resonance modes. Electrodes providing the electrical pumping may be placed near the pumping area. Figure 2.19(e) shows the transmission function of two resonators coupled by a waveguide of length $d = 490$ nm (solid line). The dashed line in the same figure represents the transmission function of the tuned modulator. The tuned parts of the waveguide are shown in the inset of Figure 2.19(e) by the gray rectangles. The refractive index of the tuned parts of the waveguide equals to $n = 2.96$. The length of each rectangle equals to four-unit cell size of the PC waveguide lattice (4×350 nm). The position of the tuned "shifted" resonant mode on Figure 2.19(e) corresponds to the one on Figure 2.17(a) ($d = 489$ nm). The position of the tuned "basic" resonant mode on Figure 2.19(e) is shifted to the short wavelengths due to the PC waveguide refractive index decreasing near the central part of the device.

In our following numerical investigation, we model phase shift θ by adjusting d, the length of the connecting waveguide, instead of adjusting the PC waveguide refractive index. Let us find the correspondence between θ and d. The wavelength of the resonant mode in the waveguide is $1{,}448/2.97 = 500$ nm. Then, the additional phase shift $\theta_s = \pi/2$ corresponds to additional waveguide length $d = 500/4 = 125$ nm. The initial phase shift $\theta_0 = \pi/2$ corresponds to an initial waveguide length $d = 350$ nm. However, the results of FDTD simulation coincide with the analytical relation Eq. (2.12) with $\theta = \pi/2$ when $d = 370$ nm. This discrepancy may be because the Bragg frequency of the PC waveguide does not coincide with the resonance mode frequency. In this case, to realize the condition $\theta_0 = \pi/2$, it is necessary to add some additional waveguide fragment between the resonators [44].

Figure 2.20(a) shows the transmission function of two resonators coupled by a waveguide of length $d = 490$ nm (solid line) or $d = 489$ nm (dashed line). These parameters correspond approximately to the parameters of the analytic curves in Figure 2.17(a). The magnitude of the resonance mode splitting in the FDTD model was also about 10 nm. FDTD modeling confirms that the displacement of the spectral peak width at half-maximum occurs with a phase shift $\theta_s = \pi/2/125 = \pi/250$. Figure 2.20(b) demonstrates the transmission function of two resonators coupled by a waveguide of length $d = 484$ nm (solid line) or $d = 480$ nm (dashed line). This matches the parameters of Figure 2.17(b). Resonance mode splitting is about 5 nm.

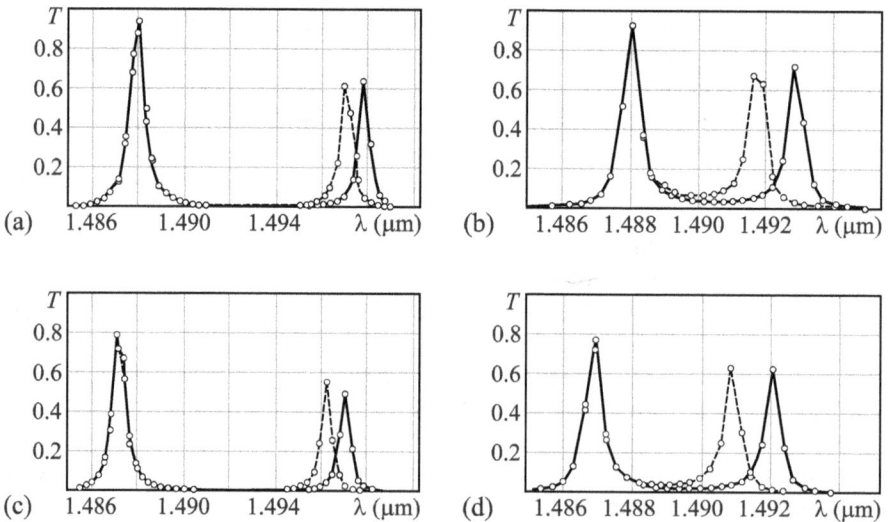

FIGURE 2.20 (a) The FDTD calculation of the transmission function of two resonators with connecting waveguide lengths of $d = 490$ nm (solid line) and $d = 489$ nm (dashed line). (b) The FDTD calculation of the transmission function of two resonators with connecting waveguide lengths of $d = 484$ nm (solid line) and $d = 480$ nm (dashed line). (c) The FDTD calculation of the transmission function of two mismatched resonators (resonance wavelengths at 1.486 μm and 1.488 μm) with connecting waveguide lengths of $d = 490$ nm (solid line) and $d = 489$ nm (dashed line). (d) The FDTD calculation of the transmission function of two mismatched resonators (resonance wavelengths at 1.486 μm and 1.488 μm) with connecting waveguide lengths of $d = 484$ nm (solid line) and $d = 480$ nm (dashed line).

Decreasing θ_0 decreases the maximum of "shifted" spectral peaks, as seen in Figures 2.20(a) and (b). This is due to mismatched modes in the PC and connecting waveguide sections. This situation could be enhanced, e.g. by tapering hole radii and/or adjusting hole spacing in the central part of the FP etalon [45].

Thus, FDTD modeling confirms the ability of our designed optical modulator to scale the additional phase shift θ_s. The value of θ_s necessary to shift the spectral peak width at half-maximum depends on the initial value of θ_0. The theoretical minimum value of θ_s is determined by the free spectral zone, which is given by the size of the band gap for PC resonators. In this chapter, we assume resonators with free spectral bandwidths of about 300 nm. Therefore, to shift the resonance mode by the peak width at half-maximum requires providing θ_0 of about 10^{-5}. Practically, before operating at such small θ_0 values, the decreasing of "shifted" spectral peaks should be diminished.

Finally, we investigated the impact of manufacturing errors on the proposed design of the optical modulator. We modify the resonant wavelength of one of the two resonators by reducing the radii of the cavity central holes pair from 65 nm to 63.5 nm; the resonant wavelength redshifts by about 2 nm. Thus, the optical modulator consists of two mismatched resonators with resonant wavelengths of 1.486 μm and 1.488 μm, respectively. Figure 2.20(c) shows the transmission function of two mismatched resonators with $d = 490$ nm (solid line) and $d = 489$ nm (dashed line). The transmission function in Figure 2.20(c) is only slightly distorted in comparison with Figure 2.20(a). A similar trend can be seen comparing Figure 2.20(b) and 2.20(d) for $d = 484$ nm (solid line) and $d = 480$ nm (dashed line).

2.6   APPLICATION OF PHOTONIC CRYSTAL COUPLED CAVITIES FOR INCREASE IN SENSITIVITY OF OPTICAL SENSOR

There are examples of application of the optical cavities of different kinds in the refractive index sensors. Among them, the sensors based on the spherical cavities were suggested in [45]. The disk cavities used as the components of the optical sensors are discussed in paper [46].

Perhaps, the refractive index sensors based on the ring cavities are the most frequently used solution [47]. They are easily integrated on the crystal and the technology of their manufacturing is well developed. The optical sensors based on the PC cavities are discussed in paper [48].

As a rule, in the optical sensors the light comes to the resonant chamber through the monomode waveguide. The transmission spectrum of the cavity has the form of the Lorenz curve. The curvature of this curve depends on the value of the cavity Q-factor, and it defines the sensor sensitivity.

The optical sensors based on the ring cavities are more compact compared with, for example, the sensors based on the spherical or disk cavities. Their dimensions are of the order of some tens of micrometers along both directions in the crystal plane.

The solutions suggested in the present publication are demonstrated using the most compact optical refractive index sensor known today based on the photonic crystal cavities (PCC) [8, 44]. Usually, their dimensions do not exceed the value of several wavelengths of the light that has been used.

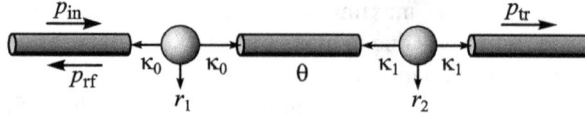

FIGURE 2.21   Scheme of optical system consisting of two coupled cavities with phase delay.

In Figure 2.21 we present the proposed scheme of the optical system consisting of two coupled cavities. The phase delay of the connection of the cavities has the value of θ. In this figure, the variable κ_j, $j = \{1,2\}$ defines the connectivity coefficient of the j-th cavity with the waveguide; $r_j = \{1,2\}$ are the coefficients of the spatial energy losses of the j-th cavity; and p_{in}, p_{rf}, p_{tr} are the amplitudes of the input, reflected, and transmitted fields, respectively.

According to the time depended coupled mode theory, the transmission function of one cavity can be written in the form:

$$T_1(s_j) = \frac{2\kappa_j}{s_j + 2\kappa_j}, \; s_j = i(\omega - \omega_j) + r_j, \; j = \{1,2\}, \tag{2.13}$$

where ω_j, $j = \{1,2\}$ is the resonant frequency of the j-th cavity. The Q-factor of the j-th cavity is defined by the relation $1/Q_j = 1/Q_{j\kappa} + 1/Q_{jr}$, where $Q_{j\kappa} = \omega_j/(4\kappa_j)$ and $Q_{jr} = \omega_j/(2r_j)$.

When $\theta = \pi/2$, the transmission function of the system of two cavities shown in Figure 2.21 is written as:

$$T_2(s) = \frac{i2\kappa_0\kappa_1}{(s + \kappa_0)^2 + \kappa_1^2}. \tag{2.14}$$

In this equation, we neglected the waveguide dispersion. The phase incursion $\theta = \pi/2$ corresponds to the situation in which the coupling waveguide is absent.

In Figure 2.22, the dashed line corresponds to the transmission function of one cavity calculated according to Eq. (2.13) and the solid line is the transmission

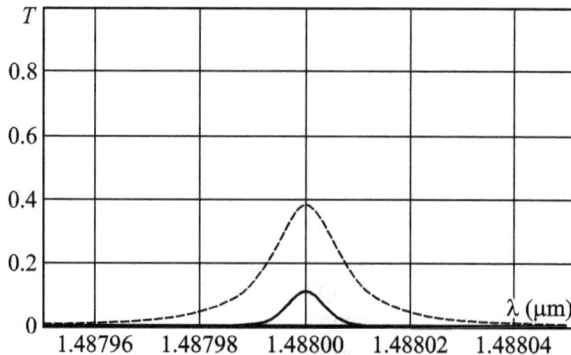

FIGURE 2.22   Transmission functions for one cavity (dashed line) and two coupled cavities for $\theta = \pi/2$ (solid line).

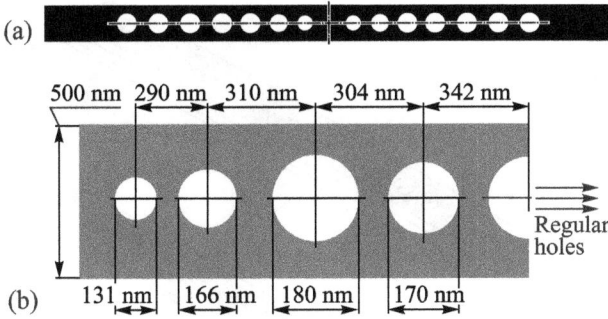

FIGURE 2.23 (a) Cavity based on ridge photonic-coupled waveguide in Si and (b) geometrical parameters of the resonant chamber.

function of the same coupled cavities defined by Eq. (2.14). We see that when two cavities are used in place of one cavity, the Q-factor of the system increases. The Q-factor of the system with one cavity and the resonant wavelength were $Q_1 = 1.5 \times 10^5$ and $\lambda_0 = 1.488$ μm, respectively. The extrinsic losses of this cavity are defined by the quantity $Q_{jr} = 2.5 \times 10^5$.

As the concrete PCC, let us calculate the cavity based on the 2D ridge PC waveguide. Such cavities have a small area, and they are integrated naturally into the waveguide geometry of the junctions on the chip. In Figure 2.23(a) the cavity based on the PC ridge waveguide is shown. The defect where the resonant mode excites is formed by reduction of the radius of the apertures in the vicinity of the resonant chamber.

In Figure 2.23(b), the geometrical parameters of the apertures in the vicinity of the resonant chamber are shown. For the given wavelength of the light, these geometrical parameters ensure the appearance of the band gap for the TE polarization mode in the waveguide.

To calculate the characteristics of the cavity we used the 2D FDTD method. The width of the waveguide was $w = 500$ nm, the refractive index of its material was equal to 2.97. This value corresponds to the effective refractive index of the Si waveguide whose width is 270 nm that is placed on the glass substrate and surrounded by the air.

In Figure 2.24(a) we show the 3D cavity based on the PC waveguide without a substrate with three regular apertures. The system consisting of two such cavities is shown in Figure 2.24(b). The apertures in the regular part of the waveguide are filled with the air and they are spaced apart at intervals of $a = 350$ nm. The wavelength of the resonant mode is equal to 1.48827 μm. We denote the distance between two cavities (that is the distance between the neighbor holes in the central part of the structure) by d. The Q-factor of one of these cavities with eight regular apertures from each side is equal to 156,413. The extrinsic loss of the given cavity is defined by the value $Q_{jr} = 2.56 \times 10^5$. These parameters are approximately equal to the ones of the cavity whose transmission function is shown in Figure 2.22.

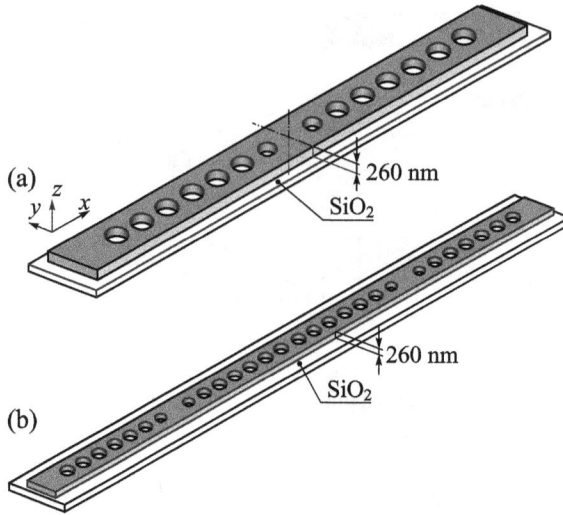

FIGURE 2.24 (a) Examples of 3D cavity based on the PC waveguide without a substrate with three regular apertures, (b) system consisting of two coupled 3D cavities.

In Figure 2.25, the readings of the transmission function for the systems consisting of one and two cavities are shown. To calculate the readings, we used the FDTD method. The parameters of the basic cavity correspond to the parameters shown in Figure 2.24(b). We see that with good accuracy the curves in Figure 2.25 correspond to the curves in Figure 2.22.

In Figure 2.26 we show the transmission functions for the systems consisting of two cavities calculated analytically (solid line) and with the aid of the FDTD method (the readings are interpolated by the dotted line). The period of the regular part of the PC waveguide is equal to 350 nm. That is the reason why the absence of

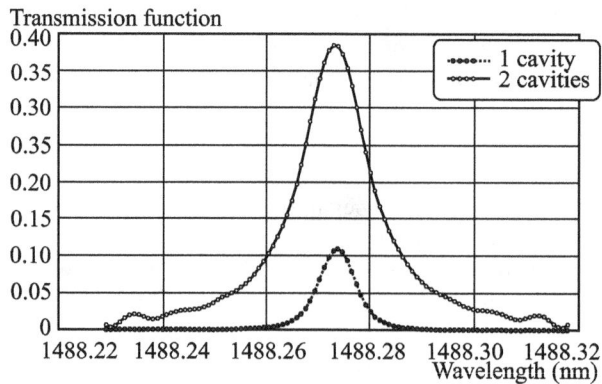

FIGURE 2.25 Readings of transmission functions for one and two cavities interpolated with the aid of FDTD method by dotted and solid lines, respectively.

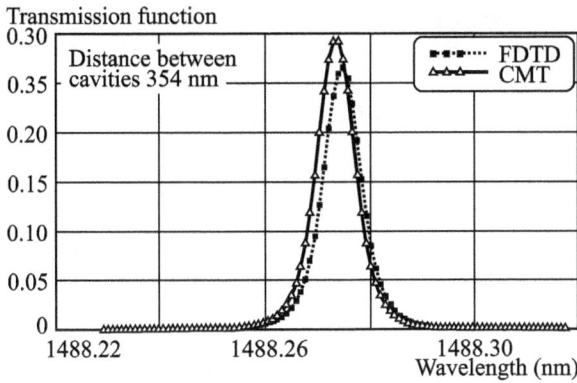

FIGURE 2.26 Transmission functions for systems consisting of two cavities for $\theta = \pi/2$ calculated analytically (solid line) and with the aid of FDTD method (readings are interpolated by dotted line).

the coupling waveguide ($\theta = \pi/2$) corresponds to the value $d = 350$. However, from Figure 2.26 we see that almost perfect correspondence between the curves defined by the analytical expression (2) for $\theta = \pi/2$ and the result of the FDTD method simulations is achieved for $d = 354$ nm. It can be explained by the fact that the Bragg frequency of the PC waveguide does not match the resonant frequency. In this case, to accomplish the condition $\theta = \pi/2$ we need to add a fragment of the waveguide between the cavities [24].

The Q-factor of the resonant system can be estimated from the slope of the spectral peak. For example, in the refractive index sensors the slope of the spectral peak corresponds to the rate of the changes of the transmission function due to the changes of the refractive index of the environment. In Figure 2.27 we show the derivatives of the transmission functions in the region near the spectral peaks calculated with the aid of the FDTD simulations for the systems consisting of one (the dotted line) and two (the solid line) cavities, respectively.

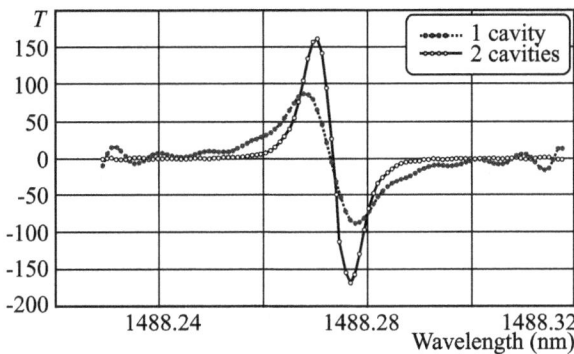

FIGURE 2.27 Derivatives of transmission functions in regions near spectral peaks for systems consisting of one (dotted line) and two (solid line) cavities, respectively.

From Figure 2.27, we see that for the system of two coupled cavities the maximal value of the derivative of the transmission function in the point of the spectral peak is approximately two times larger than the respective value in the case of one cavity. As a result, the Q-factor of the system consisting of two cavities is better compared to one of the systems with one cavity. However, as it is seen from Figure 2.25, for the parameters of the cavities in use, this improving of the Q-factor is obtained when increasing the input energy level. Nevertheless, the obvious analytical calculations show that the difference between the maximums of the transmission functions for one and two cavities can be reduced by the way of reduction of the energy losses. For example, for $Q_1 = 1.5 \times 10^5$ and $Q_{jr} = 2.5 \times 10^6$, this difference is about 3%.

With the aid of numerical simulations, let us show how the errors in manufacturing of single and coupled cavities influence the displacement of the resonant mode frequencies of the cavities and the Q-factors of resonant systems.

In Table 2.4, we present the results of computer simulations of the influence of the errors in manufacturing for one PC cavity. We varied the radii of the holes in the PC waveguide. It was assumed that the errors in the values of the radii were random numbers uniformly distributed inside the interval ±1, ±2, ±3, ±4, or ±5 nm. We used so simple a scheme when modeling the influence of the errors of the manufacturing mainly due to calculation time. Several hours of the computing time are necessary to calculate the characteristics of one cavity with the aid of the FDTD method using the computational cluster of the peak capacity equal to 1 TFLOP. From Table 2.4 we see that when the error is equal to ±5 nm the shift of the resonant frequency can reach 5 nm. At this point, when the error increases, the value of the Q-factor of the cavity can both decrease and increase.

Table 2.5 demonstrates the results of computer simulation of the influence of the manufacturing errors for the system consisting of two coupled PCC. In this case, the changes in the transmission function depend mainly on the mismatching of these two cavities in the resonant wavelengths. In the presence of a mismatching, the resonant modes of the cavities are no longer the degenerated ones, and that results in a broadening of the spectral peak of the transmission function.

TABLE 2.4
Values of Resonant Wavelengths and
Q-factors for Different Values of Errors
for the System Consisting of One PCC

Error, nm	Wavelength, nm	Q-factor
±0	1,488.2	156,413
±1	1,488.2	156,435
±2	1,488.3	153,050
±3	1,489.2	183,318
±4	1,490.7	110,346
±5	1,491.2	97,482

TABLE 2.5

Values of the Resonant Wavelengths and Q-factors for Different Values of Errors for System Consisting of Two PCC

Error, nm	Wavelength, nm	Q-factor
±0	1,488.2	252,739
±1	1,487.1	172,461
±2	1,484.8	110,686
±3	1,483.6	312,560
±4	1,482.8	180,881
±5	1,482.5	297,793

Consequently, the individual tuning of the resonant wavelengths in each resonator allows us to compensate, at least partially, the nondegeneracy of the resonant modes. Dynamic tuning of resonant wavelength of each resonator can be done by changing the refractory index of the waveguide. For this purpose, one can use, for example, the Kerr effect, the injection of the free carriers, or the thermal effect. Among these methods, the thermal effect permits the implementation at the minimal frequency. However, for the resonant system described here the high rate of the change of the resonant wavelength is not necessary.

Recently, injection of the free carriers is most frequently used to change the refractory index in PCC. The injection of the free carriers into PCC may be done using optical or electron pumping. When comparing optical and electron pumping, only electron pumping provides the possibility of practical realization of the device proposed in this chapter.

2.7 ON A SI-BASED PHOTONIC CRYSTAL CAVITY FOR THE NEAR-INFRARED REGION: NUMERICAL SIMULATION AND FORMATION TECHNOLOGY

Si is the most used material in the complementary metal–oxide–semiconductor (CMOS) technology of constructing electronic circuits [49]. However, as a rule, as materials for optoelectronic components, III–V semiconductors, e.g. GaN, InP, or GaAs, are used [50]. The monolithic integration of Si and III–V semiconductors in a crystal presents difficulties because of the difference between the crystal-lattice periods of these materials. The use of Si in various fields, including optoelectronics, would simplify the problem of the integration of photonic and electronic components. Si is rather easily oxidized. Stable Si oxide possesses a refractive index of ~1.5, which creates the required contrast with Si, whose refractive index is ~3.5 in the tele-communication wavelength region. An important element of nanophotonic devices is the optical cavity [51, 52]. The basic parameters of the cavity are the quality factor, frequency, and mode composition. High-quality-factor cavities with a small mode volume make it possible to enhance the interaction of light with the medium, to reduce the size of the photonic element, and to optimize its dispersion characteristics. Nanocavities are formed based on PCs [53]. High-quality-factor nanocavities make possible the production, e.g. of optical switches, filters, modulators, and coherent light sources with a low threshold power [54]. The fabrication of a nanocavity within a 3D PC is a technologically difficult problem. Therefore, 2D PCs in the form of a membrane are often used. In such a membrane, light propagates because of total internal reflection (TIR) [55, 56]. Further simplification of the problem is attained by using ridge PC waveguides. In such waveguides, the TIR effect prevents the propagation of light in transverse directions. In the longitudinal direction, the localization of light in the nanocavity is provided by the PC [26, 57]. In this study, we explore a cavity based on a ridge PC waveguide. The structure of such a waveguide consists of three parts. The first part is the PC mirrors formed as identical equidistant holes in the waveguide; these holes constitute the PC lattice. The second element of the structure is the transition zone between the PC waveguide and the cavity region as such. This zone is used to reduce energy losses in the cavity. The transition zone involves several holes, whose radii and positions are optimized to make possible a reduction in the losses. The third element of the structure is the cavity proper. A rise in the quality factor of the ridge PC cavity is attained by reducing losses in both longitudinal and transverse directions. Losses in the longitudinal direction can be reduced merely by increasing the number of periods of the PC [40, 58]. To reduce losses in the transverse directions, some finer optimization of the structure of the nanocavity is required.

In this section, we report the results of numerical simulation of a PC cavity based on a Si-based ridge waveguide and demonstrate the production technology of cavities of this type. As shown by the results of technological experiments, the procedure of ion-beam etching used to produce cavities here can lead to the PC cavity having cone-shaped holes rather than cylindrical holes. In the study, we simulate the conicity of the hole shape and show the possibility for compensating the effects of the deviation in the hole shape from cylindrical. We consider the influence of the height of the ridge waveguide on the parameters of the PC cavity.

The nanocavity was fabricated at the Institute for Physics of Microstructures, Russian Academy of Sciences, Nizhny Novgorod, Russia. The PC cavity structures were formed in several stages, using electron-beam and ion-beam lithography. In the preliminary stage, we formed waveguides by means of electron-beam lithography in combination with ion etching. Then, by precision etching with a sharply focused ion beam, we formed the end faces of the waveguides and the PC cavities themselves as a sequence of holes at the waveguide surface; the holes were different in diameter and positioned at a certain distance from each other. The cavities were formed on standard silicon-on-insulator (SOI) substrates (Soitec) with a thickness of the buried Si-oxide layer of ~1 μm and a thickness of the upper "bearing" Si layer hSi ~220 nm. To create a pattern, we first deposited a two-layer positive resist (polymethyl methacrylate PMMA 495 and PMMA 950) on the substrate; the total width of the resist was ~200 nm. After exposure to an electron beam and development in a solution of methyl isobutyl ketone (MIBK) in isopropyl alcohol (IPA), MIBK:IPA (1:3), a ~20-nm-thick vanadium layer was deposited onto the resist, and the lift-off procedure was carried out in acetone with ultrasonic activation. As a result of the lift-off lithography process, a vanadium mask, formed as 20×2-μm strips, was created.

This mask was further used upon etching of the Si layer with an ion beam. Primary-ion etching was conducted through the thickness of the Si layer in an atmosphere of argon. This was done using a system with a hot cathode (a Kaufman setup, an Ar^+ ion energy of ~1,000 eV). The remains of the vanadium mask were removed by wet etching in a solution of hydrogen peroxide. As a result of the procedures just described, Si strips to be used as blanks for the waveguides were formed. The dimensions of the blanks were somewhat larger than the required dimensions. Final treatment of the waveguides, specifically, the formation of necessary-sized strips, holes in the strips, and the end faces with specified parameters was carried out by local etching with a sharply focused ion beam. Etching was conducted with Ga^+ ions with an energy of 30 keV, using a Neon 40 (ZEISS, Germany) two-beam microscope. The ion-beam current was varied from 1 to 10 pA; the beam diameter was ~7–15 nm. Figure 2.28 shows micrographs of the PC cavities fabricated by the procedure just described. The micrographs were obtained with the same scanning microscope, immediately after the final stage of etching.

The images were obtained with secondary electrons, using an electron probe at different angles of sample inclination. The waveguide length and width were 11.5 and 0.5 μm, respectively; the hole diameter was varied and increased along the direction from the center of the waveguide to its edges. As shown in Figure 2.28(b), in the image of the cavity at the angle of sample inclination of 54°, we can see a characteristic contrast ratio indicative of the conical shape of holes of the cavity. This can be clearly seen in the micrograph of the cavity cross section made with the use of a sharply focused ion beam (Figure 2.28d). In this case, we can see in detail the shape of the holes and determine their dimensions.

Simulation of the PC cavities was carried out by the finite-difference time-domain (FDTD) method. In this study, we used the parallel 3D FDTD method. Calculations for PC cavities were performed for the telecommunication wavelength region (~1.5 μm). In accordance with the perfectly matched layers (PML) method, it was assumed

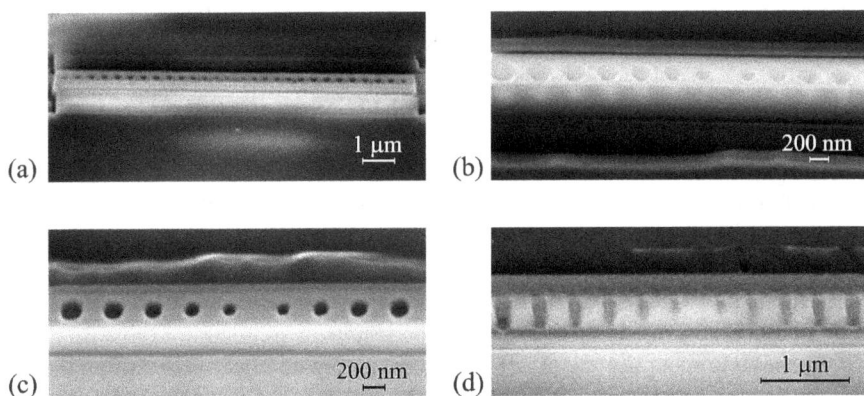

FIGURE 2.28 Micrographs of a waveguide with the photonic crystal cavity at different magnifications: (a) the whole cavity; (b) the central part of the cavity measured at an angle of sample inclination of 54°; (c) the central part of the cavity (view from above); and (d) the cross section of the cavity along waveguide central axis.

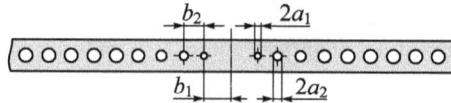

FIGURE 2.29 Geometrical parameters of the PC cavity.

TABLE 2.6

The Radii of Holes and the Spacings between Them in the PC Waveguide (in Nm)

b_1	a_1	b_2	a_2	b_3	a_3	b_4	a_4	b_5
200	65	290	82	310	90	305	85	345

that there is an absorbing layer at the boundaries of the 3D calculation region. The thickness of this layer was half the resonance wavelength. For the radiation source, we used a Gaussian pulse. The frequency width of this pulse was about 10% of the resonance frequency. The resolution of the computational mesh was chosen from the condition of convergence of the results [58].

Figure 2.29 shows the geometrical parameters of the PC cavity to be simulated. Here, a_1 and a_2 are the hole radii and b_1 and b_2 are the distance from the center and the spacing between neighboring holes, respectively. Table 2.6 lists these parameters. The parameters are close to those of the PC cavity fabricated in the study. In the PC cavity to be simulated, there are eight holes in the central transition zone and eight regular holes on each side of the central zone. The PC waveguide width and height are, correspondingly, $w = 500$ nm and $h = 260$ nm. The waveguide consists of Si with the refractive index 3.45, lies on a Si-oxide layer with the refractive index 1.45, and is surrounded by air. The holes in the regular part of the waveguide are 185 nm in diameter, filled with air, and spaced from each other by a distance of $a = 350$ nm. The wavelength of the resonance mode in the cavity with such parameters corresponds to $\lambda = 1{,}457.41$ nm (here and below, we give the wavelengths in vacuum). The quality factor of the cavity with eight regular holes on each side is 69,064.

It can be shown that a change in the PC waveguide height (h) yields a change in the band gap of the PC and, as a consequence, the frequency region of the cavity. The results of numerical simulation show that there exists an optimal height h, for which the band gap is maximal. The quality factor and cavity-mode wavelength obtained as functions of the waveguide height by theoretical calculations for the PC cavity just described are listed in Table 2.7. The quality factor of the cavity was calculated by the method described elsewhere [57]. The results of simulation (Table 2.7) show that the maximum quality factors of the PC cavity are attained on condition that the ratio of the height h to the lattice period of the PC (a) is unity. As the PC waveguide height is increased, the resonance wavelength increases and, at large values of h, is beyond the band gap of the PC. For example, the height $h = 510$ nm given in Table 2.7 corresponds to another resonance mode with a quality factor much (about an order of magnitude) lower than that of the preceding mode.

TABLE 2.7

Dependences of the Parameters of the PC Cavity on the Waveguide Height

Height h, nm	Wavelength, nm	Quality Factor
220	1,423.59	54,482
240	1,439.17	62,153
260	1,457.41	69,064
280	1,476.52	74,530
300	1,493.63	75,365
330	1,516.72	79,201
370	1,541.73	75,742
410	1,562.43	73,472
450	1,579.71	60,487
490	1,594.31	45,135
510	1,451.56	6,768

As shown in Section 2.2, some undesirable effects associated with variations in the parameters of the PC cavity can be produced during ion etching. Specifically, during etching, holes in the PC waveguide can acquire a conical shape. It is known that, during etching with an ion beam, the conicity of a hole is influenced by several factors: the relation between the diameter of the ion beam and the dimensions of the region of etching, the scanning mode, the ion energy, etc. [59]. Let us consider the influence of the hole conicity on the parameters of the PC cavity.

A conical hole is described by the cone height h equal to the waveguide height, the radii of the lower (r) and upper (R) bases, and the point of intersection of the cone with a cylinder of specified diameter (RC).

This point is defined by the distance L reckoned from the upper surface of the waveguide. The cone-shaped hole and the projection of its intersection with an ideal cylinder in the waveguide are schematically shown in Figure 2.30.

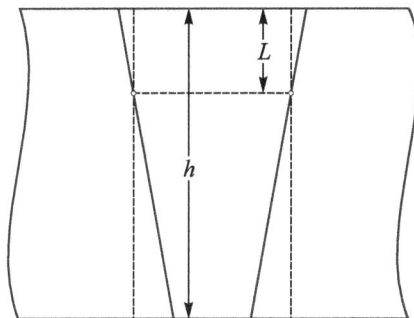

FIGURE 2.30 Projection of the intersection of a cone-shaped hole with an ideal cylinder in the waveguide.

TABLE 2.8

Calculated Parameters of the PC Cavity as Functions of the Degree of Hole Conicity

r L	5d/12	d/3	d/4
h/4	$\lambda = 1{,}461.35$ nm; $V_{cn}/V_{cl} = 0.896$; $Q = 49{,}816$	$\lambda = 1{,}476.15$ nm; $V_{cn}/V_{cl} = 0.806$; $Q = 28{,}089$	$\lambda = 1{,}490.51$ nm; $V_{cn}/V_{cl} = 0.731$; $Q = 11{,}197$
h/2	$\lambda = 1{,}484.52$ nm; $V_{cn}/V_{cl} = 1.009$; $Q = 8{,}311$	$\lambda = 1{,}466.16$ nm; $V_{cn}/V_{cl} = 1.037$; $Q = 18{,}968$	$\lambda = 1{,}483.17$ nm; $V_{cn}/V_{cl} = 1.083$; $Q = 13{,}272$
3h/4	$\lambda = 1{,}426.14$ nm; $V_{cn}/V_{cl} = 1.398$; $Q = 5{,}186$	$\lambda = 1{,}450.91$ nm; $V_{cn}/V_{cl} = 1.925$; $Q = 10{,}020$	$\lambda = 1{,}472.59$ nm; $V_{cn}/V_{cl} = 2.583$; $Q = 12{,}951$

In the numerical experiment, the values of L were chosen as fractions of the waveguide height [$3h/4$, $h/2$, $h/4$] and the radius of the lower cone base was defined in terms of its ratio to the diameter of the distorted cylindrical hole r [$5d/12$, $d/3$, $d/4$]. Thus, the value of h was 260 nm and the value of d corresponded to the diameter of each cylinder distorted. It was assumed that all the holes in the PC cavity were distorted in a similar manner. The resultant values of the resonance-mode wavelength (λ), the ratio between the volumes of the conical and cylindrical holes (V_{cn}/V_{cl}), and the quality factor of the cavity in relation to the parameters r and L are given in Table 2.8.

As follows from Table 2.8, the quality factor and the resonance-mode wavelength depend on the volume and the degree of hole conicity. In this case, the maximum quality factor corresponds to a hole with a minimum degree of conicity ($L = h/4$, $r = 5d/12$). Eventually, the hole conicity influences the effective refractive index of the medium. From the results of numerical simulations, it follows that, by optimizing the volume and conicity of holes, we can tune the cavity to the desired wavelength and, at the same time, retain a high-quality factor of the cavity.

2.8   CONCLUSIONS

We describe and numerically investigate an all-optical temporal integrator based on PC nanobeam cavities. We show that an array of PC cavities enables high-order temporal integration. This integrator is more compact than any of those previously suggested. Its dimensions depend linearly on the order of integration. The ways to increase the time-bandwidth product of the integrator by using an active cavity are discussed. For in-plane optical pumping, the nanocavities in the cross section of ridge PC waveguides are described. The structures proposed have been shown to concentrate the electromagnetic field energy in the cavity's regions that contain the material introduced to the resonance cavity. For electrical pumping, the model of two-component nanocavity with the possibility of electrical pumping is also described. In this model, the minimum details of the structure are found only in the periodic component of the resonator. The advantages of such a structure include a promising way to construct an electrically pumped photonic cavity, the ease of introducing nonlinear optical materials in the area of the nanocavity, the possibility of formation of the desired energy distribution in the far zone, and the possibility to construct dynamic systems based on nanocavities. The production technology of a PC cavity formed as a group of holes in a Si strip waveguide by ion-beam etching is described. The parasitic effect associated with hole conicity, which develops upon hole formation by the given technology, is studied. Numerical simulation shows that the hole-conicity-induced decrease in the cavity quality factor can be compensated with consideration for the hole volume. The influence of the waveguide thickness on the resonance wavelength and quality factor of the PC cavity is analyzed.

REFERENCES

1. Ngo, N. Q. 2007. Design of an optical temporal integrator based on a phase-shifted Bragg grating in transmission. *Opt. Lett.* 32(20):3020–2. https://doi.org/10.1364/OL.32.003020.
2. Ferrera, M. et al. 2010. On-chip CMOS-compatible all-optical integrator. *Nat. Commun.* 1:29. https://doi.org/10.1038/ncomms1028.
3. Ding, Y., Zhang, X., Zhang, X., and Huang, D. 2009. Active microring optical integrator associated with electro absorption modulators for high speed low light power loadable and erasable optical memory unit. *Opt. Express.* 17(15):12835–48. https://doi.org/10.1364/OE.17.012835.
4. Slavík, R. et al. 2008. Photonic temporal integrator for all-optical computing. *Opt. Express.* 16(22):18202–14. https://doi.org/10.1364/OE.16.018202.
5. Babiker, S. G., Shuai, Y., Sid-Ahmed, M. O., and Xie, M. 2014. One-dimensional Si/SiO2 photonic crystals filter for thermophotovoltaic applications. *WSEAS Trans. Appl. Theor. Mech.* 9:97–103.
6. Anagnostakis, E. A. 2010. A qualitative comprehension of nanophotonics. *Proceedings of the European Conference of Systems, and European Conference of Circuits Technology and Devices, and European Conference of Communications, and European Conference on Computer Science*, ed. V. Mladenov, K. Psarris, N. Mastorakis, A. Caballero, G. Vachtsevanos, 84–100. Wisconsin, US: World Scientific and Engineering Academy and Society (WSEAS).
7. Singh, S., and Sarin, R. K. 2007. Enhanced performance of microstrip-fed wide slot antenna using periodic gaps in dielectric substrate. *Proc 11th Conf on 11th WSEAS Int Conf on Communications (ICCOM'07)* 11:127–9.
8. Akahane, Y., Asano, T., Song, B.-S., and Noda, S. 2005. Fine-tuned high-Q photonic-crystal nanocavity. *Opt. Express.* 13(4):1202–14. https://doi.org/10.1364/OPEX.13.001202.
9. Reed, G. T., Mashanovich, G., Gardes, F. Y., and Thomson, D. J. 2010. Silicon optical modulators. *Nat. Photonics.* 4:518–26. https://doi.org/10.1038/nphoton.2010.179.
10. Takano, H., Song, B.-S., Asano, T., and Noda, S. 2006. Highly efficient multi-channel drop filter in a two-dimensional hetero photonic crystal. *Opt. Express* 14(8):3491–6. https://doi.org/10.1364/OE.14.003491.
11. Englund, D., Altug, H., Ellis, B., and Vučković, J. 2008. Ultrafast photonic crystal lasers. *Laser Photon. Rev.* 2(4):264–74. https://doi.org/10.1002/lpor.200710032.

12. Faraon, A., and Vučković, J. 2009. Local temperature control of photonic crystal devices via micron-scale electrical heaters. *Appl. Phys. Lett.* 95(4):043102. https://doi.org/10.1063/1.3189081.
13. Tanaka, Y., Upham, J., Nagashima, T., Sugiya, T., Asano, T., and Noda, S. 2007. Dynamic control of the *q* factor in a photonic crystal nanocavity. *Nat. Mater.* 6:862–5. https://doi.org/10.1038/nmat1994.
14. Fushman, I., Waks, E., Englund, D., Stoltz, N., Petroff, P., and Vučković, J. 2007. Ultrafast nonlinear optical tuning of photonic crystal cavities. *Appl. Phys. Lett.* 90(9):091118. https://doi.org/10.1063/1.2710080.
15. Englund, D., Ellis, B., Edwards, E., Sarmiento, T., Harris, J. S., Miller, D. A. B., and Vučković, J. 2009. Electrically controlled modulation in a photonic crystal nanocavity. *Opt. Express.* 17(18):15409–19. https://doi.org/10.1364/OE.17.015409.
16. Tanabe, T., Nishiguchi, K., Kuramochi, E., and Notomi, M. 2009. Low power and fast electro-optic silicon modulator with lateral *p-i-n* embedded photonic crystal nanocavity. *Opt. Express.* 17(25):22505–13. https://doi.org/10.1364/OE.17.022505.
17. Debnath, K., O'Faolain, L., Gardes, F. Y., Steffan, A. G., Reed, G. T., and Krauss, T. F. 2012. Cascaded modulator architecture for WDM applications. *Opt. Express.* 20(25):27420–8. https://doi.org/10.1364/OE.20.027420.
18. Schmidt, B., Xu, Q., Shakya, J., Manipatruni, S., and Lipson, M. 2007. Compact electro-optic modulator on silicon-on-insulator substrates using cavities with ultra-small modal volumes. *Opt. Express.* 15(6):3140–8. https://doi.org/10.1364/OE.15.003140.
19. Shakoor, A., Nozaki, K., Kuramochi, E., Nishiguchi, K., Shinya, A., and Notomi, M. 2014. Compact 1D-silicon photonic crystal electro-optic modulator operating with ultra-low switching voltage and energy. *Opt. Express.* 22(23):28623–34. https://doi.org/10.1364/OE.22.028623.
20. Quan, Q., Floyd, D. L., Burgess, I. B., Deotare, P. B., Frank, I. W., Tang, S. K., and Loncar, M. 2013. Single particle detection in CMOS compatible photonic crystal nanobeam cavities. *Opt. Express.* 21(26):32225–33. https://doi.org/10.1364/OE.21.032225.
21. Kazanskiy, N. L., Serafimovich, P. G., and Khonina, S. N. 2010. Harnessing the guided-mode resonance to design nanooptical transmission spectral filters. *Opt. Mem. Neural Netw.* 19(4):318–24. https://doi.org/10.3103/S1060992X10040090.
22. Serafimovich, P. G., and Kazanskiy, N. L. 2015. Active photonic crystal cavities for optical signal integration. *Opt. Mem. Neural Netw.* 24(4):260–71. https://doi.org/10.3103/S1060992X15040050.
23. Vollmer, F., Braun, D., Libchaber, A., Teraoka, I., and Arnold, S. 2002. Protein detection by optical shift of a resonant microcavity. *Appl. Phys. Lett.* 80(21):4057. https://doi.org/10.1063/1.1482797.
24. Robinson, J. T., Chen, L., and Lipson, M. 2008. On-chip gas detection in silicon optical microcavities. *Opt. Express.* 16(6):4296–301. https://doi.org/10.1364/OE.16.004296.
25. Liu, H. C., and Yariv, A. 2012. Designing coupled-resonator optical waveguides based on high-*q* tapered grating-defect resonators. *Opt. Express.* 20(8):9249–63. https://doi.org/10.1364/OE.20.009249.
26. Kazanskiy, N. L., and Serafimovich, P. G. 2014. Coupled-resonator optical waveguides for temporal integration of optical signals. *Opt. Express.* 22(11):14004–13. https://doi.org/10.1364/OE.22.014004.
27. Asghari, M. H., Wang, C., Yao, J., and Azaña, J. 2010. High-order passive photonic temporal integrators. *Opt. Lett.* 35(8):1191–3. https://doi.org/10.1364/OL.35.001191.
28. Quan, Q., and Loncar, M. 2011. Deterministic design of wavelength scale, ultra-high *q* photonic crystal nanobeam cavities. *Opt. Express.* 19(5):18529–42. https://doi.org/10.1364/OE.19.018529.

29. Huang, N., Li, M., Ashrafi, R., Wang, L., Wang, X., Azaña, J., and Zhu, N. 2014. Active Fabry-Perot cavity for photonic temporal integrator with ultra-long operation time window. *Opt. Express.* 22(3):3105–16. https://doi.org/10.1364/OE.22.003105.

30. Kazanskiy, N. L., Serafimovich, P. G., and Khonina, S. N. 2013. Use of photonic crystal cavities for temporal differentiation of optical signals. *Opt. Lett.* 38(7):1149–51. https://doi.org/10.1364/OL.38.001149.

31. Coccioli, R. 1998. Boroditsky, M., Kim, K. W., Rahmat-Samii, Y., and Yablonovitch, E. Smallest possible electromagnetic mode volume in a dielectric cavity. *IEE Proc., Optoelectron.* 145(6):391–7. https://doi.org/10.1049/ip-opt:19982468.

32. Zhang, Y., McCutcheon, M. W., Burgess, I. B., and Loncar, M. 2009. Ultra-high-*q* TE/TM dual-polarized photonic crystal nanocavities. *Opt. Lett.* 34(17):2694–6. https://doi.org/10.1364/OL.34.002694.

33. Rivoire, K., Buckley, S., and Vučković, J. 2011. Multiply resonant photonic crystal nanocavities for nonlinear frequency conversion. *Opt. Express.* 19(22):22198–207. https://doi.org/10.1364/OE.19.022198.

34. Schriever, C., Bohley, C., and Schilling, J. 2010. Designing the quality factor of infiltrate photonic wire slot microcavities. *Opt. Express.* 18(24):25217–24. https://doi.org/10.1364/OE.18.025217.

35. Yamamoto, T., Notomi, M., Taniyama, H., Kuramochi, E., Yoshikawa, Y., Torii, Y., and Kuga, T. 2008. Design of a high-Q air-slot cavity based on a width-modulated line-defect in a photonic crystal slab. *Opt. Express.* 16(18):13809–17. https://doi.org/10.1364/OE.16.013809.

36. Fan, S., Winn, J. N., Devenyi, A., Chen, J. C., Meade, R. D., and Joannopoulos, J. D. 1995. Guided and defect modes in periodic dielectric waveguides. *J. Opt. Soc. Am. B.* 12(7):1267–72. https://doi.org/10.1364/JOSAB.12.001267.

37. Taflove, A., and Hagness, S. C. 2005. *Computational electrodynamics: The finite-difference time-domain method.* 3rd ed. Norwood, MA: Artech House. ISBN: 978-1-58053-832-9.

38. Almeida, V. R., Xu, Q., Barrios, C. A., and Lipson, M. 2004. Guiding and confining light in void nanostructure. *Opt. Lett.* 29(11):1209–11. https://doi.org/10.1364/OL.29.001209.

39. Ahn, B. H., Kang, J. H., Kim, M. K., Song, J. H., Min, B., Kim, K. S., and Lee, Y. H. 2010. One-dimensional parabolic beam photonic crystal laser. *Opt. Express.* 18(6):5654–60. DOI: 10.1364/OE.18.005654.

40. Serafimovich, P. G., Kazanskiy, N. L., and Khonina, S. N. 2013. Two-component cavity based on a regular photonic crystal nanobeam. *Appl. Opt.* 52(23):5830–4. https://doi.org/10.1364/AO.52.005830.

41. Yang, X., Yu, M., Kwong, D. L., and Wong, C. W. 2009. All-optical analog to electromagnetically induced transparency in multiple coupled photonic crystal cavities. *Phys. Rev. Lett.* 102(17):173902. https://doi.org/10.1103/PhysRevLett.102.173902.

42. Gan, X., Pervez, N., Kymissis, I., Hatami, F., and Englund, D. 2012. A high-resolution spectrometer based on a compact planar two dimensional photonic crystal cavity array. *Appl. Phys. Lett.* 100(23):231104. https://doi.org/10.1063/1.4724177.

43. Deotare, P., Kogos, L., Quan, Q., Ilic, R., and Loncar, M. 2012. On-chip integrated spectrometer using nanobeam photonic crystal cavities. *Conf on Lasers and Electro-Optics*: CM3B.4. https://doi.org/10.1364/CLEO_SI.2012.CM3B.4.

44. Velha, P., Rodier, J. C., Lalanne, P., Hugonin, J. P., Peyrade, D., Picard, E., Charvolin, T., and Hadji, E. 2006. Ultra-high-reflectivity photonic-bandgap mirrors in a ridge SOI waveguide. *New J. Phys.* 8(9):204. https://doi.org/10.1088/1367-2630/8/9/204.

45. Arnold, S., Khoshsima, M., Teraoka, I., Holler, S., and Vollmer, F. 2003. Shift of whispering-gallery modes in micro-spheres by protein adsorption. *Opt. Lett.* 28(4):272–4. https://doi.org/10.1364/OL.28.000272.

46. Boyd, R. W., and Heebner, J. E. 2001. Sensitive disk resonator photonic biosensor. *Appl. Opt.* 40(31):5742–7. https://doi.org/10.1364/AO.40.005742.

47. White, I. M., Oveys, H., and Fan, X. 2006. Liquid-core optical ring-resonator sensors. *Opt. Lett.* 31(9):1319–21. https://doi.org/10.1364/OL.31.001319.

48. Lee, M., and Fauchet, P. M. 2007. Two-dimensional silicon photonic crystal based biosensing platform for protein detection. *Opt. Express* 15(8):4530–5. https://doi.org/10.1364/OE.15.004530.

49. Baba, T. 1997. Photonic crystals and microdisk cavities based on GaInAsP-InP system. *IEEE J. Sel. Top. Quantum Electron.* 3(3):808–30. https://doi.org/10.1109/2944.640635.

50. Pottier, P., Seassal, C., Letartre, X., Leclercq, J. L., Viktorovitch, P., Cassagne, D., and Jouanin, C. 1999. Triangular and hexagonal high Q-factor 2-D photonic bandgap cavities on III-V suspended membranes. *J. Lightw. Technol.* 17(11):2058–62. https://doi.org/10.1109/50.802995.

51. Little, B. E., Chu, S. T., Haus, H. A., Foresi, J., and Laine, J. P. 1997. Microring resonator channel dropping filters. *J. Lightw. Technol.* 15(6):998–1005. https://doi.org/10.1109/50.588673.

52. Yamamoto, Y., and Slusher, R. E. 1995. Optical processes in microcavities. In *Confined electrons and photons: New physics and applications*, ed. E. Burstein, C. Weisbuch, 871–8. New York, US: Springer. https://doi.org/10.1007/978-1-4615-1963-8_46.

53. Labilloy, D., Benisty, H., Weisbuch, C., Krauss, T. F., Bardinal, V., and Oesterle, U. 1997. Demonstration of cavity mode between two-dimensional photonic-crystal mirrors. *Electron. Lett.* 33(23):1978–80. https://doi.org/10.1049/el:19971321.

54. Beaky, M. M., Burk, J. B., Everitt, H. O., Haider, M., and Venakides, S. 1999. Two-dimensional photonic crystal Fabry Perot resonators with lossy dielectrics. *IEEE Trans. Microw. Theory Tech.* 47(11):2085–91. https://doi.org/10.1109/22.798003.

55. Painter, O., Lee, R. K., Scherer, A., Yariv, A., O'Brien, J. D., Dapkus, P. D., and Kim, I. 1999. Two-dimensional photonic band-gap defect mode laser. *Science.* 284(5421): 1819–21. https://doi.org/10.1126/science.284.5421.1819.

56. Chow, E., Lin, S. Y., Johnson, S. G., Villeneuve, P. R., Joannopoulos, J. D., Wendt, J. R., Vawter, G. A., Zubrzycki, W., Hou, H., and Alleman, A. 2000. Three-dimensional control of light in a two-dimensional photonic crystal slab. *Nature.* 407(6807):983–6. https://doi.org/10.1038/35039583.

57. Kazanskiy, N. L., Serafimovich, P. G., and Khonina, S. N. 2012. Enhancement of spatial modal overlap for photonic crystal nanocavities. *Computer Optics.* 36(2):199–204.

58. Kazanskiy, N. L., Serafimovich, P. G., and Khonina, S. N. 2012. Use of photonic crystal resonators for differentiation of optical impulses in time. *Computer Optics.* 36(4):474–8.

59. Hopman, W. C. L., Ay, F., Hu, W., Gadgil, V. J., Kuipers, L., Pollnau, M., and de Ridder, R. M. 2007. Focused ion beam scan routine, dwell time and dose optimizations for submicrometre period planar photonic crystal components and stamps in silicon. *Nanotechnology.* 18(19):195305. https://doi.org/10.1088/0957-4484/18/19/195305.

3 Nanoplasmonic Sensors

Recent Advances

Muhammad Ali Butt[1,2], Nikolay Lvovich Kazanskiy[1,3], and Svetlana Nikolaevna Khonina[1,3]
[1]Samara National Research University, Samara, Russia
[2]Warsaw University of Technology, Institute of Microelectronics and Optoelectronics, Warszawa, Poland
[3]Image Processing Systems Institute – Branch of the Federal Scientific Research Centre "Crystallography and Photonics" of Russian Academy of Sciences, Samara, Russia

3.1 INTRODUCTION

The study of light–matter interaction via surface plasmon polaritons (SPPs) or surface plasmons (SPs) is known as plasmonics. SPs are mutual vibrations of free electrons (e⁻) at a metal surface that are optically stimulated. SPPs are quanta of collective charge density oscillations that are strictly restricted at the boundary of negative and positive permittivity materials. SPs may, however, be described in classical electrodynamics without restoring to quantum physics. They represent genuine mechanical oscillations of the e⁻ gas controlled by the incident light's electromagnetic (EM) field. SP modes are enclosed in nanosized, i.e. subwavelength metal structures-localized surface plasmons (LSPs). Light interaction with nanoscopic matter has been seen for eras, such as in the stained-glass goblets of the Roman Empire or magnificent medieval cathedral windows. More than a century ago, Michael Faraday and Gustav Mie theorized about it. Our capacity to create and manipulate matter on a nanoscale has only recently given birth to what is known as nanoplasmonics.

Photonics, biological imaging, molecular spectroscopy, and sensing, to mention a few, have all benefited from the advancement of nanoplasmonic technology during the previous decade. Subwavelength confinement, a high cross section for scattering and absorption of light and, most importantly for sensing purposes, significantly increased EM fields in the immediate vicinity of nanostructures are all key characteristics that enables such progress. Through field amplification, nanoplasmonic resonances are sensitive to small changes in the surrounding refractive index (RI). This assumption is the foundation for nanoplasmonic sensing, which is essentially RI sensing with the strong addition that the investigated volumes are nanoscale. In a variety of nanoplasmonic sensing systems, even single biological or chemical molecular substances can be identified.

The metal-insulator-metal waveguide configuration is among the most frequently utilized plasmonic-based n-structures for the development of dense optical

DOI: 10.1201/9781003439165-3

circuits. Metal-insulator-metal waveguides are plasmonic structures with two metal claddings around an insulator. The essential aspects of this system are its simple construction and the ability to restrict light at the subwavelength level. Researchers have studied the construction of a wide range of devices utilizing metal-insulator-metal waveguides to build highly integrated optical circuits. Filters [1], couplers [2], sensing devices [3], demultiplexers [4], switches [5], modulators [6], and splitters [7] are among the devices included. Photonic crystals are used to create a variety of efficient devices, including optical filters, logic gates, sensing devices, and demultiplexers, which may be built in plasmonic waveguides with less than one footprint area.

RI-sensing devices offer a wide range of uses in the biological and chemical sciences, and have been extensively explored recently, including the measurement of unknown concentration of the solution and pH value via RI variations [8, 9]. As illustrated in Figure 3.1(a), an EM field is created by activating the sensor element with light, which creates extremely focused SPs at the metal's surface. When the material under investigation comes into contact with the sensor, the metal-insulator-metal waveguide's n_{eff} variations cause λ_{res} to redshift, as illustrated in Figure 3.1(b). Variations in the RI in the neighborhood of the surface are extremely sensitive to SPs. Deviations in one of the properties of the light connected to the SP, such as the λ_{res}, intensity, or phase, can be used to calculate the ΔRI.

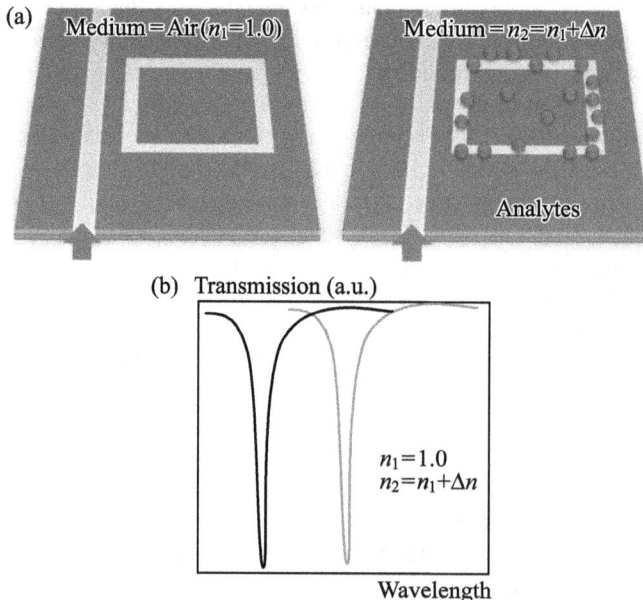

FIGURE 3.1 (a) Diagram of a sensing method in plasmonic sensing devices. (b) Redshift in λ_{res} because of the variation in RI in the adjacent medium.

3.2 PLASMONIC WAVEGUIDES

In terms of speed and data transfer rate, semiconductor-integrated circuits and recent electronic-integrated devices are swiftly nearing their basic limits. One possible answer to these issues is to employ light rather than e⁻s in the fundamental processing components. Nevertheless, due to the prevailing diffraction limit of light when the size of photonic device is near to or smaller than the λ of light in the material, the feasibility of creating nanoscale photonic devices may be constrained. Bypassing the diffraction limit, SPP EM waves linked to charge oscillation at the metal-dielectric interface can accomplish localization of EM energy in nanoscale areas that are significantly smaller than the $\lambda(s)$ of light in the material. The EM field perpendicular to the metal-dielectric boundary decays exponentially from the metal surface while preserving long-range EM energy transmission from the surface. SPPs may be utilized to make photonic components and optical signal processing devices at the subwavelength scale because of this significant feature.

In recent years, plasmonic waveguides such as metal-insulator-metal, insulator-metal-insulator, metal grooves, metal strips, metal wedge, and hybrid Bragg waveguides have been suggested. These waveguides may confine EM waves close to the boundary, well beyond the diffraction limit of light. Metal-insulator-metal waveguide formations are one of the most attractive approaches for building nanoscale photonic-integrated circuits, and they are particularly well suited to domains of optical communication and sensing. As illustrated in Figure 3.2(a), this waveguide system is made up of three layers, with a dielectric core of either air or another low RI material inserted between two metal claddings. In [10], the dispersion relationship for metal-insulator-metal configurations is analytically explained, as shown in Figure 3.2(b). Gordon [11] and Dionne et al. [12] provided thorough analytical analyses of metal-insulator-metal waveguides in 2006, while Bozhevolnyi et al. experimentally showed channel plasmon subwavelength waveguide elements such as interferometers and ring resonators (RRs) [13]. An insulator-metal-insulator is one more plasmonic waveguide construction in which metal is inserted between two insulator claddings that have been employed in several eye-catching utilizations for active [14] and passive [15] elements. In Japan, an SPP-based research article for the fast detection of COVID-19 was recently published.

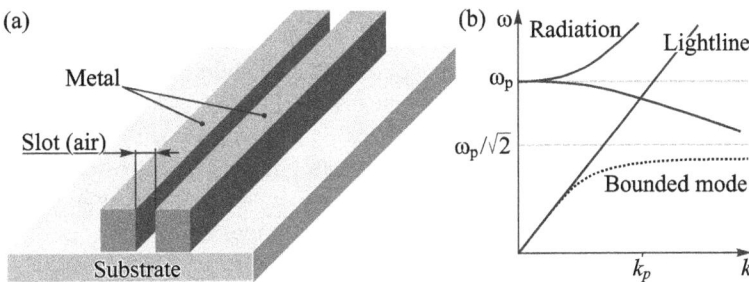

FIGURE 3.2 (a) Graphical illustration of a metal-insulator-metal waveguide and (b) dispersion plot of a metal-insulator-metal layered assembly adapted from [10].

Antibiotic-coated Au nanoparticles (NPs) experience a resonance peak shift when the viruses are captured, resulting in a unique color transition [Kurabo. (2020). Available online at: https://www.kurabo.co.jp/news/products/ (accessed July 7, 2020).]. Comparable techniques are widely used in pregnancy testing.

Several plasmonic waveguides, such as groove waveguides, multilayer metal-insulator-metal waveguides, and nanowires, have recently been proven to allow the long-distance propagation of SPPs. Fundamental and higher-order modes occur in metal groove waveguides, generally identified as channel plasmon polariton (CPP) modes [13]. Depending on the modes, the energy might be restricted to various places in the groove. The taper angle can be used to change the modes and propagation distance. To properly direct the SPPs, the groove depth must not be substantially less than the basic mode's penetration depth. [13] described the setup, manufacturing, and characterization of CPP-based subwavelength waveguide components working at telecom λ(s), such as Y-splitters, Mach–Zehnder interferometers (MZI), and RRs. [5] showed a passive imbalanced MZI established on multilayer metal-insulator-metal plasmonic waveguides, paving the path for a small EO modulator. Owing to their single crystal formations and flat surfaces, chemically produced nanowires have a low loss. The cylinder's Bessel function [16] may represent the basic modes of wires, and the transmission of SPPs on a metal wire can create a helix. When the wire is put on a substratum, the top and bottom of the wire exhibit faster and slower modes, respectively.

3.3 CRITICAL ISSUES ASSOCIATED WITH PLASMONIC SENSING DEVICES ESTABLISHED ON METAL-INSULATOR-METAL WAVEGUIDE FORMATIONS

In this section, we offer a short explanation of the most extensively utilized plasmonic materials, fabrication methods, transmission losses, modal solutions, and inquisition methods.

3.3.1 FAVORITE PLASMONIC MATERIALS

Polarization describes the interaction between a material and an EM wave. Although polarization can be electric and/or magnetic in nature, for frequencies greater than several hundred THz, the magnetic polarization of naturally occurring materials is insignificant. The electric polarization can be defined by the complex electric permittivity or dielectric function of material, which is signified by $\varepsilon(\omega)$. Whereas, in fact, the real part of the dielectric function (ε') defines the intensity of the polarization induced by an exterior electric field, the imaginary part (ε'') explains the losses faced in polarizing the material. Therefore, a material with insignificant loss is related to a low value of ε''.

Primary loss mechanism in the soft UV, visible (VIS), and near-IR frequencies may be arising from events associated with conduction e⁻ and bound e⁻ (interband effects). Losses of conduction e⁻ are caused by collisions between e⁻-e⁻ and e⁻-phonon and by scattering because of lattice defects or grain boundaries. As the conduction e⁻ have

almost a continuum of vacant states, their contact with an EM field is perfectly estimated by classical theory. Through considering conduction e⁻ as a three-dimensional
(3D) free e⁻ gas, the Drude theory explains this phenomenon. The permittivity of a
material can be expressed according to the generalized Drude principle:

$$\varepsilon(\omega) = \varepsilon(\omega)' + i\varepsilon(\omega)'' = \varepsilon_{int} - \frac{\omega_p^2}{\omega\left(\omega + \dfrac{i}{\tau}\right)}, \tag{3.1}$$

where

$$\omega_p^2 = \frac{ne^2}{\varepsilon_0 m^*} \tag{3.2}$$

In Eq. (3.1), ε_{int} is an input caused by interband transitions (it is unity for a
perfectly free e⁻-gas) and τ is the mean relaxation time of conduction e⁻. The plasma
frequency (ω_p) is assumed by Eq. (3.2), where n is the conduction e⁻ density, and m^*
is the effective optical mass of conduction e⁻. Generally, ε_{int} counts on λ, which is
usually described by adding the Lorentz oscillators terms, but for some frequency
domains, it can be estimated as constant, as presented in Table 3.1.

Plasmonic utilizations necessitate materials with negative ε'. This condition is
met by materials with a plasma frequency greater than the preferred application
frequency. Metals are frequently chosen because of their high electric conductivity
and large plasma frequencies. The material properties for high-conductivity metals,
as indicated in the references, are summarized in Table 3.1. Among the metallic
elements, Ag has the lowest size and is the best alternative at optical frequencies.
Because of its minimal imaginary part of permittivity and superlative e⁻ conduction, Ag is regarded the most promising material in the near-infrared (NIR) range.
The disadvantage of Ag is that it is easily oxidized, which limits its utilizations
and allows Au to be an acceptable option in this frequency range. At lower NIR
frequencies, Au is frequently used because it is chemically stable under a variety
of circumstances. In the VIS continuum, however, Au suffers substantial interband
losses at $\lambda(s)$ below or about 500 nm. Cu, too, suffers from substantial interband loss
over most of the VIS continuum. Al is the most suited material for UV or deep UV
(DUV) utilizations, for example, 193 nm light for DUV optical litho.

TABLE 3.1
Drude Model Parameters for Metals

Metals	ε_{int}	ω_p (eV)	Γ (eV)	ω_{int} (eV)	Reference
Silver (Ag)	3.7	9.2	0.02	3.9	[17]
Gold (Au)	6.9	8.9	0.07	2.3	[17]
Copper (Cu)	6.7	8.7	0.07	2.1	[17]
Aluminum (Al)	0.7	12.7	0.13	1.41	[18]

For plasmonic utilizations around the optical frequencies, Ag and Au were primarily the materials of choice. Future plasmonic utilizations, however, need even lower losses to make full use of their prospective. Cu is an alternative plasmonic material with almost comparable interband transition and optical damping as Au in the wavelength range 600 nm to 750 nm. Unfortunately, Cu is also susceptible to oxidation. To overcome the corrosion problem, graphene has recently been used on top of Cu or Ag. Because graphene is chemically and mechanically inert, it is kept away from fluidics. Cu-graphene and Ag-graphene optical fiber surface plasmon resonance (SPR) sensors provide long-term stability and reliability. Niobium is another one-of-a-kind plasmonic material with a high chemical resistance and high mechanical strength. The linking between niobium film and silica glass is so strong that no extra bonding layer is required. SPR sensors based on indium tin oxide have lately drawn a lot of interest because of its low bulk plasma frequency. Furthermore, it has the same optical damping as Ag and Au.

3.3.2 Fabrication Methods

Metals are more difficult to microfabricate than semiconductors (mainly noble metals). With the use of sputtering or evaporation methods, plasmonic metals may be deposited in a vacuum. The deposition rate is generally restricted to <1 nm/s to achieve adequate uniformity and quality. Because the necessary plasmonic system height is usually modest, the slow deposition rate is not a big problem. The electroplating [19] technique can produce faster growth rates, but the surface quality may deteriorate, making it unsuitable for SP waves. A prototype shift from the photoresist (PR) to noble metals is difficult due to the restricted number of efficient etching techniques for Cu, Ag, and Au. In Cl_2-based settings, however, Al may be plasma etched.

For pattern metals, two main techniques are widely employed. The first method involves image reversal and lift-off, which necessitates resist overhang after advancement. MicroChem Corp. [20] sells special PRs for the lift-off process, for example lift-off resist. Ion milling is a second method for patterning metal. A focused ion beam may be employed to mill designs directly onto Au or Ag film [21]; nevertheless, redeposition from the beam source is common. Although an Ar ion beam [22] is a viable option to etch metals, it necessitates the use of hard masks and may increase the manufacturing stages and processing complication. Several techniques, including e−-beam and ion-beam litho, may be used to fabricate n-structures for plasmonic sensing devices, enabling the cost-effective production of n-structures across vast regions. Furthermore, wet chemical etching or vapor deposition methods may be used to make metal NPs quickly and easily. The purpose of this chapter, however, is not to go through these approaches in depth. We recommend some literature [23, 24] that provides an in-depth discussion of this essential issue as an alternative.

3.3.3 Transmission Loss

Unlike dielectric waveguides, where transmission loss is unimportant, plasmonic waveguides are very lossy due to the presence of metal in the waveguide system. Major ohmic losses impact the transmission of directed SPs, limiting the maximum propagation length. Numerous structures have been created utilizing collections of nanosize dimension characteristics to offset these losses [25, 26]. A system featuring a thin lossy metal sheet resting on a dielectric substratum and enclosed by a varied dielectric superstrate was used to attain the longest propagation length of 13.6 mm [26].

In general, transmission loss and mode confinement are trade-offs in plasmonic waveguides. The bigger the transmission loss, the smaller the mode size. The modes validated by plasmonic waveguides are usually more complicated in form. As a result, determining the mode confinement of plasmonic waveguide is difficult. This problem has been debated, with the conclusion that the definition of the mode region should be determined by the specific purpose [27]. Energy dissipation occurs in both electronic and plasmonic circuits. The SP transmission length during which the strength of the SPP falls to 1/e of its initial value is used to characterize the transmission loss.

Other variables, like as the spacer and insulator medium in the slot, also influence SPP propagation in metal-insulator-metal waveguides. However, if the waveguide is tens of kilometers long and anagogic to electronic devices, the losses caused by the metal absorption are considerable. As a result, relaying the signals to configure SPP-based devices is critical.

3.3.4 INQUISITION METHOD

Various light properties are used to investigate the transmission spectra of plasmonic sensor devices. As a result, plasmonic sensing devices may be divided into three groups based on the inquisition scheme used, for instance intensity, phase, or wavelength. Lasers or superluminescent light-emitting diodes (SLEDs) with narrowband filters are mandatory to excite the plasmonic modes. When compared to broadband sources, these light sources have superior stability and power, resulting in a higher signal-to-noise ratio. Any variation in the intensity of light linked to a plasmonic mode is often monitored with a one-dimensional (1D)-PDA (personal digital assistant) or 2D detectors like CCD (continuity of care document) or CMOS (complementary metal–oxide–semiconductor) cameras.

Phase inquisition is the process of converting phase variations into physically detectable intensity signals by interacting a data-carrying light with a reference beam. When common route interferometry methods are employed, the primary advantage of phase investigation is evident, because the reference and signal beams are accessed over the same optical channel and hence are impacted by the same noise components. As the measured signal is constantly referenced, phase inquisition eliminates background and nonspecific noise from the environment. Different ways to such evaluations have recently been presented for plasmonic sensing devices that use the phase inquisition method. Retrieving phase information from intensity data is one of the most intriguing ideas. The most common method is polarimetry, which involves extracting phase information from the interference of s- and p-polarized light using a polarizer. The electric field perpendicular to the plane of incidence is denoted by s-polarization, whereas the electric field parallel to the plane of incidence is denoted by p-polarization. Their relative phase difference is determined from a series of intensity measurements using angular modulation created by a rotating analyzer or phase modulation formed by a liquid crystal variable wave plate (LCVWP) and a static output analyzer [28]. Because this is effectively the equivalent idea as ellipsometry and is a mature approach for the depiction of thin films, phase inquisition-based plasmonic sensing devices might benefit from commercially available equipment.

The most common inquisition approach employed in RRs based on plasmonic systems is the measurement of variations in the λ_{res} [29]. λ inquisition-based optical systems often employ a polychromatic light source, for instance a halogen lamp or a SLED, to span the whole continuum where λ_{res} is likely to be seen. In terms of light continuum, halogen lamp technology is superior. When using the fixed incidence angle configuration and λ inquisition approach, this light source is preferable. A CCD-, CMOS-, or PDA-based spectrometer is often used to collect continuum of light coupled to an SP, and the spectral location of the plasmonic feature is scrutinized utilizing suitable feature-tracking algorithms.

3.4 PERFORMANCE CHARACTERISTICS OF SURFACE PLASMON POLARITON RESONATOR

In this section, the most studied metal-insulator-metal sensing device configurations for RI-sensing purposes, for example circular ring resonators (CRRs), split square ring resonators (SSRRs), notched RRs, gear-shaped nanocavities, square ring resonators (SRRs), and hexagon cavities, among others, are reviewed. Scientists proposed these sensing devices employing either a side-coupled cavity or edge-coupled cavity. Because of the short coupling length, edge-coupled resonators have tinier FWHM in contrast to side-coupled configurations [30]. Most metal-insulator-metal plasmonic resonator sensing devices are currently being proposed theoretically and numerically. However, it is necessary to realize these concepts experimentally and to make them available in biomedical purposes. Now, we will discuss two major attributes of plasmonic sensing devices based on how their performance is evaluated.

3.4.1 SENSITIVITY

The capacity to detect variations in the RI is the most utilized performance parameter of plasmonic sensing devices. This is commonly expressed as the bulk RI S, which is defined as sensitivity $S = \Delta\lambda_{res}/\Delta n$, where λ_{res} is the λ at which the SP excitation arises, and Δn is the RI shift. Table 3.3 shows the S of a variety of plasmonic sensing device designs that have been published. The S varies significantly depending on the type of the EM mode, resonant λ, excitation geometry, and substratum, among other aspects. S has been studied in relation to the geometric characteristics of metal-insulator-metal waveguide RRs encased by a (homogeneous) sample by numerous researchers [31–50].

Plasmonic metamaterials (MMs) made from Au nanorods (abbreviated as NRs) are used to represent the dispersion of SPPs localized on a metal sheet [51], therefore enhancing S. The plasmonic MMs support a guided mode with resonant excitation conditions equivalent to the SPP mode of a smooth metal sheet when the spacing between the NRs is smaller than the incident λ. This MM has 3,000 nm/RIU S to RI variations in the medium between the rods. Circular dichroism can be used instead of extinction-based sensing methods for chiral nanostructures. Their spectral characteristics are sharper than chiral particles because of their POL-dependent (polarization-dependent) spectra. These characteristics offer chiral plasmonic nanohelices with astonishing RI S of 1,091 nm/RIU and FOM* = 2,800 RIU^{-1} [52].

3.4.2 The Figure of Merit

The capability of a plasmonic sensing device to detect small variations in the RI is proportionate to S and inversely related to the spectral width of the resonance feature that is being tracked. The Figure of Merit (FOM) is expressed as the product of these characteristics and is represented as FOM = S/FWHM. Even though the FOM is quite high in certain studies [34, 41, 46, 48], it is computed differently in each study and is typically referred to as FOM*. For example, the expression $\Delta R/(R\Delta n)$ at a fixed λ is used in [53] to calculate an extraordinary FOM* = 2.33 × 10^4, where R signifies the variation in reflection intensity due to variations in the refractive index (n) of the ambient medium and R is the reflection rate in the sensing device system. FOM = 4.05 × 10^4 is calculated in [46] using the equation $\Delta T/T\Delta n$, where T indicates the transmittance in the suggested formations and $\Delta T/\Delta n$ signifies the transmission variation at a constant λ caused by a RI modification. FOM is frequently utilized to evaluate diverse plasmonic sensing devices from a sensing standpoint.

Fano resonances (FRs) are a technique for improving FOM by lowering the spectral width of a plasmonic feature. The goal of mode coupling research has been to achieve a small FWHM. Cetin et al. [54] revealed that Fano resonant asymmetric ring/disk systems have spectrally sharp features with FWHM as small as 9 nm and a high S as big as 648 nm/RIU. FR is a sort of resonant scattering phenomenon that happens when a discrete state and a continuous state are coupled and interfere [55]. In metallic n-structures, FR has been explored as the ideal characteristic for preventing the diffraction limit of light [56] produced by SPPs. Consequently, several plasmonic forms have been suggested to produce FR, including rectangular cavities [57], plasmonic n-clusters [58], n-slits [59], and metal-insulator-metal waveguide structures [60].

Because of their sharp and asymmetric line structure [61], which allows the transmission continuum to be rapidly decreased from peak to trough, plasmonic sensing devices based on FR are expected to be very sensitive. The transmission continuum's FWHM is quite small, which increases the sensing device's sensing resolution substantially [62]. In 2019, several RI-sensing devices based on FR were disclosed, although only a handful are cited here [63–71]. In terms of FOM, Fano resonant systems may be able to outperform Lorentz resonant systems; nevertheless, the insufficient contrast of plasmonic spectral characteristics may restrict the system's capacity to convert advances in FOM to the capacity to identify smaller variations in the bulk RI.

3.4.3 Newly Developed Refractive Index Sensing Devices

Many metal-insulator-metal resonator models have been suggested and studied for RI-detecting purposes in recent years [30–75]. Figure 3.3(a–h) depicts a summary of such arrangements. Some display Lorentzian line forms, whereas others show FRs, which enhance FOM considerably.

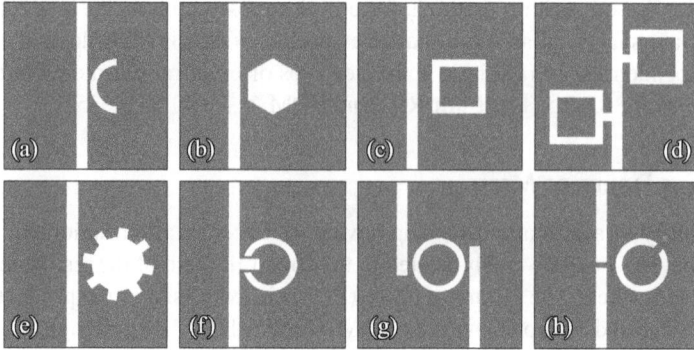

FIGURE 3.3 (a) Semi-RR, (b) side-coupled hexagonal cavity resonator, (c) side-coupled RR, (d) double-sided coupled RR, (e) gear-shaped nanocavity resonator, (f) notched RR, (g) two metal-insulator-metal waveguide-coupled circular RR, and (h) side-coupled split circular RR.

Plasmonic sensing devices are smaller and easier to incorporate on-chip than fiber sensors [72, 73]. The S, on the other hand, is not as high as that of fiber sensors. As a result, the major focus of research is on improving the S of plasmonic sensing devices. In [32], the author has proposed an RI-sensing device with an S higher than most previously described sensing devices.

A metal-insulator-metal waveguide-coupled asymmetric resonator for biosensing and slow light purposes was presented. The transmission continuum of the device had electromagnetically induced transparency (EIT)–like features. EIT denotes a quantum mechanical phenomenon involving quantum destructive interference between resonant excitation channels at the higher atomic level, which can affect an atomic system's optical properties. This sensor's S and FOM are 806 nm/RIU and 66, respectively [76]. Because of steep dispersion and strong resonance at the transparency window, many utilizations such as ultrafast switching and filtering are based on the EIT effect [77, 78].

Given the progress of highly integrated photonic circuits, the mechanism of multi FRs in single one subwavelength metal-insulator-metal formations has received the most attention. As a result, metal-insulator-metal waveguide-based composite configurations, such as groove-cavity composite formations and cascaded groove formations, have been successfully suggested and explored [34, 44, 46, 70]. In the IR wavelength region, dual Fano transmission peaks with asymmetrical line forms have been obtained. Chen et al. [79] suggested a novel method for producing double FRs by combining two equal stub resonators combined to a rectangular cavity offers S and FOM of 1,100 nm/RIU and 91, respectively. An end-coupled slot resonator with extraordinary S and FOM was used to create multiple FRs [80].

3.4.4 Examples

Here, we have presented two detailed examples for designing the plasmonic sensor for RI sensing purpose via the finite element method (FEM). In the first illustration, the multichannel metallic dual nanowall (NW) square split RR is explained, which can be employed as a filter/sensor, whereas in the second, the plasmonic sensor with nanodots is presented for enhanced S.

3.4.4.1 Numerical Modeling of a Multichannel Metallic Dual Nanowall Split Square Ring Resonator

This section numerically analyzes a new metal-insulator-metal waveguide construction consisting of a multichannel dual metallic NW SSRR side paired with a metal-insulator-metal bus waveguide. A square ring geometry has the benefit of revealing additional resonance modes that are attributed to the involvement of bends in a rectangular ring construction [81]. The suggested tool is customized by embedding a dual metallic NW within a standard metal-insulator-metal square ring, which disrupts the metal-insulator-metal RR's symmetry and disqualifies these resonators from being classified as traveling-wave resonators. By appropriately placing the NW, the resonance modes may be extremely stimulated or repressed. The effects of NW on filtering and sensing properties are quantitatively investigated using FEM. Other related characteristics are carefully chosen, such as the relative permittivity (ε_r) of Ag and the RI of the air. The relative permittivity of Ag abides by the Drude–Lorentz model as follows:

$$\varepsilon_m(\omega) = \varepsilon_\infty - \frac{\omega_p^2}{\omega(\omega + i\gamma)}, \tag{3.3}$$

where $\varepsilon_\infty = 3.7$ is the dielectric constant at an infinite frequency, $\omega_p = 9.1$ eV is the bulk plasma frequency of free conduction e$^-$, and $\gamma = 0.018$ eV is the e$^-$ collision frequency. ω is the angular frequency of the incident light in a vacuum. Because of its small imaginary part of ε_r in the NIR band, Ag is used in the arrangement. As a result, as compared to Au and Al, its power consumption is minimal. Furthermore, the Ag layer has great adherence to the substrate and can be easily patterned with strong etch selectivity in HNO_3 and H_2O.

3.4.4.1.1 *Selective Wavelength Filter Configuration*
Metal-insulator-metal waveguide filter configuration in this example comprises four dual NWs SSRR, two on each side of the bus waveguide. The outer length of rings is signified by $l1$, $l2$, $l3$, and $l4$, where $l1 > l2 > l3 > l4$. The inner length of the ring is l_i-d, where $i = 1, 2, 3,$ and 4 and d is the cavity width. The distance between bus waveguide and rings is represented by g, which is static during the examination process. The dual NWs are introduced in the ring; the width of the wall is represented by S. All the four rings are coupled with the output ports, P_i that transmits the filtered λ contingent on the resonance conditions. The graphical illustration of the filter configuration is presented in Figure 3.4. The geometric variables of the dual NW SSRR are specified in Table 3.2.

When a TM-polarized light is introduced into a metal-insulator-metal waveguide, it excites SPPs that propagate along the metal's interface. Only the fundamental TM mode can be sustained in this configuration since the waveguide width is less than

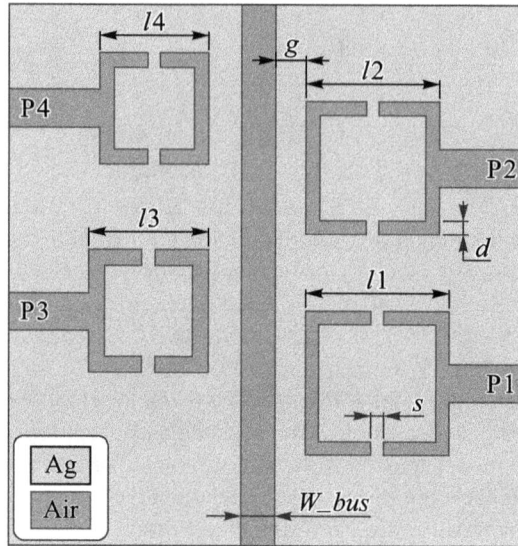

FIGURE 3.4   Graphical illustration of a multichannel λ filter.

TABLE 3.2

Geometric Variables of the Dual NW Square Split Ring Resonator (SSRR)

W_bus (nm)	l1 (nm)	l2 (nm)	l3 (nm)	l4 (nm)	s (nm)	d (nm)	g (nm)
30	100	90	80	70	5–25	5–20	20

the incident λ. Figure 3.5(a) shows the E-field mapping at the cross section of the metal-insulator-metal waveguide. The λ_{res} may be calculated by changing the RR's effective index (n_{eff}). As shown in Figure 3.5(b), the n_{eff} of the resonator is completely dependent on the size of the metal-insulator-metal waveguide. Using the Maxwell equations and boundary conditions, the n_{eff} in the waveguide may be calculated. Furthermore, by adding a material with a marginally higher RI into the surrounding medium, n_{eff} may be changed. Consequently, the resonant modes shift to longer λ(s) as the RI of the medium increases. This capability permits this configuration to be used in sensor purposes.

As the NWs are positioned in the middle of each ring, integer modes in the VIS and near-IR λ domain are stimulated. Figure 3.6(a–d) displays the excitation of modes at λ_{res} for the corresponding rings.

3.4.4.1.2 Biological-Sensing Applications

In addition to filter applications, the suggested design may be employed in biosensing utilizations [82–85]. By increasing the RI in the medium, λ_{res} can be moved to

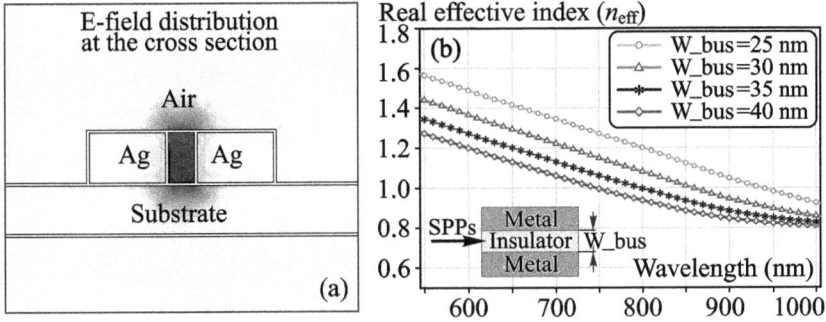

FIGURE 3.5 (a) E-field mapping at the cross section of a metal-insulator-metal waveguide.
(b) Real part of the n_{eff} of a metal-insulator-metal waveguide for bus waveguide width 25 nm,
30 nm, 35 nm, and 40 nm.

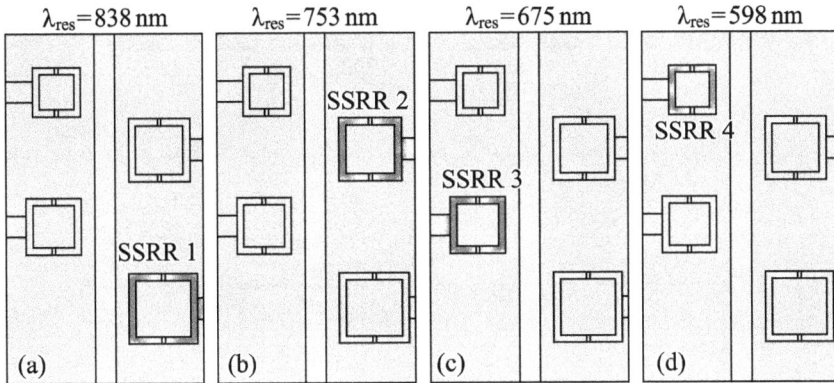

FIGURE 3.6 E-field mapping in the SSRR at λ_{res}: (a) 838 nm, (b) 753 nm, (c) 675 nm,
(d) 598 nm.

higher $\lambda(s)$. As a result, the suggested structure can simultaneously serve as a sensor
and a filter. The S, FOM, and Q-factor are utilized to determine the RI sensor's
performance quantitively. We filled the adjacent medium with RI = 1.3 to 1.5 to estimate the S of the suggested configuration and displayed the transmission continuum
attained from P1, P2, P3, and P4, as revealed in Figure 3.7. A significant redshift
in λ_{res} is seen when the RI of the medium increases. This improvement implies that
altering the RI in the cavity might provide another way to adjust the structure's
filtering properties.

S, FOM, and Q-factor all grow linearly with the size of the SSRR, as shown in
Figure 3.8(a)–(c). Dual NW SSRR 1 has the best S, FOM, and Q-factor of 793.3 nm/
RIU, 52.9, and 82.1, respectively. In Table 3.3, the sensor's best continuum charac-
teristics are listed.

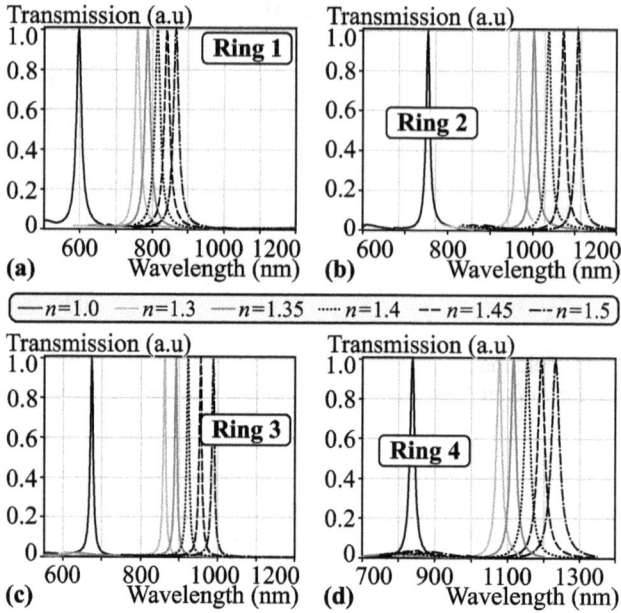

FIGURE 3.7   The transmission field of dual NW SSRR configurations for different values of n. The output from (a) P1, (b) P2, (c) P3, (d) P4.

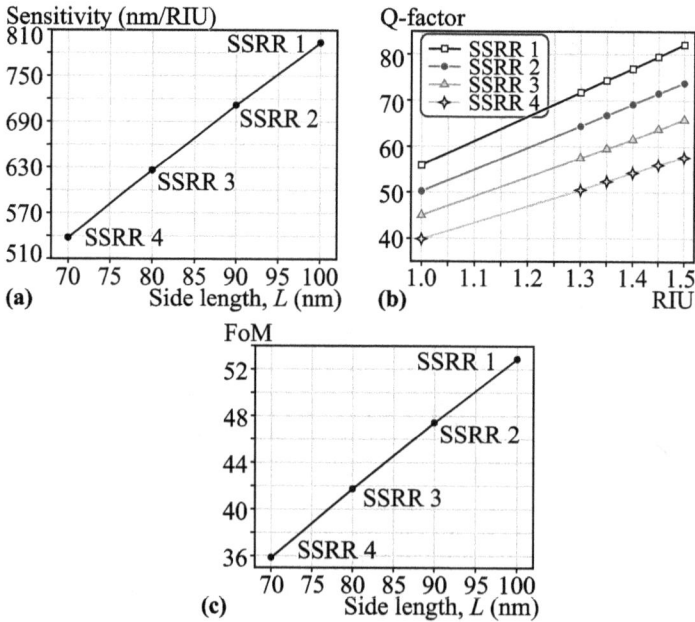

FIGURE 3.8   (a) S versus L, (b) Q-factor, (c) FOM versus L.

TABLE 3.3

Spectral Features of a Dual NW SSRR

	SSRR 1	SSRR 2	SSRR 3	SSRR 4
S (nm/RIU)	793.3	711.4	626	537.8
Q-factor	82.1	73.8	65.9	57.7
FOM	52.9	47.4	41.7	35.9

3.4.4.2 A Technique for Boosting the Sensitivity of a Conventional Plasmonic Metal-Insulator-Metal Square Ring Resonator

In this illustration, a square ring resonator (SRR) arrangement combined with Ag nanodots is presented. The width of the bus waveguide, the side length of the ring, cavity width, and the gap between bus and ring are signified by w, L, c, and g, respectively. The ring waveguide is connected to the output port, which transfers the filtered λ reliant on the resonance conditions. A surface integration of the output port yields the transmission continuum. As seen in Figure 3.9(a), there are seven nanodots on either side of the cavity. The nanodots are put in a cavity with a period of $3r$, where r is the nanodots' radius. The sensor's structural variables are listed in Table 3.4. The E-field mapping of light in the conventional ring arrangement at $\lambda_{res} = 712$ nm is shown in Figure 3.9(b).

As seen in Figure 3.9(c), an array of nanodots acts as a grating, allowing the energy in the cavity to build up. The inset illustrates the amplification of SPPs at nanodot's borders, which allows for more light contact with the surrounding medium and therefore increases the sensor's S.

Filling the surrounding medium with a variety of RIs is used to determine the resonator's continuum response. For the λ range of 600 nm to 1,200 nm, the normalized intensity at the sensor's out is computed for $n = 1.0$, 1.1, 1.15, 1.2, 1.25, 1.30, and 1.35. The S is computed in Figure 3.10(a) concerning the number of nanodots. Seven nanodots are initially inserted in one side length of the ring in the cavity, followed by the addition of the other three sides regularly. The nanodot's radius is set at 2.5 nm.

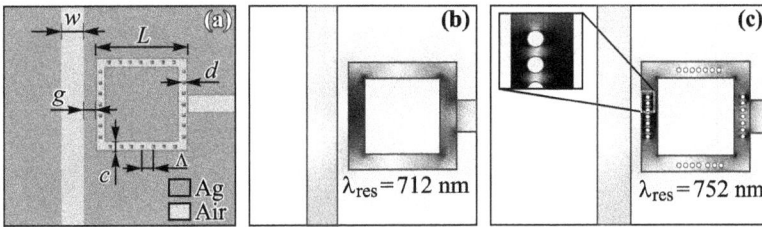

FIGURE 3.9 (a) Schematic illustration of a metal-insulator-metal SRR formation with nanodots embedded in the cavity. (b) E-field mapping in the conventional SRR at λ_{res}. (c) E-field mapping in the nanodots embedded SRR formation at λ_{res}.

TABLE 3.4

Geometric Variables of the Sensor

W (nm)	L (nm)	g (nm)	c (nm)	r (nm)
30	100	10	15	2, 2.5, 3, 3.5, 4

TABLE 3.5

The S of the Nanodots Embedded SRR

	Standard Ring	Nanodots on One Side	Nanodots on Two Sides	Nanodots on Three Sides	Nanodots on Four Sides
S_{max} (nm/RIU)	670	720	740	732	738

The inclusion of nanodots in the cavity results in a significant increase in S. Table 3.5 shows the resonator's maximum S_{max}.

The resonator's S, Q-factor, FOM, and λ_{res} are all affected by the size of the nanodots. As a result, we looked at the influence of nanodots radius on spectral competence by keeping the number of nanodots at 28 (seven on each side of the ring), as shown in Figure 3.10(b). The resonator's S_{max}, FOM, and Q-factor are shown in Table 3.6. The FOM and Q-factor of the resonator fall dramatically as the size of the nanodots increases, resulting in a substantial increase in S. The widening of the FWHM is linked to a reduction in the FOM and Q-factor. As shown in Figure 3.11(a), the highest FOM and Q-factor have been obtained for the resonator with nanodots of size 2 nm, whereas $S_{max} = 907$ nm is obtained for the nanodots of radius 4 nm due to the intensification of SPPs at the metal-insulator-metal bus waveguide boundary. Furthermore, as seen in Figure 3.11(b), λ_{res} exhibits a redshift as the size of the nanodots increases.

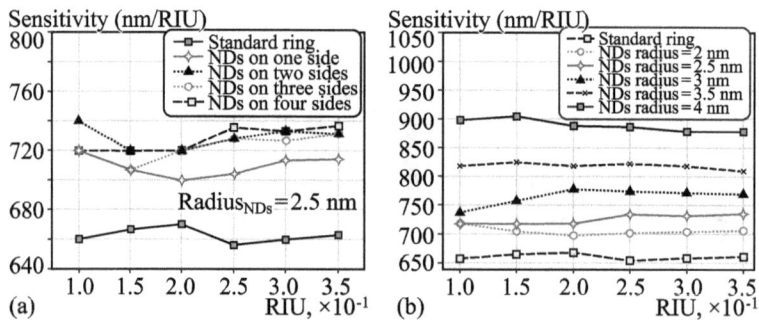

FIGURE 3.10   The S of the SRR by placing nanodots of (a) 2.5 nm radius on different sides of the cavity; (b) various sizes in all four sides of the cavity.

TABLE 3.6

Spectral Competence of the Nanodots Embedded Metal-Insulator-Metal S-R-R

Variables	Conventional RR	Nanodots $r = 2$ nm	Nanodots $r = 2.5$ nm	Nanodots $r = 3$ nm	Nanodots $r = 3.5$ nm	Nanodots $r = 4$ nm
S_{max} (nm/RIU)	670	720	738	780	827	907
Q-factor	52.4	52.6	45.9	46.6	45.7	41
FOM	49.2	50.4	43.4	44.3	43.7	40.8

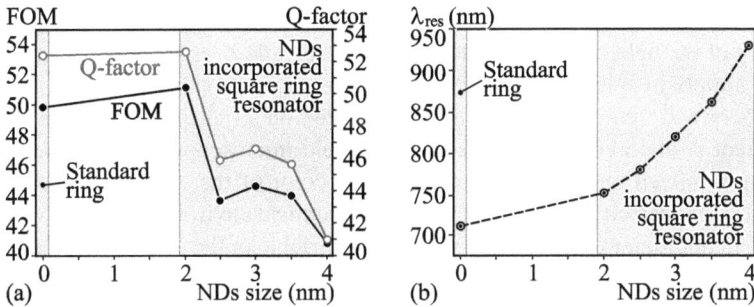

FIGURE 3.11 (a) FOM and Q-factor versus the size of nanodots; (b) λ_{res} shift versus nanodots.

3.5 SURFACE PLASMON RESONANCE SENSOR ARRANGEMENT

The resonance of SP is the secret to great performance in the design of SPR sensors. A p-polarized light can best couple the resonance state at the interface of the thin metal and dielectric medium at a given λ and incidence angle. When the SP's wave vector (β_{SP}) value equals the incident light's propagation constant, this happens. To have a better understanding of this phenomenon, the SP wave vector in the nano-metal film surface is hypothesized in the following equation:

$$\beta_{SP} = \text{Re}\left\{ \frac{2\pi}{\lambda} \sqrt{\frac{\varepsilon_M \varepsilon_D}{\varepsilon_M + \varepsilon_D}} \right\}, \tag{3.4}$$

where λ is the wavelength of the incident light, while ε_M and ε_D are the real part value of the metal and medium dielectric constant, respectively. One more important factor is the propagation length (P_{length}) of the SP wave articulated by the following formula:

$$P_{length} = \frac{\pi}{2\lambda} \frac{\varepsilon_{iM}}{\varepsilon_M^2} \left(\frac{\varepsilon_M \varepsilon_D}{\varepsilon_M + \varepsilon_D} \right)^{\frac{3}{2}}, \tag{3.5}$$

where the ε_{iM} is the imaginary part value of the metal-dielectric constant. A certain λ necessitates a specific configuration in the SP generation to meet the incident light's propagation constant at a precise angle of incidence. The availability of a light source determines the priority of investigation and modification of the overall sensor setup.

3.5.1 SURFACE PLASMON RESONANCE SENSING WITH NANOPARTICLES AND THIN FILMS

SPR-based methods make use of the optical phenomenon of a fixed λ resonance involving free e$^-$ on a metal surface. Metallic NP coloring might be the most persuasive proof of this optical phenomena. When Au particles are smaller than a nanometer, they lose their typical yellow or bulk Au appearance. For example, spherical Au NPs with a diameter of 13 nm look brilliant red. This optical phenomenon is described by the fact that light with the suitable wavevector can resonate with the metal's free e$^-$, causing a collective oscillation of the metal's free e$^-$ known as SP and instigating the intensity of the light beam to be attenuated at a specific λ, as shown in Figure 3.12.

The simplicity with which LSPR may be measured explains its rising popularity. For NPs or n-structured substratum immobilized on a transparent surface, UV-VIS equipment is often employed, whereas dark-field microscopy may also be used for NPs immobilized on solid support [86]. The color of the light absorbed is determined by the dielectric constant or RI of the chemical environment in the vicinity of the NP, in addition to the type and geometric variable of the metallic nanomaterials.

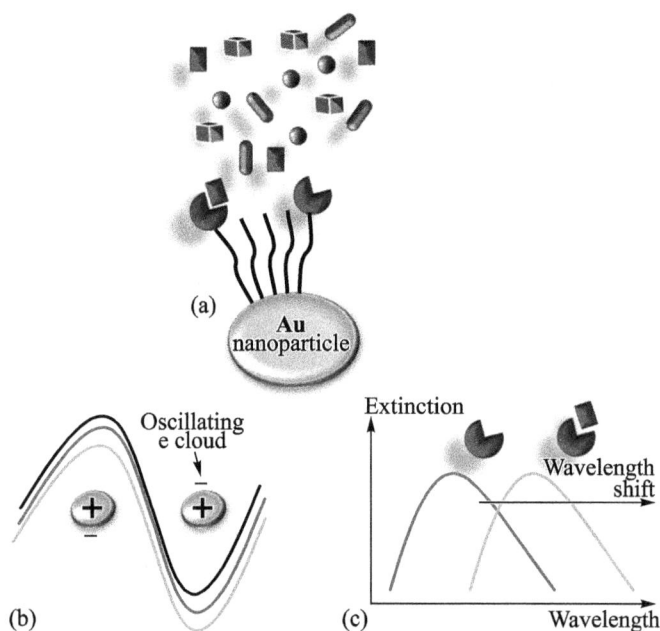

FIGURE 3.12 Graphical illustration of biosensing experiments with LSPR. (a) Biosensing scheme on an Au NP, onto which a chemical layer is placed to selectively capture a molecule in the occurrence of interfering agents. (b) Excitation in extinction spectroscopy by a beam of light entering resonance at a precise λ with the free e$^-$ cloud. (c) The extinction continuum redshifts with the binding of molecules to the Au NP.

The λ of the resonance λ performs a redshift as the RI of the solution increases, which is conventionally referred to as the bulk S, which is measured in nm/RIU. SPR has a detecting range of a few tens to hundreds of nanometers, which is restricted by the depth of penetration SPR probes have into the solution. The λ shift seen in SPR sensing is frequently determined by the quantity or size of molecules adsorbed on metallic surface, and this notion lies at the heart of SPR-sensing devices. SPR materials having a shallow penetration depth are more responsive to the development of a molecular adsorbate, and, as a result, to analyte detection. High S is a dynamic field of study in SPR sensing, as it is in other analytical methods, to further increase the detection of even lower concentrations of analytes.

3.5.1.1 Kretschmann Configuration

The pioneering work of Otto Kretschmann (KR) and Raether in the late 1960s discovered that the SPR phenomenon occurs when the light at a metal/liquid interface induces SP waves, which are linked with collective excitations of e− in the metal. The majority of commercially available SPR biosensors still employ KR and Raether's technique of focusing p-polarized light through a glass prism and then reflecting it from a thin Au layer of 50 nm placed on its surface. When the tangential x-component of the incident optical wave vector, k_x, is equal to k_{SP}, the SP wave vector, the pumping light energy is reassigned to the SPs. Under total internal reflection circumstances, the SPR effect results in the form of a strong resonant dip in the reflected light intensity at a specified angle (θ). In reality, because of the strict need for optical and evanescent wave vector matching, an alteration in the RI of the ambient medium that touches Au (within 200–300 nm) would have a significant impact. Figure 3.13 shows a graphical representation of the KR setup.

Most SPR-sensing devices monitor RI variations because of binding events near the Au layer that cause SPR to arise. More specifically, the aggregation of biological material within that thin layer, which occurs because of interactions with macromolecules

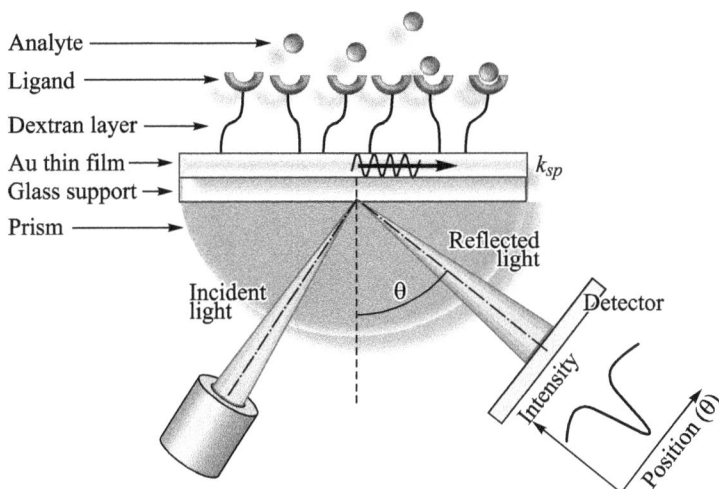

FIGURE 3.13 SPR detection mechanism in the KR setup.

found in the liquid and species formerly immobilized on the sensing device's surface, influences the RI consequently results in a change in θ observed in real time. The detection limit (DL) of existing equipment for RI change is, at best, 10^{-7} RIU. It results in a DL of around 0.1 pg mm^{-2} of biomaterial collecting on the biosensor surface. Angle scanning, λ scanning, and SPR imaging are three types of equipment used in SPR sensing investigations. Angle scanning instruments are the highly popular commercial equipment, and they are often attained by optical or mechanical means.

For analyzing samples on a few channels, such instruments offer great *S*. An SPR image-sensing device's function is based on measuring the intensity of a reflected beam incident at a certain angle and λ. An imaging camera detects a change in intensity due to variations in SPR response. The SPR-sensing device designates areas for various analytes. SPR image-sensing devices are useful because they provide multiplexing capabilities for evaluating numerous analytes. SPR sensing based on the KR arrangement has evolved into a sophisticated method for examining biomolecular interactions without the use of labels. The ability of SPR sensing to determine binding constants and binding partners has been particularly useful in biological sciences. We have tabulated a few of the sensing device configurations based on the KR configuration in Table 3.7 for the reader's convenience.

TABLE 3.7
The Summary of SPR-Sensing Device Advancement

No.	Scheme	Target Sample	Performance	Ref
1	KR, SPR, λ = 632.8 nm, the μ-array sensing membrane	Si coating SAM	Decay length ~4 μm	87
2	KR, dielectric mirror TiO₂/SiO₂ sensing formation, λ = 632 nm	Glucose solution	Res: 1.28×10^{-5} RIU Dynamic range: 1.331–1.5 RIU	88
3	KR, angular int., λ = 632.8 nm	Sugar content in carbonated drink	LOD: 0.01–0.05%	89
4	KR, Au/SnO₂ sensing film, angular int., λ = 633 nm	Ammonia gas	*S* 0.055 o/ppm (0.5–250 ppm)	90
5	KR, Au/ZnO sensing film, angular int., λ = 633 nm	DNA of N. meningitidis	LOD: 5 ng/μL	91
6	KR, intensity mod. at dual λ(s) references	DNA hybridization	LOD: 2×10^{-6} RIU	92
7	KR CaF₂ prism, TiO₂/Au layer on sensing, angular int., λ = 633 nm	CO₂	5X *S* improvement	93
8	KR, angular int., diverging laser beam, λ = 637 nm	Ethanol solution	LOD: 5×10^{-6} RIU	94
9	KR with the rotating diffuser, SPRi, CCD camera detector, λ = 633 nm, high throughput, and disposable sensing configuration	IgG, BSA	Proof of concept for multi-sample detection	95

DNA – deoxyribonucleic acid; IgG – immunoglobulin G; BSA – bovine serum albumin; LOD – limit of detection

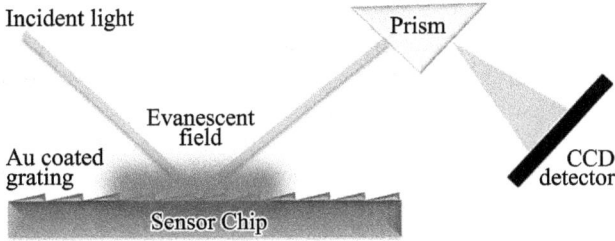

FIGURE 3.14 The configuration of the SP grating coupler.

3.5.1.2 Grating Coupled Surface Plasmon Resonance

Grating-enhanced SPR excitation has been a remarkable pioneer in SPR arrangement. Figure 3.14 demonstrates the grating configuration of SP excitation. Wood published the first SPR excitation phenomena on the grating arrangement in 1902 [96]. Enhancements were subsequently made until Rayleigh established an analytical explanation to the Wood's anomalies explaining the diffraction angles used in grating methods by means of the following formula [97]:

$$\sin(\theta p) = \sin\theta + p\frac{\lambda}{\Lambda}, \tag{3.6}$$

where θ is the incidence angle of p-polarized light, θp is the diffraction angle, λ is the λ of the incoming light, and Λ is the period of groove. This narrative permits the control of diffraction angle of any scattered order from the grating period p, the incidence angle of light (θ), and the λ. The passing-off of the order n occurs when $\sin(\theta p) = \pm 1$. Hence, from Eq. (11), the λ of a continuum producing the passing-off of a diffracted order are expressed as [98]:

$$n\frac{\lambda}{\Lambda} = -\sin(\theta m) \pm 1; m = \pm1, \pm2, \pm3, ... \tag{3.7}$$

The λ of the passing-off defined earlier is known as the Rayleigh λ. The relation of incident light vector and grating configuration in resonance condition can be articulated by [99]:

$$n_D\sin\theta + m\frac{\lambda}{\Lambda} = \pm\mathrm{Re}\left\{\sqrt{\frac{\varepsilon_M\varepsilon_D}{\varepsilon_M + \varepsilon_D}}\right\} + \Delta n_{\mathrm{ef}}, \tag{3.8}$$

where n_D is the RI of the medium and $\Delta n_{\mathrm{ef}} = \mathrm{Re}\{\Delta\beta(\lambda/2\pi)\}$, and $\Delta\beta$ is the propagation constant fluctuating in the presence of the grating formation. In the grating coupler-based SPR-sensing devices, the prism does not need to be directly coupled to the chip.

The sensing device chip substratum is embossed with a diffractive grating that separates the incident light beam into several beams that exit at different angles. These beams can be further collected and separate by prism sets, making it possible to observe the intensity of the spectra. Since a sensing device chip is not directly linked to a prism, the systems needed for chip handling can be significantly simplified. This method also eliminates the requirement that the width of the plasmon-forming metal layer be precisely regulated, although the need to emboss the diffraction grating on the chip complicates the production process. Another important consideration is that since the light must pass through the sample to enter the sensing device chip, the optical clarity of both the analyte and the buffer solution is important. We are referring to a few recently proposed grating-based SPR-sensing devices that might interest the readers [100–103].

3.5.2 BREAKTHROUGH IN LOCALIZED SURFACE PLASMON RESONATOR SENSING

Developing new plasmonic materials that are highly sensitive to bulk RI directs to greater variations in the SPR interaction and low DL. Colloidal chemistry has been advanced to a great extent over the past decade to realize metallic NPs of different forms and materials, containing spheres, cubes, rods, and pyramid-shaped NPs. The excitation λ and S are vastly dependent on the geometric variables of the NPs. Even though widespread investigation has been dedicated to improving the analytical properties of metallic NPs, yet there is no agreement on the optimal shape and size of the NP. Nanospheres are simple to fabricate but at the cost of the low S of ~76 nm/RIU [104]. Au NRs are more sensitive than nanospheres (366 nm/RIU) and are stimulated in the near-IR, a spectral range that is relatively free of interference, but long-term strength of Au NRs must be enhanced for sensing utilizations [104]. Certain formations such as cubes, hollow spheres, and pyramids are ideal substratum for enhanced Raman scattering, but they have not been commonly used in SPR sensing.

The current advancement in producing precise dimension of NP and n-structured materials have facilitated the advancement of novel analytical approaches using localized surface plasmon resonator (LSPR) sensing. Some of these analytical tactics exploit the plasmonic coupling method. Metallic NPs placed in close vicinity to an Au film or another metallic NP results in large variations in coloration, which may offer sufficient S for single molecular recognition. Proteins, enzymes, and DNA interactions can be effectively examined with the configuration of a colorimetric-sensing device based on the combination of Au NP. Analytes with multiple binding sites or the interaction of binding partners on different NPs can induce aggregation. In addition, colorimetric-sensing devices can be built into an immunochromatographic test-chip, analyzing or a paper-based detection scheme designed to generate low-priced biosensors. Various uses other than sensing are also envisaged. Au NPs are being studied for photothermal therapy, using heat dissipation around Au NP while excited by a laser beam to eradicate cancer cells. This method can result in tumor staining and marginally invasive therapy with suitable surface chemistry to identify tumors. The low penetration of light in tissues because of absorption and scattering makes this treatment inadequate for cancer therapy. Therefore, the manufacturing of n-structured surfaces has created a multitude of different plasmonic materials that have been immobilized onto a substratum. This enables the advancement of SPR-sensing devices, allowing the multiplexing of SPR-sensing devices based on NPs. For instance, e-beam or photolitho focus ion beam milling and colloidal litho techniques were used to fabricate n-structured plasmonic sensing devices, such as nanotriangle arrays, films over nanospheres, nanocrescents, nanohole arrays, and a collection of other exotic structures. Chip-based LSPR-sensing devices are being built in a test-strip with an optical reader system or in a multiplexed array of sensing devices. Such advancements in nanomaterials have had a significant impact on sensing devices established on LSPR by offering reproducible means for producing n-structures on a simple to the manageable substratum, enhancing S and creating colorimetric assays for biomolecules.

3.6 FUTURE OF PLASMONIC SENSING DEVICES

In just about a decade, n-plasmonic sensing devices made a remarkable evolution from the theoretical idea to the operational devices. It is apparent that this technology is proficient of providing high S, small size, readily accessible multiplexing competences that can be apprehended and applied in a wide range of areas, both in investigation and even consumer devices. The functional surface chemistry has progressively become the central point in the advancement of n-plasmonic sensing devices that target at point-of-care diagnostics. New conceptual configurations have been revealed to further simplify n-plasmonic sensing devices such as the recent advances in paper-based substratum for MM devices that can potentially be used for quantitative examination in biochemical sensing purposes, as shown in Figure 3.15 [105]. Experimental demonstrations are carried out by patterning μm-sized metal-insulator-metal resonators on paper substratum and measuring the resonance shift

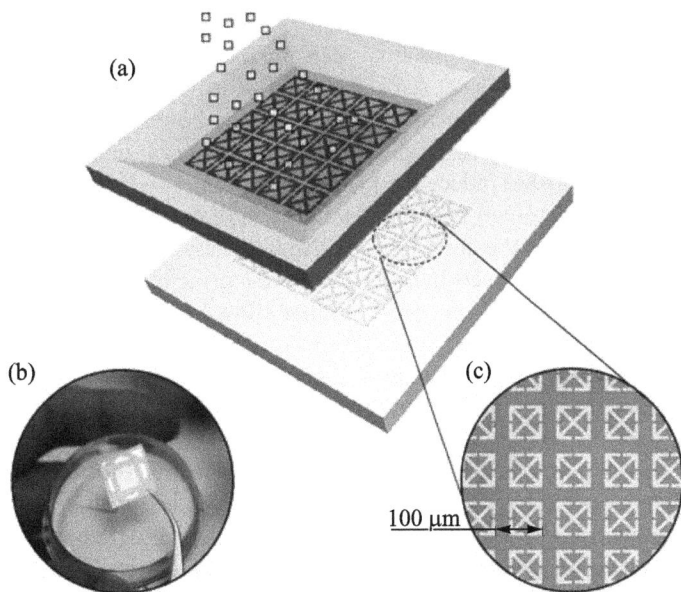

FIGURE 3.15   (a) Graphical illustration of μm-sized MM resonators sprayed on paper substratum with a predefined micron stencil. (b) Photo of a paper-based THz MM device. (c) Optical microscopy photo of one part of manufactured paper MM sample [105].

brought by engaging various concentrations of glucose solution on the metal-insulator-metal surface.

A highly responsive label-free sensing method established on asymmetric FRs in plasmonic nanoholes with comprehensive repercussions for point-of-care diagnostics is established. By utilizing exceptional light transmission phenomena through high Q-factor ~200 subradiant dark modes, FOM as great as 162 for intrinsic dynamic light scattering (DLs) exceeding that of the Au standard prism coupled surface-plasmon sensing devices (KR configuration) have been demonstrated [106]. The examples of the recent progress are the selective decoration with analyte-capturing chemistry of nanostars [107, 108], nanocrescents [109], nanotriangles [110], and others. In combination with surface chemistry, proper configuration of plasmonic n-structures can also yield significantly increased and "localized" S [111]. It is expected that the interplay of numerous n-plasmonic resonances in one structure would generate n-plasmonic sensing devices with the ability to resolve the tridimensional analyte structure [112]. It is likewise incredibly clear that applications outside biological and chemical sensing are increasingly sought by n-plasmonic detectors.

Materials science would greatly profit from the wider application of specific n-plasmonic sensing methods. Intriguingly, since the most sophisticated and commercially productive plasmonic sensing platforms are those using thin noble metal films to propagate SPPs, there is a drift to merge both [113]. This is realized, for instance, in plasmonic nanowires. Extending the sensing device's probing volume further, as with SPR technology, while retaining intense enhancement of the EM field characteristic of the n-plasmonic sensing devices, is certainly beneficial. On the side of n-plasmonic sensing device's fundamental characteristics, it is imperative that consistent predictive models emerge that explicitly link the structure with optical and sensing properties of n-plasmonic sensing platforms.

On one more functional side, scalable manufacturing methods that can deliver large arrays of n-plasmonic structures are expected to be further developed. At the same time, nanofabrication of an industrial scale that aims at multiplexed array format is essential for further reduce the price per chip in n-plasmonic-based sensing. All in all, we will most likely see a gradual growth of n-plasmonic sensing devices for different purposes in the coming years. Nevertheless, there are still fascinating fundamental research prospects in the configuration principles and function of sensing devices. Methods have been developed that allow low-cost substratum, such as plastics and polymers, to manufacture these n-structures. Plasmonics is one of the most exciting research areas in modern science, which will continue to evolve. In the years to come, faster, smaller, and cheaper sensing device architectures based on plasmonics will be reported, capitalizing on recent breakthroughs such as nanoflow through LSPR single-molecule detection and highly sensitive FRs.

Nevertheless, incorporating these innovations into biomedical devices is a challenge for the industry. Soon, biosensors will be universal as they are needed for significant transformative approaches that have been expected, such as routine point-of-care-based diagnosis and personalized medicine. Low-cost biosensors are also required for rapid diagnosis of infectious diseases in developing countries and remote locations. Therefore, plasmonic biosensors are well prepared not only to be at the center of an emerging healthcare revolution but also to propel it. Furthermore, advancements in perpetual issues in surface chemistry, such as nonspecific binding and consistent molecular immobilization schemes, have helped the growth of the field. The variety of established sensing devices and the various sensing schemes suggest that a plasmonic solution can be used for several biochemical or biomedical problems, ranging from fundamental protein–protein interaction studies and single protein dynamics to disease diagnosis by detecting DNA fragments in complex biological samples at femtomolar concentrations.

ACKNOWLEDGEMENT

"The work was supported by the Russian Science Foundation grant No. 21-79-20075"

REFERENCES

1. Chen, L., Lu, P., Tian, M., Liu, D., and Zhang, J. 2013. A subwavelength MIM waveguide filter with single-cavity and multi-cavity structures. *Optik.* 124(18):3701–4. https://doi.org/10.1016/j.ijleo.2012.11.025.
2. Pu, M., Yao, N., Hu, C., Xin, X., Zhao, Z., Wang, C., and Luo, X. 2010. Directional coupler and nonlinear Mach-Zehnder interferometer based on the metal-insulator-metal plasmonic waveguide. *Opt. Express.* 18(20):21030–7. https://doi.org/10.1364/OE.18.021030.
3. Duan, G., Lang, P., Wang, L., Yu, L., and Xiao, J. 2016. An optical pressure sensor based on π-shaped surface plasmon polariton resonator. *Mod. Phys. Lett. B.* 30(21):1650284. https://doi.org/10.1142/S0217984916502845.
4. Kou, Y., and Chen, X. 2011. Multimode interference demultiplexers and splitters in metal-insulator-metal waveguides. *Opt. Express.* 19(7):6042–7. https://doi.org/10.1364/OE.19.006042.
5. Kamada, S., Okamoto, T., El-Zohary, S. E., and Haraguchi, M. 2016. Design optimization and fabrication of Mach-Zehnder interferometer based on MIM plasmonic waveguides. *Opt. Express.* 24(15):16224–31. https://doi.org/10.1364/OE.24.016224.
6. Babicheva, V. E., and Lavrinenko, A. V. 2013. A Plasmonic modulator based on a metal-insulator-metal waveguide with a barium titanate core. *Photonics Lett. Poland.* 5(2):57–9. https://doi.org/10.4302/plp.2013.2.08.
7. Butt, M. A., Khonina, S. N., and Kazanskiy, N. L. 2020. Ultra-short lossless plasmonic power splitter design based on metal-insulator-metal waveguide. *Laser Phys.* 30:016201. https://doi.org/10.1088/1555-6611/ab5577.
8. Kuchmizhak, A., Pustovalov, E., Syubaev, S., Vitrik, O., Kulchin, Y., Porfirev, A., Khonina, S. N., Kudryashov, S. I., Danilov, P., and Ionin, A. 2016. On-fly femtosecond-laser fabrication of self-organized plasmonic nanotextures for chemo- and biosensing applications. *ACS Appl. Mater. Interfaces.* 8(37):24946–55. https://doi.org/10.1021/acsami.6b07740.
9. Kudryashov, S. I., Danilov, P. A., Porfirev, A. P., Saraeva, I. N., Nguyen, T. H. T., Rudenko, A. A., Khmelnitskii, R. A., Zayarny, D. A., Ionin, A. A., Kuchmizhak, A. A., Khonina, S. N., and Vitrik, O.B. 2019. High-throughput micropatterening of plasmonic surfaces by multiplexed femtosecond laser pulses for advanced IR-sensing applications. *Appl. Surf. Sci.* 484:948–56. https://doi.org/10.1016/j.apsusc.2019.04.048.
10. Economou, E. N. 1969. Surface plasmons in thin films. *Phys. Rev.* 182(2):539–54. https://doi.org/10.1103/PhysRev.182.539.
11. Gordon, R. 2006. Light in a subwavelength slit in a metal: Propagation and reflection. *Phys. Rev. B.* 73(15):153405. https://doi.org/10.1103/PhysRevB.73.153405.

12. Dionne, J. A., Sweatlock, L. A., Atwater, H. A., and Polman, A. 2006. Plasmon slot waveguides: Towards chip-scale propagation with subwavelength-scale localization. *Phys. Rev. B.* 73(3):035407. https://doi.org/10.1103/PhysRevB.73.035407.

13. Bozhevolnyi, S. I., Volkov, V. S., Devaux, E., Laluet, J.-Y., and Ebbesen, T. W. 2006. Channel plasmon subwavelength waveguide components including interferometers and ring resonators. *Nature.* 440(7083):508–11. https://doi.org/10.1038/nature04594.

14. Nikolajsen, T., Leosson, K., and Bozhevolnyi, S. I. 2004. Surface plasmon polariton based modulators and switches operating at telecom wavelengths. *Appl. Phys. Lett.* 85(24):5833–5. https://doi.org/10.1063/1.1835997.

15. Charbonneau, R., Scales, C., and Breukelaar, I. et al. 2006. Passive integrated optics elements based on long-range surface plasmon polaritons. *J. Lightw. Technol.* 24(1):477–94. https://doi.org/10.1109/JLT.2005.859856.

16. Ditlbacher, H. et al. 2005. Silver nanowires as surface plasmon resonators. *Phys. Rev. Lett.* 95(25):257403. https://doi.org/10.1103/PhysRevLett.95.257403.

17. Johnson, P. B., and Christy, R. W. 1972. Optical constants of the noble metals. *Phys. Rev. B.* 6(12):4370–9. https://doi.org/10.1103/PhysRevB.6.4370.

18. Ehrenreich, H., Philipp, H., and Segall, B. 1963. Optical properties of aluminum. *Phys. Rev.* 132(5): 1918–28. https://doi.org/10.1103/PhysRev.132.1918.

19. Giurlani, W. et al. 2018. Electroplating for decorative applications: Recent trends in research and development. *Coatings.* 8(8):260. https://doi.org/10.3390/coatings8080260.

20. MicroChem Corp. 2012, March. http://www.microchem.com.

21. Kannegulla, A., and Cheng, L.-J. 2016. Metal assisted focused-ion beam nanopatterning. *Nanotechnology.* 27(36):36LT01. https://doi.org/10.1088/0957-4484/27/36/36LT01.

22. Hindmarch, A. T., Parkes, D. E., and Rushforth, A. W. 2012. Fabrication of metallic magnetic nanostructures by argon ion milling using a reversed-polarity planar magnetron ion source. *Vacuum.* 86(10):1600–4. https://doi.org/10.1016/j.vacuum.2012.02.019.

23. Masson, J.-F., Murray-Methot, M.-P., and Live, L. S. 2010. Nanohole arrays in chemical analysis: Manufacturing methods and applications. *Analyst.* 135(7):1483–9. https://doi.org/10.1039/c0an00053a.

24. Cao, J., Sun, T., and Grattan, K. T. V. 2014. Gold nanorod-based localized surface plasmon resonance biosensors: A review. *Sens. Actuators B Chem.* 195:332–51. https://doi.org/10.1016/j.snb.2014.01.056.

25. Maier, S. A., Barclay, P. E., Johnson, T. J., Friedman, M. D., and Painter, O. 2004. Low-loss fiber accessible plasmon waveguide for planar energy guiding and sensing. *Appl. Phys. Lett.* 84(20):3990. https://doi.org/10.1063/1.1753060.

26. Maier, S. A., Friedman, M. D., Barclay, P. E., and Painter, O. 2005. Experimental demonstration of fibre-accessible metal nanoparticle plasmon waveguides for planar energy guiding and sensing. *Appl. Phys. Lett.* 86:071103. https://doi.org/10.1063/1.1862340.

27. Oulton, R. F., Bartal, G., Pile, D. F. P., and Zhang, X. 2008. Confinement and propagation characteristics of subwavelength plasmonic modes. *New J. Phys.* 10:105018. https://doi.org/10.1088/1367-2630/10/10/105018.

28. Otto, L. M., Mohr, D. A., Johnson, T. W., Oh, S. H., and Lindquist, N. C. 2015. Polarization interferometry for real-time spectroscopic plasmonic sensing. *Nanoscale.* 7(9):4226–33. https://doi.org/10.1039/C4NR06586G.

29. Homola, J., Sinclair, S., and Gauglitz, G. 1999. Surface plasmon resonance sensors: Review. *Sens. Actuators B.* 54(1–2):3–15. https://doi.org/10.1016/S0925-4005(98)00321-9.

30. Butt, M. A., Khonina, S. N., and Kazanskiy, N. L. 2021. Metal-insulator-metal nano square ring resonator for gas sensing applications. *Waves Random Complex Media.* 31(1):146–56. https://doi.org/10.1080/17455030.2019.1568609.

31. Zhang, Z., Yang, J., He, X., Zhang, J., Huang, J., Chen, D., and Han, Y. 2018. Plasmonic refractive index sensor with high figure of merit based on concentric-rings resonator. *Sensors.* 18(1):116. https://doi.org/10.3390/s18010116.

32. Wu, T., Liu, Y., Yu, Z., Peng, Y., Shu, C., and Ye, H. 2014. The sensing characteristics of plasmonic waveguide with a ring resonator. *Opt. Express.* 22(7):7669–77. https://doi.org/10.1364/OE.22.007669.

33. Wei, W., Zhang, X., and Ren, X. 2015. Plasmonic circular resonators for refractive index sensors and filters. *Nanoscale Res. Lett.* 10:211. https://doi.org/10.1186/s11671-015-0913-4.

34. Chen, Z., and Yu, L. 2014. Multiple Fano resonances based on different waveguide modes in a symmetry breaking plasmonic system. *IEEE Photonics J.* 6(6):4802208. https://doi.org/10.1109/JPHOT.2014.2368779.

35. Gaur, S., Zafar, R., and Somwanshi, D. 2016. Plasmonic refractive index sensor based on metal insulator metal waveguide. *IEEE International Conference on Recent Advances and Innovations in Engineering (ICRAIE-2016)*: 1–4. https://doi.org/10.1109/ICRAIE.2016.7939557.

36. Zhang, Z., Luo, L., Xue, C., Zhang, W., and Yan, S. 2016. Fano Resonance based on metal-insulator-metal waveguide-coupled double rectangular cavities for plasmonic nanosensors. *Sensors.* 16(5):642. https://doi.org/10.3390/s16050642.

37. Binfeng, Y., Hu, G., Zhang, R., and Yiping, C. 2016. Fano resonances in a plasmonic waveguide system composed of stub coupled with a square cavity resonator. *J. Opt.* 18(5):055002. https://doi.org/10.1088/2040-8978/18/5/055002.

38. Yan, S., Zhang, M., Zhao, X., Zhang, Y., Wang, J., and Jin, W. 2017. Refractive index sensor based on a metal-insulator-metal waveguide coupled with a symmetric structure. *Sensors.* 17(12):2879. https://doi.org/10.3390/s17122879.

39. Zhao, X., Zhang, Z., and Yan, S. 2017. Tunable Fano resonance in asymmetric MIM waveguide structure. *Sensors.* 17(7):1494. https://doi.org/10.3390/s17071494.

40. Butt, M. A., Khonina, S. N., and Kazanskiy, N. L. 2018. Hybrid plasmonic waveguide-assisted metal-insulator-metal ring resonator for refractive index sensing. *J. Mod. Opt.* 65(9):1135–40. https://doi.org/10.1080/09500340.2018.1427290.

41. Rakhshani, M. R., Tavousi, A., and Mansouri-Birjandi, M. A. 2018. Design of a plasmonic sensor based on a square array of nanorods and two slot cavities with a high figure of merit for glucose concentration monitoring. *Appl. Opt.* 57(27):7798–804. https://doi.org/10.1364/AO.57.007798.

42. Wang, L., Zeng, Y.-P., Wang, Z.-Y., Xia, X.-P., and Liang, Q.-Q. 2018. A Refractive index sensor based on an analogy t shaped metal-insulator-metal waveguide. *Optik.* 172:1199–204. https://doi.org/10.1016/j.ijleo.2018.07.093.

43. Butt, M. A., Khonina, S. N., and Kazanskiy, N.L. 2018. Plasmonic refractive index sensor based on M-I-M square ring resonator. *Int. Conf. on Computing, Electronic and Electrical Engineering (ICE Cube)*: 1–4. https://doi.org/10.1109/ICECUBE.2018.8610998.

44. Fang, Y., Wen, K., and Li, Z. et al. 2019. Multiple Fano resonances based on end-coupled semi-ring rectangular resonator. *IEEE Photon. J.* 11(4):4801308. https://doi.org/10.1109/JPHOT.2019.2914483.

45. Chen, Y., Xu, Y., and Cao, J. 2019. Fano Resonance sensing characteristics of MIM waveguide coupled square convex ring resonator with metallic baffle. *Results Phys.* 14:102420. https://doi.org/10.1016/j.rinp.2019.102420.

46. Yu, S., Zhao, T., Yu, J., and Pan, D. 2019. Tuning multiple Fano resonances for on-chip sensors in a plasmonic system. *Sensors.* 19(7):1559. https://doi.org/10.3390/s19071559.

47. Asgari, S., and Granpayeh, N. 2019. Tunable mid-infrared refractive index sensor composed of asymmetric double graphene layers. *IEEE Sens. J.* 19(14):5686–91. https://doi.org/10.1109/JSEN.2019.2906759.

48. Shi, Y., Zhang, G.-M., An, H.-L., Hu, N., and Gu, M.-Q. 2017. Controllable Fano resonance based on coupled square split ring resonance cavity. *Acta Photonica Sinica* 46(4):0413002. https://doi.org/10.3788/gzxb20174604.0413002.

49. Zafar, R., and Salim, M. 2015. Enhanced figure of merit in Fano resonance-based plasmonic. *IEEE Sens. J.* 15(11):6313–7. https://doi.org/10.1109/JSEN.2015.2455534.

50. Butt, M. A., Khonina, S. N., and Kazanskiy, N. L. 2019. A multichannel metallic dual nano-wall square split-ring resonator: Design analysis and applications. *Laser Phys. Lett.* 16(12):126201. https://doi.org/10.1088/1612-202X/ab5574.

51. Kabashin, A.V. et al. 2009. Plasmonic nanorod metamaterials for biosensing. *Nat. Mater.* 8(11):867–71. https://doi.org/10.1038/nmat2546.

52. Jeong, H. H. et al. 2016. Dispersion and shape engineered plasmonic nanosensors. *Nat. Commun.* 7:11331. https://doi.org/10.1038/ncomms11331.

53. Danaie, M., and Shahzadi, A. 2019. Design of a high resolution metal-insulator-metal plasmonic refractive index sensor based on a ring shaped Si resonator. *Plasmonics.* 14:1453–65. https://doi.org/10.1007/s11468-019-00926-9.

54. Cetin, A. E., and Atlug, H. 2012. Fano resonant ring/disk plasmonic nanocavities on conducting substrates for advanced biosensing. *ACS Nano.* 6(11):9989–95. https://doi.org/10.1021/nn303643w.

55. Limonov, M. F., Rybin, M. V., Poddubny, A. N., and Kivshar, Y. S. 2017. Fano Resonances in photonics. *Nat. Photonics.* 11:543–54. https://doi.org/10.1038/nphoton.2017.142.

56. Song, M. et al. 2016. Nanofocusing beyond the near-field diffraction limit via plasmonic Fano resonance. *Nanoscale.* 8:1635–41. https://doi.org/10.1039/c5nr06504f.

57. Wang, Q., Ouyang, Z., Sun, Y., Lin, M., and Liu, Q. 2019. Linearly tunable Fano resonance modes in a plasmonic nanostructure with a waveguide loaded with two rectangular cavities coupled by a circular cavity. *Nanomaterials.* 9(5):678. https://doi.org/10.3390/nano9050678.

58. Ye, J. et al. 2012. Plasmonic nanoclusters: Near field properties of the Fano resonance interrogated with SERS. *Nano Lett.* 12(3):1660–7. https://doi.org/10.1021/nl3000453.

59. Liu, J., Liu, Z., and Hu, H. 2019. Tunable multiple Fano resonance employing polarization-selective excitation of coupled surface-mode and nanoslit antenna resonance in plasmonic nanostructures. *Sci. Rep.* 9(1):2414. https://doi.org/10.1038/s41598-019-38708-2.

60. Zhan, S., Peng, Y., He, Z., Li, B., Chen, Z., Xu, H., and Li, H. 2016. Tunable nanoplasmonic sensor based on the asymmetric degree of Fano resonance in MDM waveguide. *Sci. Rep.* 6(1):22428. https://doi.org/10.1038/srep22428.

61. Deng, Y., Cao, G., Yang, H., Li, G., Chen, X., and Lu, W. 2017. Tunable and high-sensitivity sensing based on Fano resonance with coupled plasmonic cavities. *Sci. Rep.* 7:10639. https://doi.org/10.1038/s41598-017-10626-1.

62. Wen, Y. et al. 2019. High sensitivity and FOM refractive index sensing based on Fano resonance in all-grating racetrack resonators. *Opt. Commun.* 446:141–6. https://doi.org/10.1016/j.optcom.2019.04.068.

63. Wei, G., Tian, J., and Yang, R. 2019. Fano resonance in MDM plasmonic waveguides coupled with split ring resonator. *Optik.* 193(5):162990. https://doi.org/10.1016/j.ijleo.2019.162990.

64. Chen, Y., Xu, Y., and Cao, J. 2019. Fano resonance sensing characteristics of MIM waveguide coupled square convex ring resonator with metallic baffle. *Results Phys.* 14:102420. https://doi.org/10.1016/j.rinp.2019.102420.

65. Chen, F., Zhang, H., Sun, L., Li, J., and Yu, C. 2019. Temperature tunable Fano resonance based on ring resonator side coupled with a MIM waveguide. *Opt. Laser Technol.* 116:293–9. https://doi.org/10.1016/j.optlastec.2019.03.044.

66. Eisorbagy, M. H. et al. 2019. Performance improvement of refractometric sensors through hybrid plasmonic Fano resonances. *J. Lightw. Technol.* 37(13):2905–13. https://doi.org/10.1109/JLT.2019.2906933.
67. Chen, F., Zhang, H., Sun, L., Li, J., and Yu, C. 2019. Electrically tunable Fano resonance based on ring resonator coupled with a stub. *Optik.* 185:585–91. https://doi.org/10.1016/j.ijleo.2019.03.161.
68. Liu, H. et al. 2019. Metasurface generated polarization insensitive Fano resonance for high performance refractive index sensing. *Opt. Express.* 27(9):13252–62. https://doi.org/10.1364/OE.27.013252.
69. Zhang, Y., and Cui, M. 2019. Refractive index sensor based on the symmetric MIM waveguide structure. *J. Electron. Mater.* 48(2):1005–10. https://doi.org/10.1007/s11664-018-6823-3.
70. Li, Z. et al. 2019. Refractive index sensor based on multiple Fano resonances in a plasmonic MIM structure. *Appl. Opt.* 58(18):4878–83. https://doi.org/10.1364/AO.58.004878.
71. Chen, F., and Li, J. 2019. Refractive index and temperature sensing based on defect resonator coupled with a MIM waveguide. *Mod. Phys. Lett. B.* 33(3):1950017. https://doi.org/10.1142/S0217984919500179.
72. Qi, L., Zhao, C. L., Yuan, J. Y., Ye, M. P., Wang, J., Zhang, Z., and Jin, S. 2014. Highly reflective long period fibre grating sensor and its application in refractive index sensing. *Sens. Actuators B Chem.* 193:185–9. https://doi.org/10.1016/j.snb.2013.11.063.
73. Wu, D. K. C., Kuhlmey, B. T., and Eggleton, B. J. 2009. Ultrasensitive photonic crystal fibre refractive index sensor. *Opt. Lett.* 34(3):322–4. https://doi.org/10.1364/OL.34.000322.
74. Shen, Y., Zhou, J. H., and Liu, T. R. et al. 2013. Plasmonic gold mushroom arrays with refractive index sensing figures of merit approaching the theoretical limit. *Nat. Commun.* 4(4):2381. https://doi.org/10.1038/ncomms3381.
75. Yue, T., Zhang, Z., Wang, R., Hai, Z., Xue, C., Zhang, W., and Yan, S. 2017. Refractive index sensor based on Fano resonances in metal-insulator-metal waveguides coupled with resonators. *Sensors.* 17(4):784. https://doi.org/10.3390/s17040784.
76. Ali, A., Ghafoorifard, H., Abdolhosseini, S., and Habibiyan, H.. 2018. Metal-insulator-metal waveguide-coupled asymmetric resonators for sensing and slow light applications. *IET Optoelectron.* 12(5):220–7. https://doi.org/10.1049/iet-opt.2018.0028.
77. Lu, H., Liu, X., and Mao, D. 2012. Plasmonic analog of electromagnetically induced transparency in multinanoresonator-coupled waveguide systems. *Phys. Rev. A.* 85(5):053803. https://doi.org/10.1103/PhysRevA.85.053803.
78. Boller, K., Imamoglu, A., and Harris, S. E. 1991. Observation of electromagnetically induced transparency. *Phys. Rev. Lett.* 66(20):2593–6. https://doi.org/10.1103/PhysRevLett.66.2593.
79. Chen, Y., Chen, L., Wen, K., Hu, Y., and Lin, W. 2019. Double Fano resonances based on different mechanisms in a MIM plasmonic system. *Photonics Nanostruct.* 36:100714. https://doi.org/10.1016/j.photonics.2019.100714.
80. Wen, K., Chen, L., Zhou, J., Lei, L., and Fang, Y. 2018. A plasmonic chip-scale refractive index sensor design based on multiple Fano resonances. *Sensors.* 18(10):3181. https://doi.org/10.3390/s18103181.
81. Zand, I., Abrishamian, M. S., and Berini, P. 2013. Highly tunable nanoscale metal-insulator-metal split ring core ring resonators (SRCRRs). *Opt. Express.* 21(1):79–86. https://doi.org/10.1364/OE.21.000079.
82. Hirsch, L. R., Jackson, J. B., Lee, A., Halas, N. J., and West, J. L. 2003. A whole blood immunoassay using gold nanoshells. *Anal. Chem.* 75(10):2377–81. https://doi.org/10.1021/ac0262210.

83. Guler, U., Suslov, S., Kildishev, A. V., Boltasseva, A., and Shalaev, V. M. 2015. Colloidal plasmonic titanium nitride nanoparticles: Properties and applications. *Nanophotonics.* 4(3):269–76. https://doi.org/10.1515/nanoph-2015-0017.
84. Behnam, M. A., and Emami, F. et al. 2018. Novel combination of silver nanoparticles and carbon nanotubes for plasmonic photothermal therapy in a melanoma cancer model. *Adv. Pharm. Bull.* 8(1):94–5. https://doi.org/10.15171/apb.2018.006.
85. Zand, I., Mahigir, A., Pakizeh, T., and Abrishamian, M. S. 2012. Selective-mode optical nanofilters based on plasmonic complementary split-ring resonators. *Opt. Express* 20(7):7516–25. https://doi.org/10.1364/OE.20.007516.
86. Willets, K. A., and Van Duyne, R. P. 2007. Localized surface plasmon resonance spectroscopy and sensing. *Annu. Rev. Phys. Chem.* 58:267–97. https://doi.org/10.1146/annurev.physchem.58.032806.104607.
87. Grigorenko, A. N., Beloglazov, A. A., and Nikitin, P. I. 2000. Dark-field surface plasmon resonance microscopy. *Opt. Commun.* 174(1–4):151–5. https://doi.org/10.1016/S0030-4018(99)00676-8.
88. Lin, C.-W., Chen, K.-P., Hsiao, C.-N., Lin, S., and Lee, C.-K. 2006. Design and fabrication of an alternating dielectric multi-layer device for surface plasmon resonance sensor. *Sens. Actuators B.* 113(1):169–76. https://doi.org/10.1016/j.snb.2005.02.044.
89. Wan, Y. W. Y., Chuah, H. P., and Mahmood, M. Y. W. 2007. Optical properties and sugar content determination of commercial carbonated drinks using surface plasmon resonance. *Am. J. Appl. Sci.* 4(1):1–4. https://thescipub.com/abstract/ajassp.2007.1.4.
90. Paliwal, A., Sharma, A., Tomar, M., and Gupta, V. 2016. Surface plasmon resonance study on the optical sensing properties of tin oxide (SnO_2) films to NH_3 gas. *J. Appl. Phys.* 119:164502. https://doi.org/10.1063/1.4948332.
91. Kaur, G., Paliwal, A., Tomar, M., and Gupta, V. 2016. Detection of Neisseria meningitidis using surface plasmon resonance based DNA biosensor. *Biosens. Bioelectron.* 78:106–10. https://doi.org/10.1016/j.bios.2015.11.025.
92. Zybin, A., Grunwald, C., Mirsky, V. M., Wolfbeis, O. S., and Niemax, K.. 2005. Double-wavelength technique for surface plasmon resonance measurements: Basic concept and applications for single sensors and two-dimensional sensor arrays. *Anal. Chem.* 77(8):2393–9. https://doi.org/10.1021/ac048156v.
93. Herminjard, S., Sirigu, L., Herzig, H. P., Studemann, E., Crottini, A., Pellaux, J.-P., Gresch, T., Fischer, M., and Faist, J. 2009. Surface plasmon resonance sensor showing enhanced sensitivity for CO_2 detection in the mid-infrared range. *Opt. Express* 17(1):293–303. https://doi.org/10.1364/OE.17.000293.
94. Karabchevsky, A., Karabchevsky, S., and Abdulhalim, I. 2011. Fast surface plasmon resonance imaging sensor using radon transform. *Sens. Actuators B Chem.* 155(1):361–5. https://doi.org/10.1016/j.snb.2010.12.012.
95. O'Brien, M. J. 2nd, Pérez-Luna, V. H., Brueck, S. R. J., and López, G. P. 2001. A surface plasmon resonance array biosensor based on spectroscopic imaging. *Biosens. Bioelectron.* 16(1–2):97–108. https://doi.org/10.1016/s0956-5663(00)00137-8.
96. Wood, R. W. 1902. On a remarkable case of uneven distribution of light in a diffraction grating spectrum. *Proc. Phys. Soc. London.* 4:396–402. https://doi.org/10.1088/1478-7814/18/1/325.
97. Kooyman, R. P. H. 2008. Physics of surface plasmon resonance. In *Handbook of surface plasmon resonance*, Ch 2, ed. R. B. M. Schasfoort, and A. J. Tudos, 403. Cambridge, UK: The Royal Society of Chemistry. https://doi.org/10.1039/9781847558220-00015.
98. Maystre, D. 2012. Theory of wood's anomalies. In *Plasmonics: From basics to advanced topics*, ed. S. Enoch, and N. Bonod, 39–83. Berlin, Heidelberg: Springer-Verlag. https://doi.org/10.1007/978-3-642-28079-5_2.
99. Homola, J., ed. 2006. *Surface plasmon resonance based sensors*. Berlin, Heidelberg: Springer-Verlag. ISBN: 978-3-540-33918-2.

100. Rossi, S., Gazzola, E., Capaldo, P., Borile, G., and Romanato, F. 2018. Grating-coupled surface plasmon resonance (GC-SPR) optimization for phase-interrogation biosensing in a microfluidic chamber. *Sensors* 18(5):1621. https://doi.org/10.3390/s18051621.
101. Dai, Y., Xu, H., Lu, Y., and Wang, P. 2018. Experimental demonstration of high sensitivity for silver rectangular grating-coupled surface plasmon resonance (SPR) sensing. *Opt. Commun.* 416:66–70. https://doi.org/10.1016/j.optcom.2018.02.010.
102. Bonod, N., Popov, E., and McPhedran, R. C. 2008. Increased surface plasmon resonance sensitivity with the use of double Fourier harmonic gratings. *Opt. Express.* 16(16):11691–702. https://doi.org/10.1364/OE.16.011691.
103. Popov, E. K., Bonod, N., and Enoch, S. 2007. Comparison of plasmon surface waves on shallow and deep metallic 1D and 2D gratings. *Opt. Express.* 15(7):4224–37. https://doi.org/10.1364/OE.15.004224.
104. Sepulveda, B., Angelome, P. C., Lechuga, L. M., and Liz-Marzan, L. M. 2009. LSPR-based nanobiosensors. *Nano Today.* 4(3):244–51. https://doi.org/10.1016/j.nantod.2009.04.001.
105. Tao, H. et al. 2011. Metamaterials on paper as a sensing platform. *Adv. Mater.* 23:3197–201. https://doi.org/10.1002/adma.201100163.
106. Yanik, A. A., Cetin, A. E., Huang, M., Artar, A., Mousavi, S. H., Khanikaev, A., Connor, J. H., Shvets, G., and Altug, H. 2011. Seeing protein monolayers with naked eye through plasmonic Fano resonances. *Proc. Natl. Acad. Sci. USA.* 108(29):11784–9. https://doi.org/10.1073/pnas.1101910108.
107. Hrelescu, C. et al. 2011. Selective excitation of individual plasmonic hotspots at the tips of single gold nanostars. *Nano Lett.* 11(2):402–7. https://doi.org/10.1021/nl103007m.
108. Dondapati, S. K. et al. 2010. Label-free biosensing based on single gold nanostars as plasmonic transducers. *ACS Nano.* 4(11):6318–22. https://doi.org/10.1021/nn100760f.
109. Unger, A. et al. 2009. Sensitivity of crescent-shaped metal nanoparticles to attachment of dielectric colloids. *Nano Lett.* 9(6):2311–15. https://doi.org/10.1021/nl900505a.
110. Beeram, S. R. et al. 2009. Selective attachment of antibodies to the edges of gold nanostructures for enhanced localized surface plasmon resonance biosensing. *J. Am. Chem. Soc.* 131(33):11689–91. https://doi.org/10.1021/ja904387j.
111. Feuz, L. et al. 2010. Improving the limit of detection of nanoscale sensors by directed binding to high-sensitivity areas. *ACS Nano* 4(4):2167–77. https://doi.org/10.1021/nn901457f.
112. Liu, N. et al. 2011. Three-dimensional plasmon rulers. *Science* 332(6036):1407–10. https://doi.org/10.1126/science.1199958.
113. Bolduc, O. R. et al. 2011. Advances in surface plasmon resonance sensing with nanoparticles and thin films: Nanomaterials, surface chemistry, and hybrid plasmonic techniques. *Anal. Chem.* 83(21):8057–62. https://doi.org/10.1021/ac2012976.

4 Plasmonic Nanolasers

Van Duong Ta[1] and Hanh Hong Mai[2]
[1] Department of Optical Devices, Le Quy Don
Technical University, Hanoi, Vietnam
[2] Department of Quantum Optics, Faculty of
Physics, VNU University of Science, Vietnam
National University, Hanoi, Vietnam

4.1 INTRODUCTION

Since 1960 when Theodore Maiman demonstrated the first laser device [1], laser technology has had a great impact on our society. Nowadays, lasers play a crucial role in optical communications, medical therapy, precise manufacturing, scientific research, and everyday life applications. For example, laser emission spectra have been used for long-distance transmission of data over optical fibers [2]. Micrometer-sized lasers, the so-called microlasers, can be integrated into a single cell for cell-tracking and intracellular sensing [3]. The sensitivity of a microlaser-based biosensor can be down to the level of a single nanoparticle and single virus [4].

Since invention of the laser, enormous effort has been spent on laser miniaturization to create a more compact size and lower power consumption. Great success was achieved in shrinking the laser size following the introduction of the semiconductor as a gain material in 1962 [5]. By 1989, the first vertical-cavity single-quantum well microlaser was reported with a minimum size of 3 μm (about three times the wavelength of the emitted light) [6]. In the 1990s, the size of lasers was reduced to wavelength scales as demonstrated by microdisc lasers [7], microsphere lasers [8], photonic crystal lasers [9], nanowire lasers [10], and so on. However, further miniaturization of laser size, beyond the wavelength scales, was a challenging issue; conventional laser cavities confine light by using differences in the refractive index of dielectrics so the minimum optical mode size and consequently the size of the laser is determined by the diffraction limit [11].

Recently, the diffraction barrier has been resolved in metallic-based nanostructures via the surface plasmonic effect [12]. Metallic nanoplasmonics is promising in terms of faster speed, smaller size, and more efficient electronics (Figure 4.1) [13]. Owing to ultra-high optical confinement and ultrafast relaxation processes, plasmonics can overcome the speed limit of semiconductor electronics (caused by heat generation) and the size limit of dielectric photonics (determined by the diffraction law).

Owing to the unique properties of plasmonics, the size of lasers that rely on plasmonic cavities (the so-called plasmonic nanolasers) can be shrunk to less than the wavelength of the emitted light [14–16]. These novel kinds of light sources have a size of ten to hundreds of nanometers and the mode dimension is below the diffraction limit of emission. This breakthrough in laser miniaturization is promising for

DOI: 10.1201/9781003439165-4

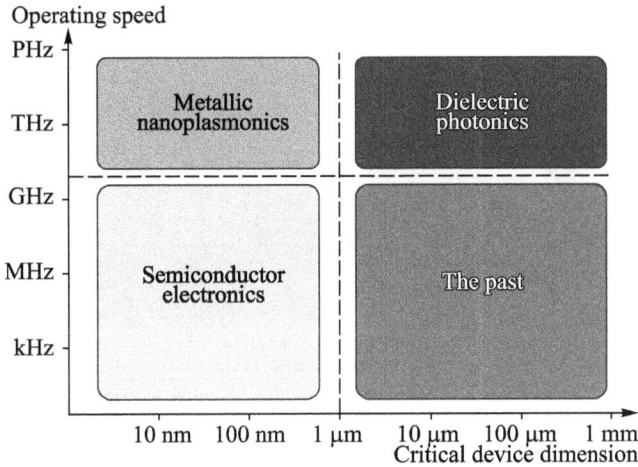

FIGURE 4.1 Comparison of operating speed and device sizes rely on typical material properties including semiconductors, insulators, and metals. The dashed lines represent the physical limitations of different technologies. (Redrawn from [13].)

high-volume data storage, ultra-high resolution imaging, and sensitive sensing. In the literature, there have been several review articles covering different aspects of nanolasers, such as theory and experiment [17–23], semiconductor nanowire-based plasmonic lasers [24], structural engineering [25], and applications [26].

In this chapter, we review the recent progress on plasmonic nanolasers, and cover various aspects of nanolasers. We first introduce surface plasmons, typical gain materials, and the working principle of plasmonic nanolasers. We then describe experimental demonstrations of these devices classified on the confinement method used for light amplification, including single particle, Fabry–Perot cavity, whispering gallery mode cavity, and metallic particle/hole array, followed by a summary of important parameters of nanolasers. After that, we discuss applications of plasmonic nanolasers in integrated photonic circuits, sensing, and biology. Finally, a summary and future prospects of plasmonic nanolasers are given.

4.2 OVERVIEW OF A PLASMONIC NANOLASER

4.2.1 SURFACE PLASMONS AND PLASMONIC MATERIALS

Surface plasmons (SPs) or plasmonics are waves that exist at the interface between a conductor (generally a metal) and a dielectric material (Figure 4.2a) [12]. These waves are formed by the interaction between the external electromagnetic field of light and the free electrons of the conductor. In principle, when the incident light comes to the metallic surface, free electrons in the metal respond by oscillating in resonance with the incident light. This electron oscillation creates electromagnetic fields outside (as well as inside) the metal. As a result, the incident light wave creates charge associated electromagnetic waves, and these two waves can interact with each

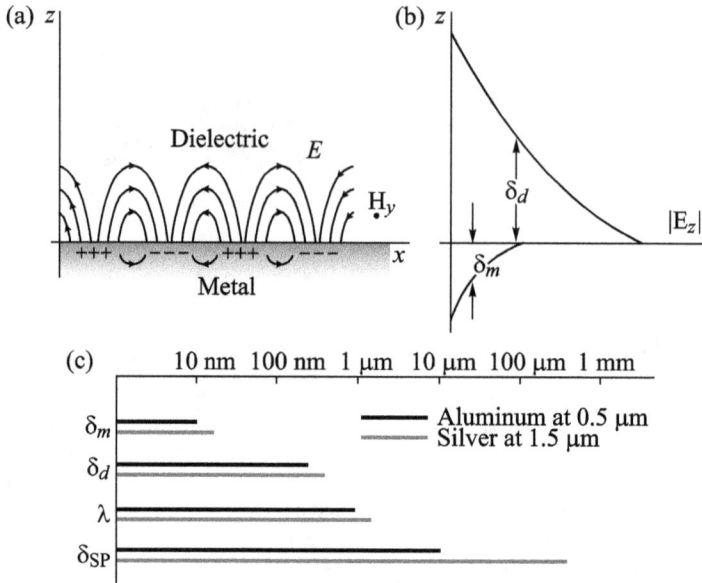

FIGURE 4.2 (a) Surface plasmon at the interface between a dielectric material and a metal. (b) The field component (E_z) perpendicular to the surface is enhanced near the surface but its strength reduces exponentially with distance away from it. The decay length of the field in the dielectric medium and in the metal is characterized by δ_d and δ_m, respectively. (c) Typical values of δ_d and δ_m in comparison with light wavelength (λ) and the propagation length of the SP mode (δ_{SP}) for aluminum and silver surfaces. (Redrawn from [12].)

other. The resonant interactions between these two waves constitute the SP and give it unique properties.

The properties of SP strongly depend on the metallic structure. Recent advanced technology allows the metal surface to be structured at the nanoscale, which enables the control of SP properties for a wide range of applications, for example in subwavelength optics, biophotonics, data storage, and microscopy [12].

SPs have two different properties in comparison with light waves. First, the momentum of the SP mode is greater than that of a free-space photon of the same frequency. This mismatch momentum is characterized by the SP dispersion relation [27]:

$$k_{SP} = k_0 \sqrt{\frac{\varepsilon_d \varepsilon_m}{\varepsilon_d + \varepsilon_m}}, \tag{4.1}$$

where, k_{SP} and k_0 are the SP and the free-space photon wavevector, respectively, and ε_m and ε_d are the permittivity of the metal and dielectric material, respectively. The mismatch momentum is needed to consider when using light for generating SPs. Second, the SP field perpendicular to the surface reduces exponentially with distance away from it [Figure 4.2(b)]. That means SPs are bound to the surface, and, therefore, their energy cannot propagate away from the surface.

SP can propagate but will quickly diminish due to the absorption in the metal. The propagation length, δ_{SP}, can be calculated as follows [12]:

$$\delta_{SP} = \frac{1}{2k''_{SP}} = \frac{c}{\omega}\left(\frac{\varepsilon'_m + \varepsilon_d}{\varepsilon'_m \varepsilon_d}\right)^{\frac{3}{2}}\frac{(\varepsilon'_m)^2}{\varepsilon''_m}, \tag{4.2}$$

where, k''_{SP} is the imaginary part of the complex SP wavevector $k_{SP} = k'_{SP} + ik''_{SP}$; ε'_m and ε''_m are the real and imaginary parts of the dielectric function of the metal $\varepsilon_m = \varepsilon'_m + i\varepsilon''_m$. In free-space, the propagation distance of the SP can be found from the real and imaginary parts of the dielectric function of the metal. The real part characterizes the electron oscillation to the external electromagnetic field, and the imaginary part determines optical loss.

Figure 4.2(c) shows the propagation distance of SP on silver and aluminum surfaces at two wavelengths. Due to low losses, the propagation distance of SP on a silver surface is about 10–100 µm in the visible range. At a wavelength of 1.5 µm, it can reach around 0.5 mm. For aluminum, the propagation distance of SP is much shorter, only 10 µm at a wavelength of 0.5 µm.

Plasmonic materials (generally metals) play an important role in a plasmonic cavity; therefore, choosing a suitable material is crucial for plasmonic lasing. Adapted data from [28, 29], Wu et al. plotted the real and imaginary parts of the dielectric function of four common metals as a function of wavelength [23]. Generally, for the wavelength ranging from 200 to 1,200 nm, aluminum (Al) has the smallest ε', followed by silver (Ag) with a medium ε', and gold (Au) and copper (Cu) have the highest ε'. Concerning the imaginary part, Ag has the smallest ε'', thus Ag exhibits the lowest loss in the visible to near-infrared (NIR) region. This is the advantage of Ag. However, Ag is easily oxidized in ambient conditions so coating a thin dielectric layer on the Ag surface is necessary to avoid degradation. In contrast to Ag, Au is chemically stable. Au also has a relatively small ε''. The disadvantage of Au is its high cost. Despite that, Au is a widely used metal in plasmonic waveguides and nanolasers. Cu and Al are more cost-effective compared with Au and Ag. Both Cu and Al have high ε'' in the visible and NIR regions. Interestingly, Al possesses a relatively small ε'', which is lower than that of Au in the wavelength from 200–400 nm, thus it may be suitable for use in plasmonic waveguides and nanolasers at this wavelength range.

Recently, alternative plasmonic materials (besides the four common metals just mentioned), such as transparent conducting oxides, metal nitrides, perovskite oxides, and two-dimensional (2D) materials have been investigated [30, 31]. For instance, indium tin oxide (ITO) exhibits a low loss, which is comparable to Ag, in the wavelength range from 500–1,000 nm [32]. Titanium nitride (TiN) demonstrates a lower loss than Au in the violet region [33].

4.2.2 Gain Materials Used in Plasmonic Nanolasers

Active materials with high optical gain are desired in a plasmonic laser because the optical losses of a plasmonic cavity are high (due to the absorption of metallic material). In general, inorganic semiconductors and organic dyes are the most widely used materials.

Inorganic semiconductors provide high gain with wavelength ranging from ultraviolet (UV) to NIR region. Typical II–IV semiconductors such as ZnO, GaN, and ZnS (for UV) and CdS, CdSe (for visible) have been attracting great research interest [34]. For the NIR region, GaSb, GaAs, and InP are important laser materials [34]. In some cases, the emission wavelength can be tuned. For example, material $In_xGa_{1-x}N$ ($0 \leq x \leq 1$) offers not only high gain but also is wavelength tunable from NIR (InN, 0.65 eV) to near UV (GaN, 3.4 eV) [35]. Recently, semiconductor colloidal quantum dots have been demonstrated as a promising laser material with a tunable spectral range [36].

Organic dyes are conventional laser materials that provide emission from near UV to NIR [37]. In general, organic dyes have a broad gain spectral range, thus they can be employed for wavelength-tunable or short-pulsed lasers. Another advantage of dyes is their high solubility in various solvents. Consequently, they can be easily incorporated into a range of matrix materials, including polymers and liquid crystals, to create compact laser configurations such as microspheres, microdiscs, and microfibers [38]. The disadvantages of dyes are their toxicity and photobleaching (due to the oxidation of dye molecules and the thermal effect caused by high-density excitation).

Recently, perovskite has emerged as a new type of promising gain material, with flexible bandgap engineering, a large absorption coefficient, and low defect state density [39, 40]. Generally, the chemical formula of perovskite is AMX_3 or A_2MX_4, where A is a cation such as Cs^+, $CH_3NH_3^+$, $NH_2CH=NH_2^+$; M is a divalent metallic cation including Pb^{2+}, Sn^{2+}, Mn^{2+}, Fe^{2+}; and X is halogen anion, Cl^-, I^-, Br^- [41]. Especially, perovskite can be synthesized into reduced-dimensional structures such as nanoplatelets, nanowires, and quantum dots, which are crucial for the development of plasmonic nanolasers.

4.2.3 Theoretical Model and Structure of a Plasmonic Nanolaser

Similar to a conventional laser, a plasmonic nanolaser consists of three main components: a plasmonic cavity (instead of an optical cavity in a conventional laser), an active medium, and a pumping source. Radiative emission is provided from active material via the pumping process. The emission is then amplified in the plasmonic cavity. When the optical gain is larger than the total optical losses, the system starts to lase. The cavity feedback can be provided by either the resonant recirculation or localized plasmonic mode.

The working principle of a plasmonic laser can be described by a three-level gain system (Figure 4.3) [42]. The external energy source (electrically or optically) excites electrons in the active medium from the ground state (N_0) to the excited level (N_2). These excited atoms then quickly relax to the lower level (N_1) and form excitons. Next, the excitation recombines and transfers its energy (nonradiative) to the plasmonic cavity mode (dash arrows). The interaction between the cavity mode and excitations at excited states provides stimulated emission for the generation of the laser. The laser operation can be described by rate equations, in which the excited

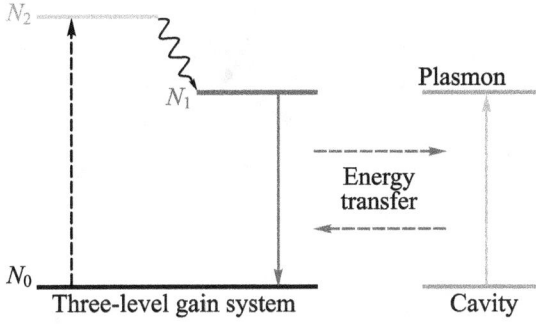

FIGURE 4.3 Schematic diagram showing energy transition in a typical plasmonic nanolaser.

carrier number N and the photon number s in the cavity mode are time-dependent functions [17, 23]:

$$\frac{dN}{dt} = R - \Gamma F_P \beta_0 A (N - N_{tr}) s - \Gamma F_P \beta_0 AN - \frac{N}{\tau_{nr}}, \qquad (4.3)$$

$$\frac{ds}{dt} = \Gamma F_P \beta_0 A (N - N_{tr}) s + \Gamma F_P \beta_0 AN - \gamma_p s, \qquad (4.4)$$

where R is the pumping rate, Γ is the confinement factor, F_P is the Purcell factor, β_0 is the spontaneous emission factor that contributes to the lasing mode without the Purcell effect, A is the spontaneous emission rate, N_{tr} is the transparent excited carrier density, τ_{nr} is the nonradiative transition lifetime, and γ_p is the damping rate of the cavity.

The Purcell factor can be evaluated as follows [43]:

$$F_p = \frac{3}{4\pi^2} \frac{Q}{V} \left(\frac{\lambda}{n}\right)^3, \qquad (4.5)$$

where Q is the quality factor of the cavity, V is the cavity mode volume, and n is the refractive index of the medium.

The lasing threshold can be characterized as [23, 26]:

$$R_{th} = \frac{\gamma_p}{2\Gamma} \left(1 + \frac{1}{\beta}\right) \left(1 + \frac{1}{\zeta}\right), \qquad (4.6)$$

where β is the spontaneous emission factor that contributes to the lasing mode with the Purcell effect, and $\zeta = \gamma_p / (\Gamma F_p \beta_0 A N_{tr})$ is the ratio of cavity loss γ_p to material loss $\Gamma F_p \beta_0 A N_{tr}$.

Plasmonic nanolasers can be realized on various architectures regarding the mechanism for light amplification. In a single metal nanoparticle structure [Figure 4.4(a)], the localized cavity mode is confined in all three dimensions (3D), thus the mode volume and the laser size is very small. However, this structure suffers

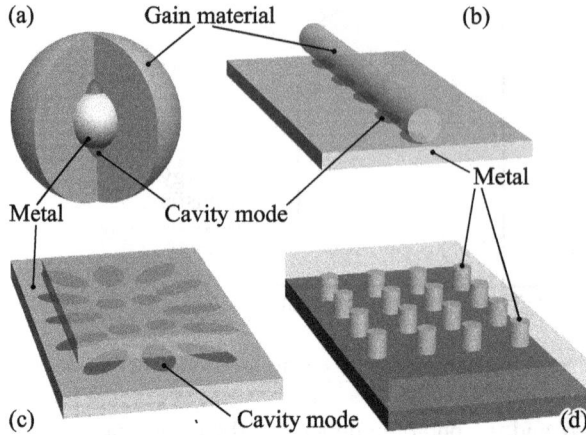

FIGURE 4.4 Typical configurations of plasmonic nanolasers according to the feedback method used for light amplification: (a) localized SP, (b) Fabry–Perot, (c) whispering gallery mode, and (d) metal particle array.

a high optical loss, making it challenging to realize lasing. For structures with lower confinement such as in 2D, the cavity mode propagating is along a waveguide where light is amplified by the Fabry–Perot cavity [Figure 4.4(b)]. In other structures, the cavity mode can be trapped in a surface where light is enhanced by whispering gallery mode [Figure 4.4(c)]. In this case, the light is confined only in 1D at the direction perpendicular to the surface plane. Recently, as shown in Figure 4.4(d), a metal particle array has demonstrated potential for arrays of nanoscale light sources with unique properties such as directional emission [25].

In general, the electric-field intensity of a cavity mode of plasmonic nanolasers can be simulated using COMSOL Multiphysics [44, 45]. Figure 4.5(a) shows a

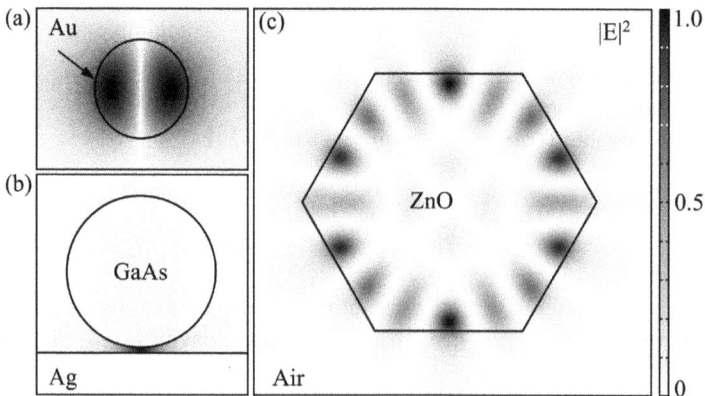

FIGURE 4.5 Simulation of electric-field intensity (in grayscale) for different structures: (a) a gold nanoparticle, (b) a GaAs nanowire separated from a metallic surface by a nanoscale dielectric gap, (c) whispering gallery mode in a ZnO hexagonal cavity.

simulation of the electric-field intensity of a gold nanoparticle. Figure 4.5(b) demonstrates the cavity mode in the gap area between a GaAs nanowire and a silver surface (using data in [46]). For a surface structure such as a hexagonal ZnO disc, light is trapped by total internal reflection at the interface of the cavity and the air, and is amplified by whispering gallery mode (Figure 4.5c).

4.3 EXPERIMENTAL DEMONSTRATIONS OF THE PLASMONIC NANOLASERS

4.3.1 SINGLE NANOPARTICLE-BASED PLASMONIC NANOLASERS

A single nanoparticle-based plasmonic nanolaser, or a spaser, is a nanoplasmonic device generating surface plasmons. The term "spaser" is short for *surface plasmon amplification by stimulated emission of radiation*, which was first proposed by Stockman and Bergman in 2003 [47]. In a spaser, a metal nanoparticle plays the role of a resonator with the replacement of its dielectric environment with an amplifying material, and localized plasmons play the role of photons. The main difference between spasers and lasers is that in a spaser plasmonic modes are nonradiatively generated, amplified, and localized on the nanoparticles, whereas in a conventional laser, wave modes are traveled within the cavity. That means conventional laser mode contains photons while spaser mode contains plasmons. However, similar to conventional lasers, when the supplied optical gain is larger than total optical losses, surface plasmons generated in metallic nanostructures are amplified. In comparison to a conventional laser, a spaser holds two unique properties. First, the localization length of spacing modes is only restricted by the inhomogeneity of the metal and the nonlocality radius (~1 nm). This provides an exclusive possibility to achieve nanoscopic optical sources [48]. Second, regardless of how small a resonator can be reached, it is still able to provide optical feedback due to local field confinement. Thus, in the spaser configuration, it is not necessary to implement high reflective mirrors like in a conventional laser [20, 42]. As a result, the spaser function is based on two main factors: (1) the intrinsic electromagnetic properties of the nanoparticle resonator (e.g. the nanoparticle size, the Q-factor, and other parameters) and (2) the interaction conditions of the amplifying medium with the nanoparticle plasmonic mode. The two simplest and most studied spaser configurations are (1) *nanoshell configuration*, with the amplifying medium at the center of a metal nanoparticle [49] and (2) *nanoparticle configuration*, with the active medium outside of a metal nanoparticle [48, 50]. In these configurations, the presence of metal near the active medium makes the possibility of transferring its excitation energy to the plasmonic mode, thus, plasmonic excitation occurs.

Spasers were first experimentally demonstrated by Noginov with nanoparticle configuration [15]. In this work, metal nanoparticles were 14-nm diameter gold nanospheres synthesized covered with a dielectric silicon shell containing Oregon green 488 dye molecules. The dye molecules have an emission wavelength of 510 nm, which coincided with the plasmonic resonance wavelength of the as-synthesized gold nanoparticles. The silicon dioxide layer thickness was 15 nm. The total size of the 3D spaser was 44 nm in diameter. The spasers were optically pumped by nanosecond pulse 488 nm laser; the lasing emission was observed at 525 nm with a Q-factor of 14.8. Since the nanolaser had a small size in comparison with the lasing wavelengths, the loss was comparably minor and the major loss was due to the gold core. The ratio between the number of dye molecules and the number of nanoparticles was $2:7 \times 10^3$. All the measurements were carried out with an ensemble of spasers at a concentration of 3×10^{11} cm^{-3} in aqueous suspension.

An example of a nanoshell spaser configuration is a dielectric core doped with an optical gain material that was covered by a metallic shell [51–53]. It was proved theoretically and experimentally that light can be amplified through a plasmonic structure containing gain material [15, 54]. The emission properties were numerically examined by finite element method, and frequency domain nonlinear solver (Wave optics module, COMSOL Multiphysics). The core was chosen to be either a mesoporous silica sphere or polystyrene that can load dye molecules; the shell was set as Ag or Au. It was shown that with this configuration an extremely high local field intensity can be achieved with an enhancement factor that reached 10^8. Additionally, this configuration could also create an enhancement factor of surface-enhanced Raman spectroscopy (SERsp) with the order of 10^{16}–10^{17} [55]. The proposed nanoshell spaser configuration has several advantages. First, the surface plasmon resonance emission can be variable from the visible to the near-infrared regions by adjusting the edge length and wall thickness of the spasers. Second, the surface plasmon resonance is strong due to the energy transfer from the gain material, which will compensate the radiative and nonradiative damping that normally occurs in surface plasmon-based devices. Thus, the amplitude and energy of the spaser are amplified, and lasing can be attained [55]. As a result, coupling among surface plasmon polariton, active atom states and light is facilitated. Third, nanoshell spaser configuration can overcome a drawback of nanoparticle configuration. Since the semiconductor cell in nanoparticle configuration must contain dye molecules, the dye molecules must be able to bind covalently with semiconductor material via a characteristic group. Besides, due to the damping, the concentration of dye molecules must be sufficient for compensating plasmonic losses [21].

4.3.2 FABRY–PEROT CAVITY-BASED PLASMONIC NANOLASERS

Technically speaking, in Fabry–Perot (FP)-based plasmonic nanolasers, optical feedback will be obtained by an FP cavity formed by the two ends of a waveguide while surface plasmonic waves are amplified by the gain medium.

In 2009, Hill et al. first demonstrated a plasmonic nanolaser-based waveguide configuration consisting of a stack of metal-dielectric-metal layers. Their waveguide consisted of a rectangular cross section InP/InGaAs/InP nanopillar surrounded by a 20-nm-thick layer of silicon nitride (SiN). The pillar was then encapsulated within two Ag walls. The InGaAs semiconductor layer formed the gain materials, while the SiN layer was used to separate gain materials. The two Ag walls were used to reduce quenching, as demonstrated in Figure 4.6(a) [56]. Due to the difference in refractive index between the InGaAs and InP, light can be confined vertically in the waveguide, and the two ends of the waveguide formed an FP cavity. The devices [SEM image shown in Figure 4.6(b)] were fabricated by using several techniques, such

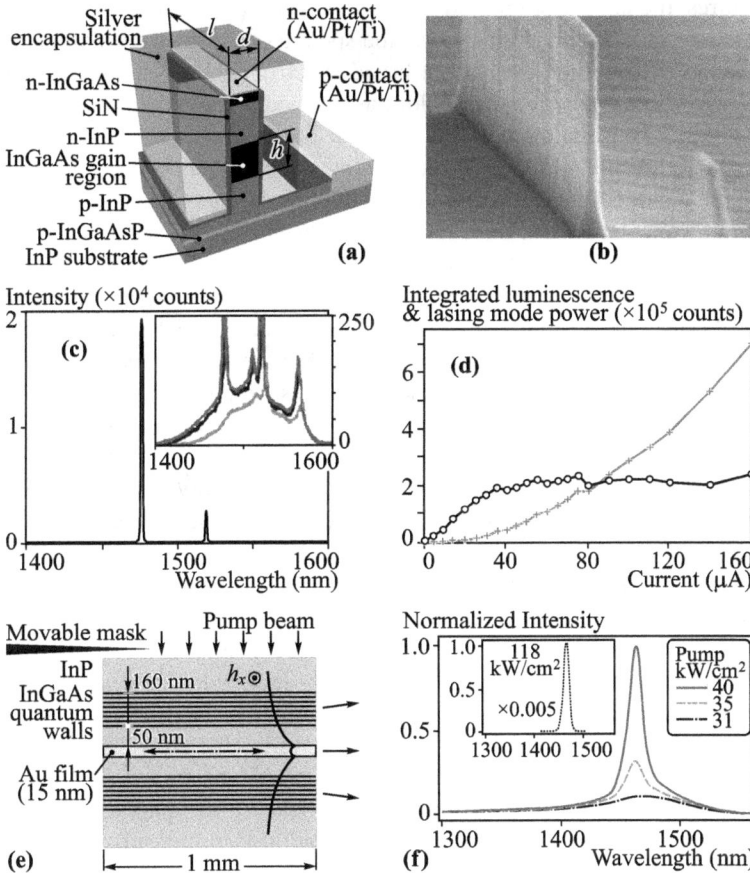

FIGURE 4.6 Fabry–Perot based plasmonic nanolasers. (a) Schematic diagram showing the structure of cavity formed by a rectangular semiconductor pillar encapsulated in Ag. (b) Scanning electron microscope image showing the semiconductor core of one of the devices. The scale bar is 1 micron. (c) Above threshold emission spectra with pump current 180 μA at 78 K. Inset: emission spectra for 20 (the lowest intensity), 40 (the middle), and 60 (the highest intensity) μA, respectively, at 78 K. (d) Lasing mode light output (crosses), integrated luminescence (circles), as a function of pump current at 78 K. Reprinted with permission [56]. Copyright © 2009, Optical Society of America. (e) Side view of an Au-film plasmonic waveguide-based nanolaser; the Au waveguide is sandwiched between two dies containing the QW stacks. (f) Emission spectra are narrower with increasing pump intensity. Inset: Spectrum at 118 kW/cm² far above the threshold. (Reprinted with permission from [58]. Copyright © 2011, Optical Society of America.)

as employing epitaxy, electron beam lithography, dry etching, and various material deposition techniques. The thickness of the InGaAs semiconductor layer could be reached down to 90 nm, which is about half of the diffraction limit in width. Such small thickness can support the high confinement of TM_{01} mode. The nanolaser was electrically pumped by injecting electrons via the top of the pillar and injecting holes via the p-InGaAsP layer and the lateral contact area. Under cryogenic temperatures, the nanolaser emitted light at telecommunication wavelength [Figure 4.6(c)]. The inset in Figure 4.6(c) shows the spectra of the device around and above the threshold. Below the threshold, a broad spectrum from spontaneous emission was observed. Just below the threshold, the full width half maximum (FWHM) of the resonance was 4 nm, indicating a cavity Q-factor on the order of 370 at a cryogenic temperature of 78 K and Q-factor ~140 at room temperature. Lasing mode light output and integrated luminescence as a function of pump current are shown in Figure 4.6(d). When the laser threshold was reached (~40 μA), a narrow lasing mode was obtained at 180 μA, and the FWHM was reduced to 0.7 nm. Later, a plasmonic laser with a similar structure of InP/InGaAsP/InGaAs layer stack was introduced whichcould function at room temperature with a Q-factor of 300. The laser included a semiconductor core of about 140 nm [57]. Within the laser, a distributed feedback (DFB) was implemented. The DFB structure was produced by applying electron beam lithography and reactive ion etching together with surface oxidation and wet chemical method. The metallic Bragg gratings provided a broad stopband of ~500 nm, and the grating coupling coefficient was more than 5,000/cm. Due to the grating, spontaneous emission was strongly suppressed; thus, the lasing threshold was significantly reduced in comparison with that of similar length FP waveguide cavities. Furthermore, the reduced threshold pumping allowed achieving room temperature lasing of the plasmonic nanolaser.

Another configuration for FP-based plasmonic nanolaser is demonstrated in Figure 4.6(e) [58]. The nanolaser included an Au waveguide thin film located inside InGaAs quantum well active medium. The waveguide length and width were 1 mm and 100 μm, respectively. The InGaAs quantum wells were grown by metal-organic chemical vapor deposition; Au thin film was formed by thin film deposition. A 50-nm thick InP spacer layer was implemented between the quantum wells and the Au film to avoid coupling to the lossy mode, and also to minimize the quenching of photoexcited electron-hole pair modes in the Au waveguide. The two ends of the Au waveguide formed an FP resonator. The InGaAs active medium was pumped by 1.06-μm laser pulses and the nanolaser threshold was ~60 kW/cm^2, which is in good agreement with calculations [Figure 4.6(f)]. Lasing was observed at room temperature at the telecommunication wavelength of 1.46 μm. It was also shown that by improving the Au film quality and bonding, the plasmonic losses can be reduced to an intrinsic ohmic level, thus a threshold of ~500 A/cm^2 could be obtained, which is slightly higher than modern laser diodes in the laser wavelength range [58].

A new configuration of an FP-based plasmonic nanolaser implementing an active medium nanowire on top of a metal substrate was first demonstrated by Oulton et al. in 2009 [16]. In this work, a semiconductor CdS nanowire was placed on top of an Ag substrate. The CdS dielectric nanowire and the Ag substrate were separated by a nano thickness MgF$_2$ layer. The CdS nanowires were used as the active medium. The feedback was provided by wave reflection from the nanowire ends due to the large difference in refractive index between the semiconductor and the air, creating an FP resonator along the nanowire. The laser devices were optically pumped by an fs pulse laser at a wavelength of 405 nm. The lasing output observed at the nanowire ends was around 489 nm wavelength at below 10 K. In this configuration, the close proximity of the semiconductor and metal interfaces confined light into an extremely small area, which is a hundred times smaller than a diffraction-limited spot. The transverse size of mode localization was extremely small, about 10 nm. However, the tight confinement also led to a high cavity loss, which consequently resulted in a high lasing threshold of ~60 MW cm^{-2}. In comparison with photonic lasing of a CdS nanowire placed on a MgF_2 substrate, which showed a cutoff frequency when the nanowire diameter reached down to 140 nm, the hybrid plasmonic lasing mode was not limited in this dimension. In fact, the lasing from the plasmonic mode was still observable with a nanowire diameter of only 52 nm.

Another approach based on a similar configuration was successfully demonstrated by Wu et al. [59]. Their plasmonic nanolaser consisted of a bundle of InGaN/ GaN nanorods placed onto a 50-nm-thick colloidal Au triangular plate, separated by a 5-nm-thick SiO_2 gap layer. The nanolaser emitted in the green region with a lower threshold of 300 kW cm^{-2}. Afterward, several works with similar confinement configurations were successfully developed to reduce the lasing threshold, operating the lasing at room temperature, broadening the lasing spectrum, or pursuing even tighter optical confinement of the lasing mode [60–63].

To obtain multiple color plasmonic nanolasers, several approaches have been applied. One of the approaches is to use a semiconductor gain media with multiple compounds [34, 35]. Since the operation band of plasmonic nanolasers mostly depends on the surface plasmon frequency, and on the electronic bandgap of the gain media, therefore, by controlling the elemental composition of semiconductor gain media, lasing emission can be tuned. For example, Lu et al. showed remarkable work in developing a tunable InGaN/GaN core-shell hexagonal nanowire nanolaser [Figure 4.7(a)–(b)] [35]. The nanolaser exhibited a single-mode, continuous, multicolor subwavelength plasmon, operating at 7 K. The nanocavity included a core-shell nanowire of InGaN/GaN with a diameter of ~30–50 nm and a length of 100–250 nm placed on a 28-nm Ag film. A 5-nm Al_2O_3 film was implemented between the nanowire and the Ag film. By tuning the percentage of In in the gain medium, varying from ~27% to 53%, the electronic bandgap was altered from ~2.65 to 1.93 eV, leading to a tunable plasmonic lasing in the range of 474–627 nm [35].

A Fabry–Perot-based plasmonic nanolaser could also be obtained by coupling an active medium nanowire with a metal nanowire. In 2013, Wu et al. reported a plasmonic lasing structure based on a longitudinally hybridized cavity [64]. As shown in Figure 4.7(c)–(d), a curved CdSe nanowire was coupled to a small point at the side of a curved Ag nanowire with a coupling efficiency of 20%. This configuration produced a longitudinal hybrid cavity, in which the lasing mode circulating was alternately changed between a photonic mode in the CdSe nanowire and a plasmonic mode in the Ag nanowire. Furthermore, it also provided the ability to separate spatial contribution from plasmonic and photonic components at the emission output [24]. By optically pumping the CdSe nanowire at room temperature, lasing emission from the plasmonic mode supported by the Ag nanowire was observed around 723 nm wavelength, with an estimated mode area of $0.008\lambda^2$ [23]. The lasing emission also exhibited a strong polarization dependence, indicating the electromagnetic nature of the surface plasmon polariton waves. Although the nanolaser still has challenges in minimizing metal loss and in improving the overall gain efficiency, the nanowire lasers offer significant advantages, such as improved thermal management.

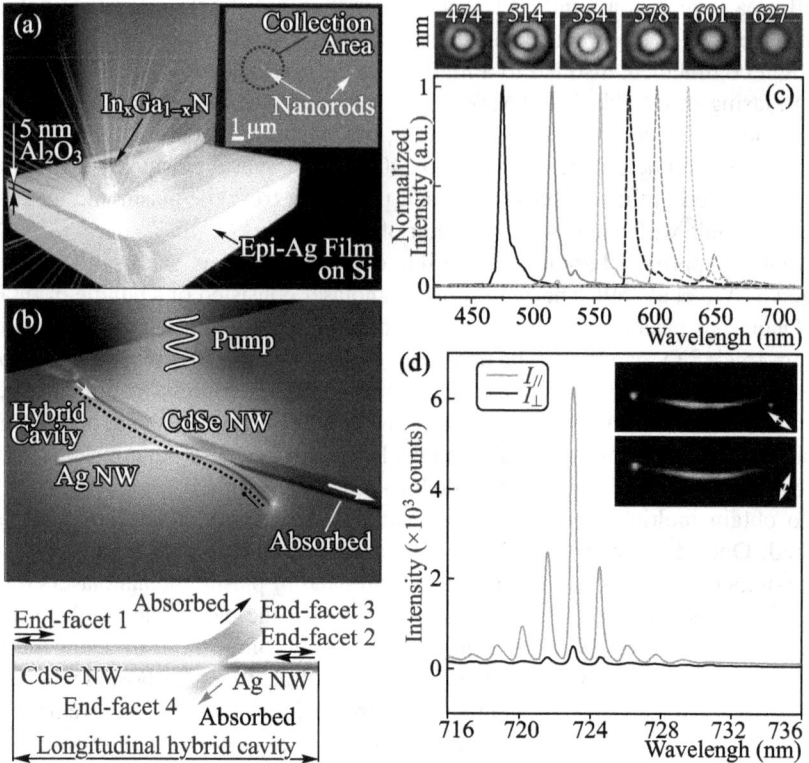

FIGURE 4.7 Fabry–Perot-based plasmonic nanowire lasers. (a) Scheme of a multicolor InGaN/GaN core-shell nanowire plasmonic laser. Inset on the upper right is the SEM image, which indicates that the laser emissions are from an individual single nanorod. (b) All-color, single-mode lasing images observed from single nanorods with an emission line width ~4 nm. Reprinted with permission [35]. Copyright © 2014, American Chemical Society. (c) Schematic diagram of a hybrid photon–plasmon nanowire laser composed of a Ag NW and a CdSe NW coupled into X-shape. (d) Polarization-sensitive lasing spectra from the Ag nanowire end-facet. The emission polarizations oriented parallel and perpendicular are demonstrated as light gray and dark gray, respectively. The insets are dark-field microscope images showing the dependence of polarization on lasing outputs. The arrows indicate the directions of the polarization. (Reprinted with permission from [64]. Copyright © 2013, American Chemical Society.)

4.3.3 Whispering Gallery Mode Cavity-Based Plasmonic Nanolasers

Besides Fabry–Perot configuration, whispering gallery mode configuration for plasmonic nanolasers also attracts a lot of attention. In this configuration, the feedback along wave propagation is obtained by the total internal reflection effect. This plasmonic nanolaser configuration was first demonstrated in a planar structure shape by Ma et al. [65]. The nanolasers had a planar nanosquare with a 45-nm CdS semiconductor on top of a Ag substrate, isolated by a 5-nm-thick MgF$_2$ layer. Here, the CdS semiconductor functioned as an active material. Due to the proximity of the CdS nanosquare and Ag surface, the modes of the CdS square were able to hybridize with the surface plasmon polariton of the metal-dielectric interface. This resulted in strong optical confinement of λ/20 in the MgF_2 gap region with a relatively low metal loss. The hybrid surface plasmon polariton mode was internally reflected on the CdS nanosquare's side borders. Since the device is ultrathin, the total internal reflection regime of the device is only satisfied by plasmonic waves, not for photon waves. Thus, only plasmonic modes were formed in this device. The total internal reflection regime can also mitigate radiation loss. With this construction, the proposed nanolaser had a strong field localization with low losses in the metal, and low radiation loss. As a result, a high Q-factor approaching 100 and a Purcell factor of 18 were obtained. By controlling the geometry of the laser's structure, a single-mode lasing was also achieved at room temperature. Similarly, a whispering gallery mode-based nanolaser was demonstrated in the form of a nanodisc [66]. The nanolaser consisted of a 235-nm InP disc with four InAsP quantum wells embedded in the middle of the disc. A glass substrate was placed on top of the disc for lasing emission collection. The bottom and the surrounding areas of the disc were covered by Ag to construct a nanopan [see Figure 4.8(a)–(d)] [66]. The Ag nanopan served as the laser cavity for plasmonic mode confinement, and the four InP quantum wells act as an active medium. The nanolasers were optically pumped by a 980-nm pulse laser and exhibited a lasing mode at 1,308 nm at a cryogenic temperature of 8 K, and the lasing threshold was ~120 kW·cm^{-2} [Figure 4.8(e)] [66].

Also based on a similar configuration, Khajavikhan et al. represented a continuous-wave telecommunications-frequency nanolaser that can function at room temperature [67]. The laser cavity is based on a nanoscale coaxial-shaped cavity design which is well known to electrical engineers and widely used in transmission lines

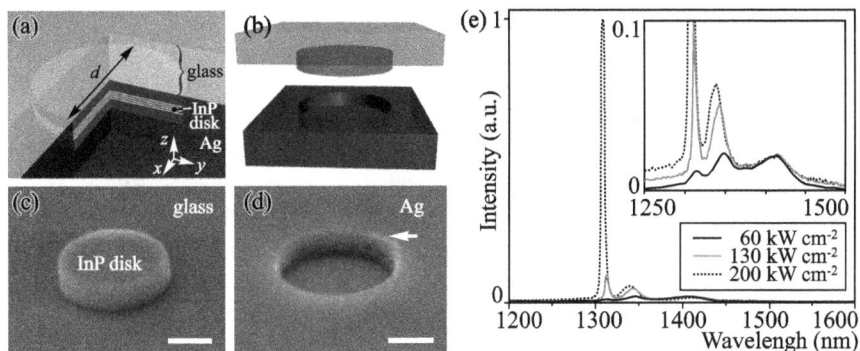

FIGURE 4.8 Whispering gallery mode-based plasmonic nanolasers. (a) Schematic diagram of the plasmonic nanolaser-based nanodisc/nanopan structure including an InP nanodisc with four InAsP quantum wells in the middle. A transparent glass substrate was placed on top of the nanodisc, and the bottom and the sidewalls of the disc were covered by Ag creating a nanopan. (b) Schematic illustration of the removal of the silver nanopan. (c) SEM image of the InP disc on glass before silver deposition. (d) SEM image of the silver film without the disc. The arrow shows damage by the separation of the disc from the silver film. Scale bars in (c) and (d) are 400 nm. (e) Lasing emission at 8 K. (Reprinted with permission from [66]. Copyright © 2010, American Chemical Society.)

in the microwave regime. The cavity had a metallic rod at its center and was surrounded by air, six InGaAsP-based quantum walls, and a thin layer of SiO_2. The whole assembly was then coated with Ag/Al alloy. The metal coverage of the devices and the air and silica "plugs" served to enforce the mode confinement.

Additionally, the metallic coating functioned as a heat sink that allowed the nanolaser to work with continuous-wave operation at room temperature. The lasers were then optically pumped using a commercial 1064 nm laser. The output room-temperature and continuous lasing modes operating at telecommunication wavelengths of 1260 nm and 1590 nm were obtained with practically no threshold. Due to the small size of the coaxial structures, the laser also exhibited a direct couple of ~99% of spontaneous emission into lasing mode. In fact, by tailoring the nanolaser's geometry, such as for the core radius, the ring width, the low index plugs, and the gain medium's height, the laser's modal content can be further modified [65]. It is noted that the advantage of a whispering gallery mode cavity is low optical loss, which is significant for obtaining a low lasing threshold. As a result, it is also widely employed for the realization of microlasers [68].

An interesting approach for obtaining nanolasers multiple colors is to vary the plasmonic nanocavity's morphology and dimension such as the substrate type, the width, and the length of semiconductors. Ma et al. demonstrated a room-temperature, multicolor plasmonic laser that consisted of a single CdS nanobelt and five Ag strips [69]. The nanobelt was crosswise placed onto the Ag strips. The nanobelt's width and thickness were ~620 nm and 100 nm, respectively. Each Ag strip had a width of 1 μm and a thickness of 250 nm. A 5-nm-thick MgF_2 gap was also implemented as an isolator. Additionally, six In/Au (10/120 nm) ohmic contact electrodes were integrated into the laser structure for additional electrical modulation, as shown in Figure 4.9(a)–(b) [69].

The overlap areas of Ag and CdS were optically pumped by an fs laser. It was shown that the square-shaped overlap areas formed a local plasmonic laser cavity by the total internal surface plasmonic reflection. It was attributed to the high refractive index contrast between the surface plasmonic mode and the surroundings. As shown in Figure 4.9(c), the evolution from broad spontaneous emission to lasing can be seen clearly and lasing threshold is determined to be around 6 GW·cm^{-2}. By varying the width of the CdS waveguide, the lasing emission was tuned from 490.2 to 502.7 nm [Figure 4.9(d)]. Interestingly, a small linear peak shift (below 0.3 nm) was observed by applying an electric field of 4V. It was referred to as the density enhancement of excited carriers. This helped to vary the wavelength in a very precise and dynamic way [69].

4.3.4 Particle and Hole Arrays-Based Plasmonic Nanolasers

A single particle-based nanolaser or spaser can be extremely small and can contain only several hundred atoms, thus the energy emission from such a small volume is also very small. The use of a periodic structure of those separate nanoparticles or nanoholes can significantly increase the generated energy due to the constructive interference between adjacent nanoparticles/nanoholes. This can support direct lasing emission and enhance the efficient out-coupling of plasmonic nanolasers.

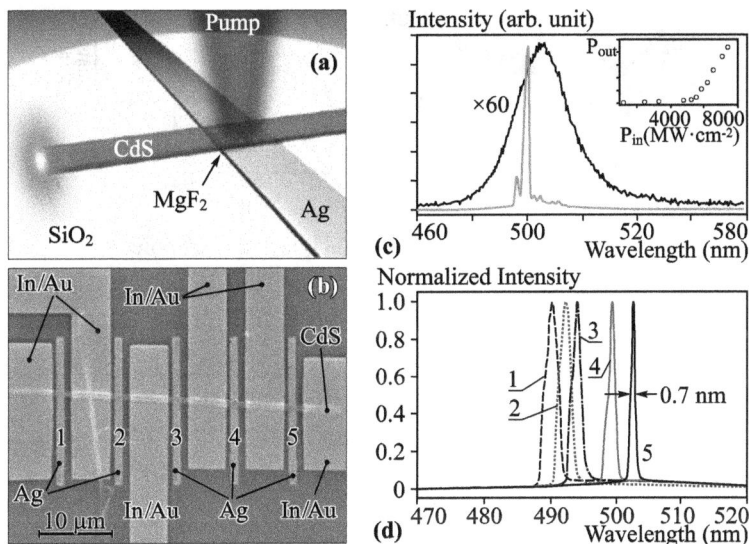

FIGURE 4.9 Multiple color plasmonic nanolasers based Whispering Gallery Mode. (a) Schematic diagram of a waveguide-embedded plasmonic laser, including a CdS nanobelt waveguide crossing over a Ag strip and thin MgF_2 gap layer. (b) SEM image of a plasmonic nanolaser including a CdS nanobelt crossing five Ag strips. (c) Lasing spectra with different pump intensities from 3.3 GW·cm^{-2} (black) to 7.4 GW·cm^{-2} (gray). The inset demonstrates integrated light output as a function of pump intensity. (d) Lasing emission was tunable in the range of 490.2–502.7 nm. (Reprinted with permission from [69]. Copyright © 2012, American Chemical Society.)

4.3.4.1 Plasmonic Nanolasers Based on Nanoparticle Arrays

The possibility of obtaining lasing from a periodic structure formed by metal nanoparticle arrays was first demonstrated by Stehr et al. [70]. Herein, arrays of Au nanodiscs with a diameter of 110 nm and a height of 30 nm were embedded into a 460-nm organic polymer film (methyl-substituted ladder-type poly para phenylene). The organic polymer functioned as a gain medium while the nanodisc arrays acted as a DFB grating. The laser was optically pumped by a 130-fs laser at a wavelength of 400 nm. In this work, the laser did not require two end mirrors for cavity feedback since the periodic structure of the nanodisc array provided optical feedback for the laser. As a result, directional lasing emission was supported at a very low threshold of only 2 nJ, and the emission line width was narrow of 0.44 nm at the wavelength of 492 nm. The constructive interference between adjacent nanodisc arrays also enhanced the efficient out-coupling.

The IR lasing emission was also obtained with a similar nanoparticle array-based nanolaser configuration. The nanolasers contained either Ag or Au nanoparticles with an average diameter of 130 nm surrounded by IR – 140 dye molecules. The laser could function at room temperature with directional emission at a wavelength of 913 nm and a spectral width of 1.3 nm [71]. Interestingly, the plasmonic character of lasing was also confirmed by performing measurements with periodic nanoparticle arrays of non-plasmon materials such as Ti and TiO_2. No lasing was observed in this case.

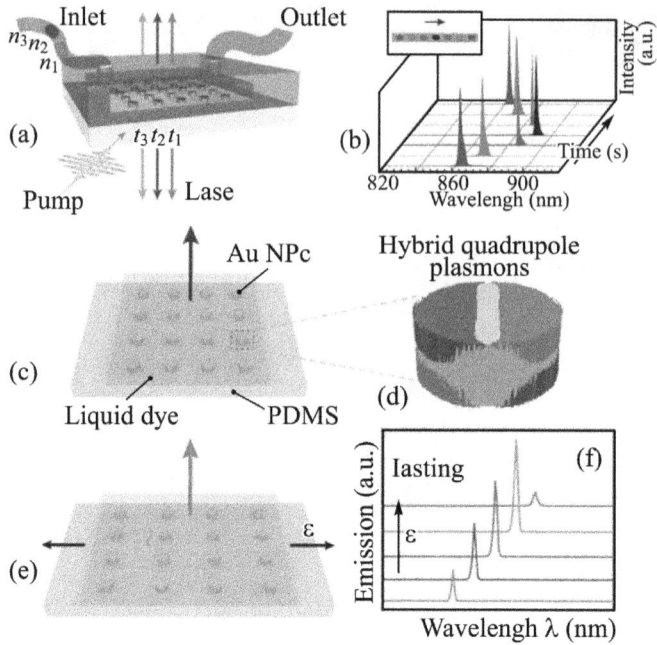

FIGURE 4.10 Tunable nanoparticle arrays based on plasmonic nanolasers. (a) The schematic diagram demonstrates a tunable laser with a microfluidic channel. (b) Continuous tuning of the lasing emission was obtained due to the difference in the refractive index of liquid gain materials. Reprinted with permission [73]. Copyright © 2015, The Authors. (c) and (d) Schematic diagram of the plasmonic nanolasers consisting of Au nanoparticle arrays based embedded within a stretchable dye-doped PDMS gain material. (e) and (f) Schematic of a tunable nanolaser based on the difference of applied strain ε on the elastomeric substrate. (Reprinted with permission from [74]. Copyright © 2018, American Chemical Society.)

Later, tunable lasing of nanoparticle array-based nanolasers was observed in [72–74]. The laser was constructed by embedding Au nanoparticle arrays within microfluidic channels [Figure 4.10(a)–(b)]. When liquid gain materials with different refractive indices flew in those channels, a plasmonic lasing emission was tuned dynamically in a wavelength range of 862 to 891 nm with an emission line width of 1.5 nm [72, 73]. The laser tuning could also be obtained by changing lattice spacing between the metal nanoparticle arrays [Figure 4.10(c)–(f)] [74]. In this approach, Au nanoparticle arrays were patterned onto a flexible substrate that contained an elastomeric slab of polydimethylsiloxane (PDMS) and liquid dye molecules. By changing the interparticle distances, the cavity resonance was modulated, and thus a continuous tuning emission over 31 nm was obtained by stretching and releasing the flexible substrate.

4.3.4.2 Plasmonic Lasers Based on Nanohole Arrays

Lasing in a periodic structure of nanoholes on a thin film embedded within an active medium was also demonstrated [75–78]. Theoretical study of a system consisting of a metal film perforated by a periodic array of nanoholes predicted that when the spasers were pumped, they exhibited a mutual synchronization between each other, which led to the generation of a superradiant lasing. The radiation intensity of the periodic nanohole arrays-based nanolasers was considerably higher than that of a single spaser, by two orders of magnitude [78].

Experimental studies on the plasmonic lasing from metal hole arrays are shown in Figure 4.11 (a) and (b) [75]. A stacked layer of 5 nm SiN, 15 nm InP, and 105 nm InGaAs was subsequently deposited on an InP substrate. Afterward, a 100-nm Au film was coated on the stack layer, and the nanohole arrays were fabricated by lithography. The nanoholes had a diameter of 160 nm each. The InGaAs acted as an active medium. The nanolaser was pumped at the wavelength of 1,064 nm at cryogenic temperatures. Below the threshold, emission at three frequencies was observed, which was attributed to the resonances of the hole arrays. Above the threshold, the resonance at 1,480 nm became dominant, with emission power increased by more than three orders of magnitude. The dominant resonance was denoted as lasing.

Later, lasing from periodic hole arrays-based nanolasers was also observed at room temperature [77, 79]. The nanolasers consisted of an array of nanoholes on

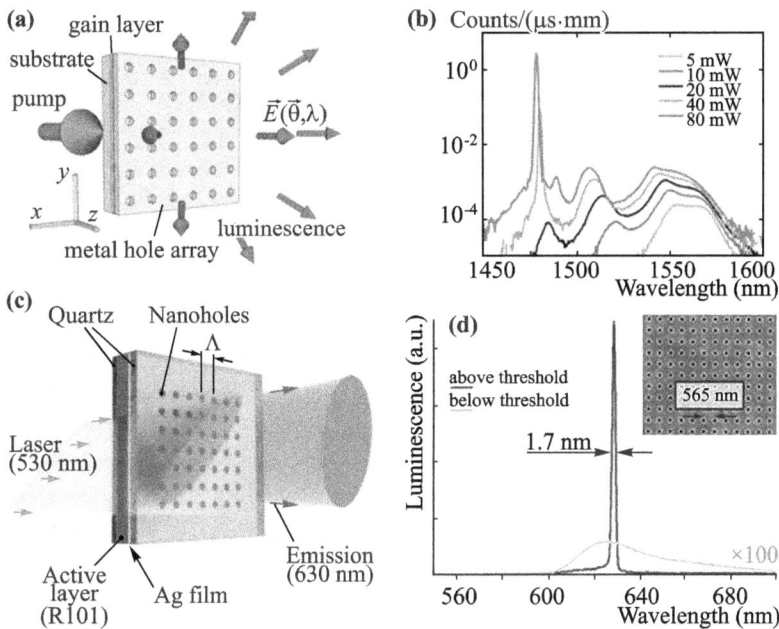

FIGURE 4.11 Nanohole arrays-based plasmonic nanolasers. (a) Schematic diagram of a nanohole arrays-based plasmonic nanolaser including a Au metal hole array deposited on an InGaAs gain medium. (b) Below the threshold, emission spectra exhibited three frequencies, and above the threshold, the wavelength at 1,480 nm became dominant and denoted as lasing. Reprinted with permission [75]. Copyright © 2013, American Physical Society. (c) Planar plasmonic laser. The laser apparatus uses a perforated silver film. (d) The spectra of a plasmonic laser used a liquid medium as an active medium at below and above the lasing threshold. The insets show the SEM image of the nanohole arrays. (Reprinted with permission from [77]. Copyright © 2017, AIP Publishing.)

Ag thin film deposited on a silica substrate [Figure 4.11(c)] [77]. An active medium could be either dye molecules in a liquid solution [77] or a polymer film [79]. A single mode at 610 nm with a narrow width of 1.5 nm was obtained when a rhodamine 101 dye-polymer film was implemented as a gain medium [79]. The nanolasers utilized a solution of R101 dye mixed with dimethyl sulfoxide (DMSO) as an active medium. The liquid active gain medium was supposed to limit the photobleaching of the nanolasers. The nanolaser exhibited a lasing emission at 628-nm wavelength with a spectral line width of 1.7 nm [Figure 4.11(d)] [77].

4.3.5 SUMMARY OF LASER CHARACTERISTICS

Summaries of some important characteristics of typical nanolasers based on the mechanism for light amplification are shown in Table 4.1.

4.4 APPLICATIONS OF NANOLASERS

Plasmonic nanolasers have demonstrated very interesting properties, such as small physical volume and the ability to deliver optical energy to even smaller modes with simultaneously spectral and temporal optical localization. These unique properties open up a wide range of applications for plasmonic nanolasers, from integrated photonic circuits, to sensing, to biological applications.

4.4.1 PLASMONIC NANOLASERS FOR INTEGRATED PHOTONIC CIRCUITS

The main challenge in the development of photonic integration is the generation of efficient radiation sources on a chip. Plasmonic nanolasers can be considered as potential sources for photonic integration due to their unique properties such as their ability to be integrated with other electronic and optical components. Their subwavelength size and low-quality cavities can offer frequency modulations with bandwidths of over hundreds of gigahertz. Furthermore, in photonic devices the capacitance load effect inherent, which is a typical characteristic of electric circuits, is absent. Therefore, photonic integration systems based on plasmonic nanolasers can significantly overcome the current problems of modern electronic devices.

In 1996, Miller et al. first proposed the possible application of nanolasers in photonic integrated circuits [83]. Based on this proposal, Kim et al. simultaneously demonstrated a metal-clad plasmonic nanolaser integrated onto a bidirectional or unidirectional, single-mode Si on an insulator waveguide [84, 85]. The actual nanocavity consists of an InGaAsP bulk semiconductor cuboid cladded by Si, SiO_2, and Ag (Figure 4.12) [84].

By adjusting the nanocavity size (l parameter), and the cladding thickness (a, g parameters), the optimum coupling efficiency of the plasmonic laser with a single-mode on an insulator Si waveguide was obtained. The maximum coupling efficiency of 90% could be achieved with a unidirectional waveguide [85]. Depending on the structure, the Q-factor of the nanolasers varied from 600 to 1,200 [84, 85].

Recently, Ma et al. experimentally demonstrated subwave guide-embedded plasmonic nanolasers with over 70% coupling efficiency into a semiconductor nanobelt

TABLE 4.1
Summaries of Some Important Characteristics of Typical Nanolasers Based on the Mechanism for Light Amplification

	Cavity Structure	Size	Gain Medium	Emission	Pumping and Lasing Conditions	Threshold	Refs
Localized SP	Spaser (core/shell) Core: Au nanoparticle/Sheel: dye-doped silica	Diameter: 44 nm	Dye	531 nm	Optical, pulsed Room temperature	~4 nJ	[15]
	Spaser (core/shell) Core: Au nanoparticle/Shell: dye-doped silica	Diameter: 22 nm	Dye	550 nm	Optical, pulsed Room temperature	~26 mJ·cm⁻²	[80]
Fabry–Perot	Metal-Insulator-Metal waveguides	Thickness: ~310 nm Length: 6 µm	InGaAs	1500 nm	Electrical, continuous wave Room temperature	~5 mA	[56]
	Metallic-coated nanopillar	Diameter: 260 nm High: 300 nm	InGaAs	1400 nm	Electrical, continuous wave 77 K	~5 µA	[14]
Fabry–Perot	Plasmonic coaxial	Diameter: 200 nm Height: 210 nm	InGaAsP MQWs	1380 nm	Optical, continuous wave 4.5 K	Thresholdless	[67]
	Plasmonic nanowire	Diameter: 60 nm Length: 480 nm	InGaN	510 nm	Optical, continuous wave 78 K	~3 kW·cm⁻²	[62]
	Plasmonic nanowire	Diameter: 100 nm Length: ~10 µm	CdS	490 nm	Optical, pulsed 10 K	~100 MW·cm⁻²	[16]
	Plasmonic nanowire	Diameter: ~100 nm Length: ~15 µm	GaN	375 nm	Optical, pulsed Room temperature	~3.5 MW·cm⁻²	[61]
	Plasmonic pseudo-wedge	Wedge length: 80 nm	ZnO	370 nm	Optical, pulsed 77 K	~55 MW·cm⁻²	[81]
Whispering Gallery Mode	Metallic-coated disc	Thickness: 480 nm Diameter: 490 nm	InGaAsP	1430 nm	Optical, pulsed Room temperature	~70 kW·cm⁻²	[82]
	Plasmonic nanodisc	Thickness: 235 nm Diameter: ~1 µm	InAsP MQWs	1300 nm	Optical, pulsed 8 K	~120 kW·cm⁻²	[66]
Whispering Gallery Mode	Plasmonic nanosquare	Thickness: 45 nm Length: 1 µm	CdS	495 nm	Optical, pulsed Room temperature	~2 GW·cm⁻²	[65]
Particle and Hole Array	Arrays of Au nanoparticles	–	Dye	913 nm	Optical, pulsed Room temperature	~0.23 mJ·cm⁻²	[71]
	Array of holes on an ITO substrate	–	Dye	620 nm	Optical, pulsed Room temperature	12 mW	[79]

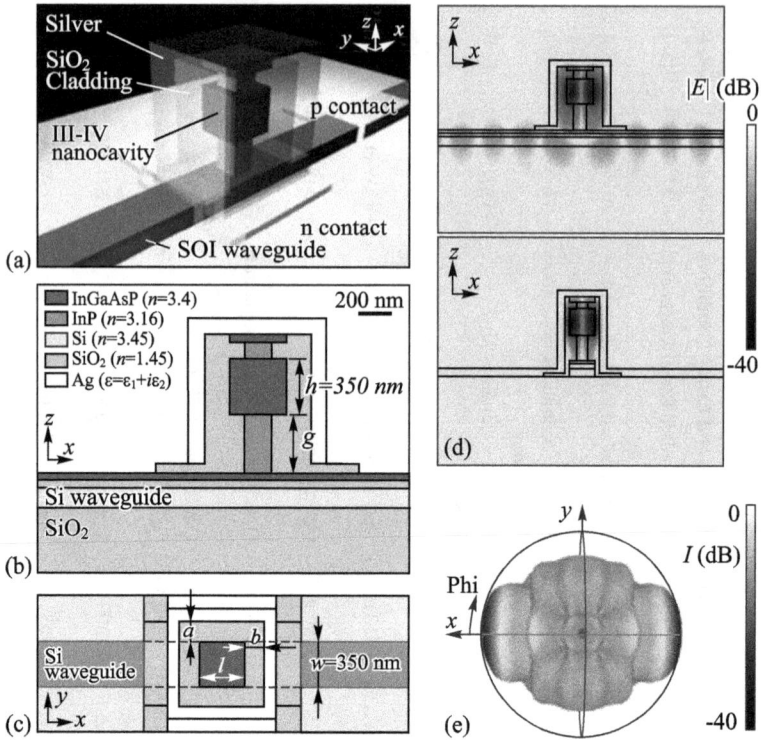

FIGURE 4.12 Plasmonic nanolasers for integrated photonic circuits: (a) schematic diagram, (b) the side view, and (c) the top view of a Si-waveguide coupled metal-clad nanolaser cavity. (d) Electric field |E| distribution in the x-z and y-z plane. (e) Far field radiation patterns to the substrate direction of the plasmonic nanolaser. (Reprinted with permission from [84]. Copyright © 2011, Optical Society of America.)

waveguide [69]. The plasmonic nanolaser-based integrated circuit had four key properties: multicolored plasmonic light source, wavelength multiplexing, efficient waveguide collection and out-coupling, and direct electrical modulation. The compact and novel configuration of the nanolaser opens up a promising application in large-scale, on-chip, integrated hybrid optoelectronic circuitry [24, 28].

4.4.2 Plasmonic Nanolasers for Sensing

Surface plasmon resonance sensors are now widely used in biochemical sensing and detection [86]. However, their strong radiative and nonradiative damping, which leads to the weakness of the plasmonic resonance, are their main drawbacks. As a result, the plasmonic resonances in the visible and near-infrared range exhibit broad line widths of tens to hundreds of nanometers. This strongly limits the sensor's sensitivity, and increase the sensor's limit of detection. However, this drawback can be fully solved by injecting a gain medium in a plasmonic nanolaser cavity. Due to the amplification of the plasmonic resonance, the line width can be narrowed by two orders of magnitude, thus, the sensor's sensitivity can be significantly enhanced [87].

The application of a spaser in surface-enhanced Raman scattering (SERS) was numerically investigated [55]. It was shown that the spaser exhibited an extremely high local field intensity with an enhancement factor reached 10^8, while the enhancement factor of SERS was on the order of 10^{16}–10^{17}. These enhancement factors are sufficient for single-molecule detection [55]. The surface plasmon resonance peak in the spasers can be tuned from visible to NIR range by controlling the edge length and the thickness of the shell. This can facilitate single-molecule SERS detection using different laser wavelengths. The strong coherent high-intensity lasing emission from the spaser will lead to the sufficiently high efficiency of single-molecule SERS signal observation [55].

Significant enhancement in sensor's performance with ultrahigh sensitivity has been obtained with plasmonic nanolaser-based sensors. Ma et al. demonstrated a nanolaser that consists of a single crystalline semiconductor CdS nanoslab (50 nm in thickness, and 600 nm in length) placed on top of an Ag surface, and an 8-nm gap layer of MgF_2 was inserted in between [88]. Here, the CdS semiconductor slab with atomically smooth surface functions at the same time as an active medium, a feedback geometry, and a sensing medium. The sensor's working principle is based on lasing-enhanced surface plasmon resonance in which the electromagnetic field was localized at the interface between the metal and the semiconductor by the surface plasmon. This allowed both the device's physical size and mode confinement to shrink down to the nm scale in the vertical dimension. As a proof of concept, the plasmonic nanolaser-based sensor was utilized for the detection of three different explosive molecules: 2,4-dinitrotoluene, ammonium nitrate, and nitrobenzene. It was shown that when the concentration of the analyte of interest was increased, it led to a decrease in laser emission intensity, while there was no noticeable change in the peak wavelength. The sensing mechanism is referred to as the surface recombination velocity modification of the gas molecules. Due to the large surface area to volume ratio, the sensor's sensitivity was ultrahigh, which could reach down to subpart per billion. It ensured that any modulation on the atomically smooth surface would strongly influence the laser emission intensity.

A similar configuration of plasmonic nanolasers with CdSe semiconductor layer on top of Au substrate separated by a few nanometers of MgF_2 was also implemented for the refractive index sensor [87, 89]. Wang et al. demonstrated that the sensor's stability was improved by thin film deposition of Al_2O_3 on top of CdSe layer [89]. Here, the sensors represented a significant wavelength shift of about 22 nm per refractive index unit at the wavelength of 718 nm. Due to the high spectral coherence of lasing surface plasmon (with a line width of less than 0.3 nm at $\lambda = 700$ nm), the sensor's sensitivity and the Figure of Merit were greatly improved. Additionally, the high signal-to-noise ratio, which was due to the considerably suppressed background emission, and the sharp tails of the Gaussian shaped laser spectrum also contributed to the significant enhancement of the sensors' Figure of Merit. When sensing the refractive index of a mixed solution of water and glycerin, the nanolasers presented a Figure of Merit of about 40 times higher than state-of-the-art surface plasmon resonance-based sensor [89]. In addition, the Figure of Merit reached up to 400 times when sensing the refractive index change of ethanol and propyl alcohol mixed solution [88]. For a demonstration, as shown in Figure 4.13a,

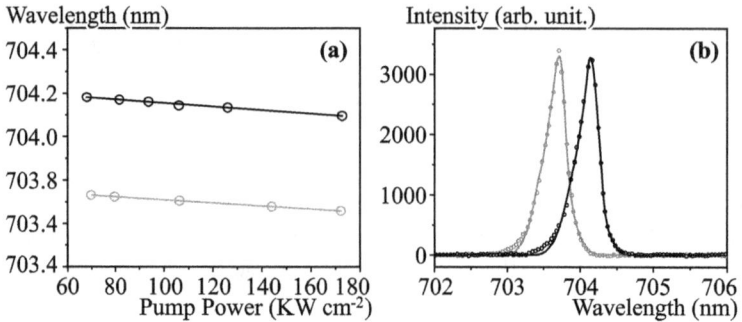

FIGURE 4.13 Plasmonic nanolaser for sensing. (a) A laser device's emission peak wavelength evolution in varied refractive index environments. From top to down, the device is in propyl alcohol and ethanol, respectively. (b) The spectrum shifts with the refractive index at the lasing state. From left to right, the device in ethanol and propyl alcohol, respectively. (Reprinted with permission from from [87]. Copyright © 2017, the Author(s).)

the lasing peak wavelength of a plasmonic nanolaser shifts with the refractive index in two different environments [87]. Interestingly, the performance of the device is quite stable because the value of the wavelength shift is independent of the pumping power. The spontaneous emission and lasing spectra in ethanol and propyl alcohol are plotted in Figure 4.13b, which indicates a clear redshift with the increase of the refractive index of the solution [87].

4.4.3 PLASMONIC NANOLASERS FOR BIOLOGICAL APPLICATIONS

Plasmonic nanolasers have been attracting a lot of attention for biological applications, including biological probing, cellular labeling and tracking, and super resolution imaging.

E.I. Galanzha et al. demonstrated an application of a biocompatible spaser as a biological probe [80]. The spaser with nanoparticle configuration had a gold nanosphere core and a silica shell doped with dye molecules (see Figure 4.14). The spasers were then loaded into complex biological backgrounds from cellular cytoplasm in vitro to a mouse tissue in vivo. The proposed spasers can be considered as one of the smallest nanolasers, with an average size of only 22 nm (including a 10-nm diameter Au core and a 6-nm-thick silica shell embedded with organic dye molecules), which is approximately 100-fold smaller than microlasers. Under the excitation of a single ns laser pulse, the spasers functioned as low-toxicity ultrafast probes with narrow emission spectra of ~1 nm and exhibited a super bright stimulated emission. This can be clearly seen above the threshold. The image of human cancer cells with the integrated spasers has a high image contrast exhibiting many individual "hot spots." Furthermore, the spasers also exhibited strong photoacoustic effects when the optical pump was increased. This opened up a promising possibility of using spasers as a useful tool for theranostic diagnosis and photothermal destruction of cancer cells.

Due to the sub-resolution size, nanolasers can be essentially used for superresolution nanoscopy. Cho et. al. represented a super high resolution optical

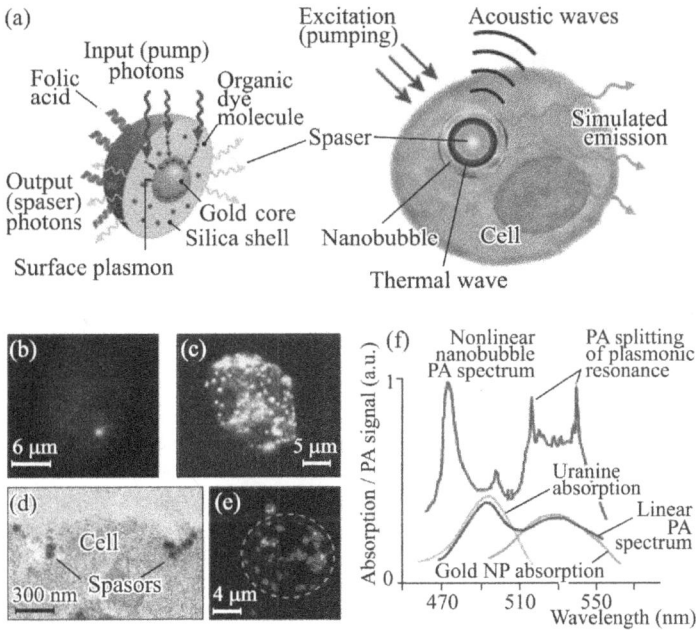

FIGURE 4.14 Spaser for biomedical application. (a) Schematic of a spaser as a multimodal nanoprobe loaded in a single cell. (b) Fluorescence image of a breast cancer cell loaded with a single spacer. (c) Fluorescent image of a breast cancer cell loaded with multiple spacers. (d) Transverse electromagnetic mode (TEM) image of spacers within a breast cancer cell. (e) Photothermal image of a breast cancer cell with spacers. (f) Absorption spectra of uranine and Au NPs and normalized photoacoustic (PhA) spectra of spasers in linear and nonlinear modes with respect to the increased optical pump. (Reprinted with permission from [80]. Copyright © 2017, the authors.)

microscopy technique, known as Laser Particle Stimulated Emission (LASE), that uses the nonlinear effect of nanolasers [90]. The technique utilized iodide perovskite nanowires laser embedded in a sample as an imaging probe. At the focal point, the excitation beam was focused into a small region which is defined by the diffraction limit.

Around the lasing threshold, the nanolasers exhibited nonlinear light output, indicating that at the focal point, lasing was achieved at a very narrow region below the diffraction limit (as seen as the beam profile along the z-axis in Figure 4.15). Outside the focal point, the stimulated emission intensity can be considered as negligible, which yielded a low out-of-focus background, and thus led to an increase in signal to background contrast. This enabled the enhancement in imaging depths in scattering samples. Since the nonlinear effect just appeared at the focal point, the technique does not require a tight pinhole to remain high resolution as it does in confocal microscopy. The stimulated emission from the nanolasers led to an enhancement of optical resolution that is around five times higher than that of the same device operated below lasing threshold [90].

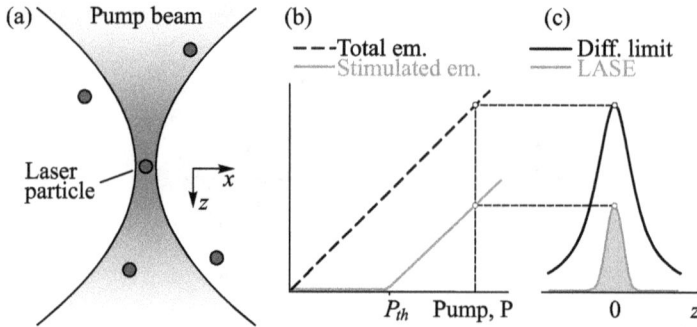

FIGURE 4.15 The principle of LASE microscopy. (a) Laser particles were excited by a tightly focused optical pump beam. (b) Emission lasing energy is a function of optical pump. The lasing emission increased sharply when the optical pump energy was above the threshold. (c) The comparison of point spread function between the laser emission (gray line) and the traditional diffraction-limited fluorescence detection (black line). (Reprinted with permission from [90]. Copyright © 2016, American Physical Society.)

4.5   SUMMARY AND OUTLOOK

We have shown the overview and advances of nanolasers. In the last decade, research works on nanolasers have obtained great achievements in both theoretical studies and experimental demonstrations. Nowadays, nanolasers are exhibited in many forms, such as single nanoparticles, an array of nanoparticles, nanowires, nanopillars, and nanodiscs.

With unique properties such as small size and low power consumption, nanolasers can be important for a wide range of applications. For instance, optically pumped nanolasers have already shown great potential for biosensing and cell tracking. More importantly, nanolasers can be employed in photonic-integrated circuits, which provide ultrafast optical communications. However, to attain the level of performance required for this application, two issues need to be addressed. First, the performance of plasmonic nanolasers has to improve to achieve electrically injected nanolasers. To solve this problem, both cavity configuration and metal structure are crucial, thus novel fabrication techniques and materials should be developed. Second, the compatibility of nanolasers within currently existing systems that mainly rely on a much larger size of nanolasers must be considered [26]. It is expected that this issue will be solved in the near future by research investment and research collaboration between industry and academic communities.

REFERENCES

1. Maiman, T. H. 1960. Stimulated optical radiation in ruby. *Nature.* 187(4736):493–4. https://doi.org/10.1038/187493a0.
2. Vahala, K. J. 2003. Optical microcavities. *Nature.* 424(6950):839–46. https://doi.org/10.1038/nature01939.
3. Humar, M., and Yun, S. H. 2015. Intracellular microlasers. *Nat. Photon.* 9(9):572–6. https://doi.org/10.1038/nphoton.2015.129.

4. He, L., Ozdemir, S. K., Zhu, J., Kim, W., and Yang, L. 2011. Detecting single vi-ruses and nanoparticles using whispering gallery microlasers. *Nat. Nanotech.* 6(7):428–32. https://doi.org/10.1038/nnano.2011.99.

5. Hall, R. N., Fenner, G. E., Kingsley, J. D., Soltys, T. J., and Carlson, R. O. 1962. Coherent light emission from GaAs junctions. *Phys. Rev. Lett.* 9(9):366–8. https://doi.org/10.1103/PhysRevLett.9.366.

6. Lee, Y. H., Jewell, J. L., Scherer, A., McCall, S. L., Harbison, J. P., and Florez, L. T. 1989. Room-temperature continuous-wave vertical-cavity single-quantum-well micro-laser diodes. *Electron. Lett.* 25(20):1377–8. https://doi.org/10.1049/el:19890921.

7. McCall, S. L., Levi, A. F. J., Slusher, R. E., Pearton, S. J., and Logan, R. A. 1992. Whispering-gallery mode microdisk lasers. *Appl. Phys. Lett.* 60(3):289–91. https://doi.org/10.1063/1.106688.

8. Spillane, S. M., Kippenberg, T. J., and Vahala, K. J. 2002. Ultralow-threshold Raman laser using a spherical dielectric microcavity. *Nature.* 415(6872):621–3. https://doi.org/10.1038/415621a.

9. Painter, O., Lee, R. K., Scherer, A., Yariv, A., O'Brien, J. D., Dapkus, P. D., and Kim, I. 1999. Two-dimensional photonic band-gap defect mode laser. *Science.* 284(5421):1819–21.

10. Huang, M. H., Mao, S., Feick, H., Yan, H., Wu, Y., Kind, H., Weber, E., Russo, R., and Yang, P. 2001. Room-temperature ultraviolet nanowire nanolasers. *Science.* 292(5523):1897–9. https://doi.org/10.1126/science.1060367.

11. Hill, M. T., and Gather, M. C. 2014. Advances in small lasers. *Nat. Photon.* 8(12): 908–18. https://doi.org/10.1038/nphoton.2014.239.

12. Barnes, W. L., Dereux, A., and Ebbesen, T. W. 2003. Surface plasmon subwavelength optics. *Nature.* 424(6950):824–30. https://doi.org/10.1038/nature01937.

13. Brongersma, M. L., and Shalaev, V. M. 2010. The case for plasmonics. *Science.* 328(5977):440–1. https://doi.org/10.1126/science.1186905.

14. Hill, M. T., Oei, Y.-S., Smalbrugge, B., Zhu, Y., de Vries, T., van Veldhoven, P. J., van Otten, F. W. M., Eijkemans, T. J., Turkiewicz, J. P., de Waardt, H., Geluk, E. J., Kwon, S.-H., Lee, Y.-H., Nötzel, R., and Smit, M. K. 2007. Lasing in metallic-coated nano-cavities. *Nat. Photon.* 1:589–94. https://doi.org/10.1038/nphoton.2007.171.

15. Noginov, M. A., Zhu, G., Belgrave, A. M., Bakker, R., Shalaev, V. M., Narimanov, E. E., Stout, S., Herz, E., Suteewong, T., and Wiesner, U. 2009. Demonstration of a spaser-based nanolaser. *Nature.* 460(7259):1110–2. https://doi.org/10.1038/nature08318.

16. Oulton, R. F., Sorger, V. J., Zentgraf, T., Ma, R.-M., Gladden, C., Dai, L., Bartal, G., and Zhang, X. 2009. Plasmon lasers at deep subwavelength scale. *Nature.* 461(7264):629–32. https://doi.org/10.1038/nature08364.

17. Ding, K., and Ning, C. Z. 2012. Metallic subwavelength-cavity semiconductor nanolas-ers. *Light Sci. Appl.* 1(7):e20. https://doi.org/10.1038/lsa.2012.20.

18. Ma, R.-M., Oulton, R. F., Sorger, V. J., and Zhang, X. 2013. Plasmon lasers: Coherent light source at molecular scales. *Laser Photon. Rev.* 7(1):1–21. https://doi.org/10.1002/lpor.201100040.

19. Premaratne, M., and Stockman, M. I. 2017. Theory and technology of SPASERs. *Adv. Opt. Photon.* 9(1):79–128. https://doi.org/10.1364/AOP.9.000079.

20. Wang, Z., Meng, X., Kildishev, A. V., Boltasseva, A., and Shalaev, V. M. 2017. Nanolasers enabled by metallic nanoparticles: From spasers to random lasers. *Laser Photon. Rev.* 11(6):1700212. https://doi.org/10.1002/lpor.201700212.

21. Balykin, V. I. 2018. Plasmon nanolaser: Current state and prospects. *Physics-Uspekhi.* 61(9):846–70. https://doi.org/10.3367/UFNe.2017.09.038206.

22. Xu, L., Li, F., Liu, Y., Yao, F., and Liu, S. 2019. Surface plasmon nanolaser: Principle, structure, characteristics and applications. *Appl. Sci.* 9(5):861. https://doi.org/10.3390/app9050861.

23. Wu, H., Gao, Y., Xu, P., Guo, X., Wang, P., Dai, D., and Tong, L. 2019. Plasmonic nanolasers: Pursuing extreme lasing conditions on nanoscale. *Adv. Optical Mater.* 7(17):1900334. https://doi.org/10.1002/adom.201900334.
24. Chun, L., Liu, Z., Chen, J., Gao, Y., Li, M., and Zhang, Q. 2019. Semiconductor nanowire plasmonic lasers. *Nanophotonics.* 8(12):2091–110. https://doi.org/10.1515/nanoph-2019-0206.
25. Wang, D., Wang, W., Knudson, M. P., Schatz, G. C., and Odom, T. W. 2018. Structural engineering in plasmon nanolasers. *Chem. Rev.* 118(6):2865–81. https://doi.org/10.1021/acs.chemrev.7b00424.
26. Ma, R.-M., and Oulton, R. F. 2019. Applications of nanolasers. *Nat. Nanotech.* 14(1): 12–22. https://doi.org/10.1038/s41565-018-0320-y.
27. Sambles, J. R., Bradbery, G. W., and Yang, F. 1991. Optical excitation of surface plasmons: An introduction. *Contemp. Phys.* 32(3):173–83. https://doi.org/10.1080/00107519108211048.
28. Johnson, P. B., and Christy, R. W. 1972. Optical constants of the noble metals. *Phys. Rev. B.* 6(12):4370–9. https://doi.org/10.1103/PhysRevB.6.4370.
29. McPeak, K. M., Jayanti, S. V., Kress, S. J. P., Meyer, S., Iotti, S., Rossinelli, A., and Norris, D. J. 2015. Plasmonic films can easily be better: Rules and recipes. *ACS Photonics.* 2(3):326–33. https://doi.org/10.1021/ph5004237.
30. West, P. R., Ishii, S., Naik, G. V., Emani, N. K., Shalaev, V. M., and Boltasseva, A. 2010. Searching for better plasmonic materials. *Laser Photon. Rev.* 4(6):795–808. https://doi.org/10.1002/lpor.200900055.
31. Naik, G. V., Shalaev, V. M., and Boltasseva, A. 2013. Alternative plasmonic materials: Beyond gold and silver. *Adv. Mater.* 25(24):3264–94. https://doi.org/10.1002/adma.201205076.
32. Naik, G. V., Kim, J., and Boltasseva, A. 2011. Oxides and nitrides as alternative plasmonic materials in the optical range [Invited]. *Opt. Mater. Express.* 1(6):1090–9. https://doi.org/10.1364/OME.1.001090.
33. Naik, G. V., Schroeder, J. L., Ni, X., Kildishev, A. V., Sands, T. D., and Boltasseva, A. 2012. Titanium nitride as a plasmonic material for visible and near-infrared wavelengths. *Opt. Mater. Express.* 2(4):478–89. https://doi.org/10.1364/OME.2.000478.
34. Ma, Y., Guo, X., Wu, X., Dai, L., and Tong, L. 2013. Semiconductor nanowire lasers. *Adv. Opt. Photon.* 5(3):216–73. https://doi.org/10.1364/AOP.5.000216.
35. Lu, Y.-J., Wang, C.-Y., Kim, J., Chen, H.-Y., Lu, M.-Y., Chen, Y.-C., Chang, W.-H., Chen, L.-J., Stockman, M. I., Shih, C.-K., and Gwo, S. 2014. All-color plasmonic nanolasers with ultralow thresholds: Autotuning mechanism for single-mode lasing. *Nano Lett.* 14(8):4381–8. https://doi.org/10.1021/nl501273u.
36. Klimov, V. I., Mikhailovsky, A. A., Xu, S., Malko, A., Hollingsworth, J. A., Leatherdale, C. A., Eisler, H.-J., and Bawendi, M. G. 2000. Optical gain and stimulated emission in nanocrystal quantum dots. *Science.* 290(5490):314–17.
37. Duarte, F. J., and Hillman, L. W., eds. 1990. *Dye laser principles: With applications.* New York: Academic Press. ISBN: 978-0-12-396073-3.
38. Ta, V. D., Wang, Y., and Sun, H. 2019. Microlasers enabled by soft-matter technology. *Adv. Optical Mater.* 7(17):1900057. https://doi.org/10.1002/adom.201900057.
39. Liao, Q., Jin, X., and Fu, H. 2019. Tunable halide perovskites for miniaturized solid-state laser applications. *Adv. Optical Mater.* 7(17):1900099. https://doi.org/10.1002/ADOM.201900099.
40. Wei, Q., Li, X., Liang, C., Zhang, Z., Guo, J., Hong, G., Xing, G., and Huang, W. 2019. Recent progress in metal halide perovskite micro- and nanolasers. *Adv. Optical Mater.* 7(17):1900080. https://doi.org/10.1002/adom.201900080.
41. Zhang, Q., Su, R., Du, W., Liu, X., Zhao, L., Ha, S. T., and Xiong, Q. 2017. Advances in small perovskite-based lasers. *Small Methods.* 1(9):1700163. https://doi.org/10.1002/smtd.201700163.

42. Stockman, M. 2008. Spasers explained. *Nat. Photon.* 2:327–9. https://doi.org/10.1038/nphoton.2008.85.
43. Purcell, E. M. 1995. Spontaneous emission probabilities at radio frequencies. In *Confined electrons and photons: New physics and applications*, ed. E. Burstein and C. Weisbuch, 839–839. Boston, MA: Springer Science+Business Media LLC. https://doi.org/10.1007/978-1-4615-1963-8_40.
44. Oxborrow, M. 2007. Traceable 2-d finite-element simulation of the whispering-gallery modes of axisymmetric electromagnetic resonators. *IEEE Trans. Microw. Theory Tech.* 55(6):1209–18. https://doi.org/10.1109/TMTT.2007.897850.
45. Ta, V. D., Chen, R., and Sun, H. D. 2011. Wide-range coupling between surface plasmon polariton and cylindrical dielectric waveguide mode. *Opt. Express.* 19(14):13598–603. https://doi.org/10.1364/OE.19.013598.
46. Oulton, R. F., Sorger, V. J., Genov, D. A., Pile, D. F. P., and Zhang, X. 2008. A Hybrid plasmonic waveguide for subwavelength confinement and long-range propagation. *Nat. Photon.* 2(8):496–500. https://doi.org/10.1038/nphoton.2008.131.
47. Bergman, D. J., and Stockman, M. I. 2003. Surface plasmon amplification by stimulated emission of radiation: Quantum generation of coherent surface plasmons in nanosystems. *Phys. Rev. Lett.* 90(2):027402. https://doi.org/10.1103/PhysRevLett.90.027402.
48. Stockman, M. I. 2010. The spaser as a nanoscale quantum generator and ultrafast amplifier. *J. Opt.* 12(2):024004. https://doi.org/10.1088/2040-8978/12/2/024004.
49. Baranov, D. G., Andrianov, E. S., Vinogradov, A. P., and Lisyansky, A. A. 2013. Exactly solvable toy model for surface plasmon amplification by stimulated emission of radiation. *Opt. Express.* 21(9):10779–91. https://doi.org/10.1364/OE.21.010779.
50. Gordon, J. A., and Ziolkowski, R. W. 2007. The design and simulated performance of a coated nano-particle laser. *Opt. Express.* 15(5):2622–53. https://doi.org/10.1364/OE.15.002622.
51. Meng, X., Guler, U., Kildishev, A. V., Fujita, K., Tanaka, K., and Shalaev, V. M. 2013. Unidirectional spaser in symmetry-broken plasmonic core-shell nanocavity. *Sci. Rep.* 3(1):1241. https://doi.org/10.1038/srep01241.
52. Bordo, V. G. 2017. Cooperative effects in spherical spasers: *ab initio* analytical model. *Phys. Rev. B.* 95(23):235412. https://doi.org/10.1103/PhysRevB.95.235412.
53. Arnold, N., Piglmayer, K., Kildishev, A. V., and Klar, T. A. 2015. Spasers with retardation and gain saturation: Electrodynamic description of fields and optical cross-sections. *Opt. Mater. Express.* 5(11):2546–77. https://doi.org/10.1364/OME.5.002546.
54. Plum, E., Fedotov, V. A., Kuo, P., Tsai, D. P., and Zheludev, N. I. 2009. Towards the lasing spaser: Controlling metamaterial optical response with semiconductor quantum dots. *Opt. Express.* 17(10):8548–51. https://doi.org/10.1364/OE.17.008548.
55. Li, Z.-Y., and Xia, Y. 2010. Metal nanoparticles with gain toward single-molecule detection by surface-enhanced Raman scattering. *Nano Lett.* 10(1):243–9. https://doi.org/10.1021/nl903409x.
56. Hill, M. T., Marell, M., Leong, E. S. P., Smalbrugge, B., Zhu, Y., Sun, M., van Veldhoven, P. J., Geluk, E. J., Karouta, F., Oei, Y.-S., Nötzel, R., Ning, C.-Z., and Smit, M. K. 2009. Lasing in metal-insulator-metal sub-wavelength plasmonic waveguides. *Opt. Express.* 17(13):11107–12. https://doi.org/10.1364/OE.17.011107.
57. Marell, M. J. H., Smalbrugge, B., Geluk, E. J., van Veldhoven, P. J., Barcones, B., Koopmans, B., Nötzel, R., Smit, M. K., and Hill, M. T. 2011. Plasmonic distributed feedback lasers at telecommunications wavelengths. *Opt. Express.* 19(16):15109–18. https://doi.org/10.1364/OE.19.015109.
58. Flynn, R. A., Kim, C. S., Vurgaftman, I., Kim, M., Meyer, J. R., Mäkinen, A. J., Bussmann, K., Cheng, L., Choa, F. S., and Long, J. P. 2011. A room-temperature semiconductor spaser operating near 1.5 μm. *Opt. Express* 19(9):8954–61. https://doi.org/10.1364/OE.19.008954.

59. Wu, C.-Y., Kuo, C.-T., Wang, C.-Y., He, C.-L., Lin, M.-H., Ahn, H., and Gwo, S. 2011. Plasmonic green nanolaser based on a metal–oxide–semiconductor structure. *Nano Lett.* 11(10):4256–60. https://doi.org/10.1021/nl2022477.

60. Ho, J., Tatebayashi, J., Sergent, S., Fong, C. F., Iwamoto, S., and Arakawa, Y. 2015. Low-threshold near-infrared GaAs–AlGaAs core–shell nanowire plasmon laser. *ACS Photonics.* 2(1):165–71. https://doi.org/10.1021/ph5003945.

61. Zhang, Q., Li, G., Liu, X., Qian, F., Li, Y., Sum, T. C., Lieber, C. M., and Xiong, Q. 2014. A Room temperature low-threshold ultraviolet plasmonic nanolaser. *Nat. Commun.* 5(1):4953. https://doi.org/10.1038/ncomms5953.

62. Lu, Y.-J., Kim, J., Chen, H.-Y., Wu, C., Dabidian, N., Sanders, C. E., Wang, C.-Y., Lu, M.-Y., Li, B.-H., Qiu, X., Chang, W.-H., Chen, L.-J., Shvets, G., Shih, C.-K., and Gwo, S. 2012. Plasmonic nanolaser using epitaxially grown silver film. *Science.* 337(6093):450–3. https://doi.org/10.1126/science.1223504.

63. Liao, Y.-J., Cheng, C.-W., Wu, B.-H., Wang, C.-Y., Chen, C.-Y., Gwo, S., and Chen, L.-J. 2019. Low threshold room-temperature UV surface plasmon polariton lasers with ZnO nanowires on single-crystal aluminum films with Al$_2$O$_3$ interlayers. *RSC Advances.* 9(24):13600–07. https://doi.org/10.1039/C9RA01484E.

64. Wu, X., Xiao, Y., Meng, C., Zhang, X., Yu, S., Wang, Y., Yang, C., Guo, X., Ning, C. Z., and Tong, L. 2013. Hybrid photon-plasmon nanowire lasers. *Nano Lett.* 13(11):5654–9. https://doi.org/10.1021/nl403325j.

65. Ma, R.-M., Oulton, R. F., Sorger, V. J., Bartal, G., and Zhang, X. 2011. Room-temperature sub-diffraction-limited plasmon laser by total internal reflection. *Nat. Mater.* 10(2):110–3. https://doi.org/10.1038/nmat2919.

66. Kwon, S.-H., Kang, J.-H., Seassal, C., Kim, S.-K., Regreny, P., Lee, Y.-H., Lieber, C. M., and Park, H.-G. 2010. Subwavelength plasmonic lasing from a semiconductor nanodisk with silver nanopan cavity. *Nano Lett.* 10(9):3679–83. https://doi.org/10.1021/nl1021706.

67. Khajavikhan, M., Simic, A., Katz, M., Lee, J. H., Slutsky, B., Mizrahi, A., Lomakin, V., and Fainman, Y. 2012. Thresholdless nanoscale coaxial lasers. *Nature.* 482(7384):204–7. https://doi.org/10.1038/nature10840.

68. Nguyen, T. V., Pham, N. V., Mai, H. H., Duong, D. C., Le, H. H., Sapienza, R., and Ta, V.-D. 2019. Protein-based microsphere biolasers fabricated by dehydration. *Soft Matter.* 15(47):9721–6. https://doi.org/10.1039/c9sm01610d.

69. Ma, R.-M., Yin, X., Oulton, R. F., Sorger, V. J., and Zhang, X. 2012. Multiplexed and electrically modulated plasmon laser circuit. *Nano Lett.* 12(10):5396–402. https://doi.org/10.1021/nl302809a.

70. Stehr, J., Crewett, J., Schindler, F., Sperling, R., von Plessen, G., Lemmer, U., Lupton, J. M., Klar, T. A., Feldmann, J., Holleitner, A. W., Forster, M., and Scherf, U. 2003. A low threshold polymer laser based on metallic nanoparticle gratings. *Adv. Mater.* 15(20):1726–9. https://doi.org/10.1002/adma.200305221.

71. Zhou, W., Dridi, M., Suh, J. Y., Kim, C. H., Co, D. T., Wasielewski, M. R., Schatz, G. C., and Odom, T. W. 2013. Lasing action in strongly coupled plasmonic nanocavity arrays. *Nat. Nanotech.* 8(7):506–11.

72. Yang, A., Li, Z., Knudson, M. P., Hryn, A. J., Wang, W., Aydin, K., and Odom, T. W. 2015. Unidirectional lasing from template-stripped two-dimensional plasmonic crystals. *ACS Nano.* 9(12):11582–8. https://doi.org/10.1021/acsnano.5b05419.

73. Yang, A., Hoang, T. B., Dridi, M., Deeb, C., Mikkelsen, M. H., Schatz, G. C., and Odom, T. W. 2015. Real-time tunable lasing from plasmonic nanocavity arrays. *Nat. Commun.* 6(1):6939. https://doi.org/10.1038/ncomms7939.

74. Wang, D., Bourgeois, M. R., Lee, W.-K., Li, R., Trivedi, D., Knudson, M. P., Wang, W., Schatz, G. C., and Odom, T. W. 2018. Stretchable nanolasing from hybrid quadrupole plasmons. *Nano Lett.* 18(7):4549–55. https://doi.org/10.1021/acs.nanolett.8b01774.

75. Beijnum, F., Veldhoven, P. J., Geluk, E., de Dood, M., Hooft, G., and Exter, M. 2013. Surface plasmon lasing observed in metal hole arrays. *Phys. Rew. Lett.* 110(20):206802. https://doi.org/10.1103/PhysRevLett.110.206802.

76. Wu, C., Khanal, S., Reno, J., and Kumar, S. 2016. Terahertz plasmonic laser radiating in an ultra-narrow beam. *Optica* 3(7):734–40. https://doi.org/10.1364/OPTICA.3.000734.

77. Melentiev, P., Kalmykov, A., Gritchenko, A., Afanasiev, A., Balykin, V., Baburin, A. S., Ryzhova, E., Filippov, I., Rodionov, I. A., Nechepurenko, I. A., Dorofeenko, A. V., Ryzhikov, I., Vinogradov, A. P., Zyablovsky, A. A., Andrianov, E. S., and Lisyansky, A. A. 2017. Plasmonic nanolaser for intracavity spectroscopy and sensorics. *Appl. Phys. Lett.* 111(21):213104. https://doi.org/10.1063/1.5003655.

78. Dorofeenko, A. V., Zyablovsky, A. A., Vinogradov, A. P., Andrianov, E. S., Pukhov, A. A., and Lisyansky, A. A. 2013. Steady state superradiance of a 2D-spaser array. *Opt. Express.* 21(12):14539–47. https://doi.org/10.1364/OE.21.014539.

79. Meng, X., Liu, J., Kildishev, A. V., and Shalaev, V. M. 2014. Highly directional spaser array for the red wavelength region. *Laser Photon. Rev.* 8(6):896–903. https://doi.org/10.1002/lpor.201400056.

80. Galanzha, E. I., Weingold, R., Nedosekin, D. A., Sarimollaoglu, M., Nolan, J., Harrington, W., Kuchyanov, A. S., Parkhomenko, R. G., Watanabe, F., Nima, Z., Biris, A. S., Plekhanov, A. I., Stockman, M. I., and Zharov, V. P. 2017. Spaser as a biological probe. *Nat. Commun.* 8(1):15528. https://doi.org/10.1038/ncomms15528.

81. Chou, Y.-H., Hong, K.-B., Chang, C.-T., Chang, T.-C., Huang, Z.-T., Cheng, P.-J., Yang, J.-H., Lin, M.-H., Lin, T.-R., Chen, K.-P., Gwo, S., and Lu, T.-C. 2018. Ultracompact pseudowedge plasmonic lasers and laser arrays. *Nano Lett.* 18(2):747–53. https://doi.org/10.1021/acs.nanolett.7b03956.

82. Nezhad, M. P., Simic, A., Bondarenko, O., Slutsky, B., Mizrahi, A., Feng, L., Lomakin, V., and Fainman, Y. 2010. Room-temperature subwavelength metallo-dielectric lasers. *Nat. Photon.* 4(6):395–9. https://doi.org/10.1038/nphoton.2010.88.

83. Miller, S. E. 1969. Integrated optics: An introduction. *The Bell System Technical Journal* 48(7):2059–69. https://doi.org/10.1002/j.1538-7305.1969.tb01165.x.

84. Kim, M.-K., Lakhani, A. M., and Wu, M. C. 2011. Efficient waveguide-coupling of metal-clad nanolaser cavities. *Opt. Express.* 19(23):23504–12. https://doi.org/10.1364/OE.19.023504.

85. Kim, M.-K., Li, Z., Huang, K., Going, R., Wu, M. C., and Choo, H. 2013. Engineering of metal-clad optical nanocavity to optimize coupling with integrated waveguides. *Opt. Express.* 21(22):25796–804. https://doi.org/10.1364/OE.21.025796.

86. Prabowo, B. A., Purwidyantri, A., and Liu, K.-C. 2018. Surface plasmon resonance optical sensor: A review on light source technology. *Biosensors* 8(3):80. https://doi.org/10.3390/bios8030080.

87. Wang, X.-Y., Wang, Y.-L., Wang, S., Li, B., Zhang, X.-W., Dai, L., and Ma, R.-M. 2017. Lasing enhanced surface plasmon resonance sensing. *Nanophotonics.* 6(2):6. https://doi.org/10.1515/nanoph-2016-0006.

88. Ma, R.-M., Ota, S., Li, Y., Yang, S., and Zhang, X. 2014. Explosives detection in a lasing plasmon nanocavity. *Nat. Nanotech.* 9(8):600–4. https://doi.org/10.1038/nnano.2014.135.

89. Wang, S., Li, B., Wang, X.-Y., Chen, H.-Z., Wang, Y.-L., Zhang, X.-W., Dai, L., and Ma, R.-M. 2017. High-yield plasmonic nanolasers with superior stability for sensing in aqueous solution. *ACS Photonics* 4(6):1355–60. https://doi.org/10.1021/acsphotonics.7b00438.

90. Cho, S., Humar, M., Martino, N., and Yun, S. H. 2016. Laser particle stimulated emission microscopy. *Phys. Rev. Lett.* 117(19):193902. https://doi.org/10.1103/PhysRevLett.117.193902.

5 Metasurfaces and Several Well-Studied Applications

Muhammad Ali Butt[1,2], Nikolay Lvovich Kazanskiy[1,3], and Svetlana Nikolaevna Khonina[1,3]
[1]Samara National Research University, Samara, Russia
[2]Warsaw University of Technology, Institute of Microelectronics and Optoelectronics, Warszawa, Poland
[3]Image Processing Systems Institute – Branch of the Federal Scientific Research Centre "Crystallography and Photonics" of Russian Academy of Sciences, Samara, Russia

5.1 INTRODUCTION

Metasurfaces (hereafter represented as MSs) are a major research subject and are used in different utilizations because of their exceptional capability to manipulate electromagnetic (EM) waves at microwave (MWv) and optical frequencies. Since the theoretical calculation of their fascinating capabilities such as super-radiance and perfect absorbance or invisibility cloak configuration, they have received a lot of attention. Due to negative diffraction, these can be obtained experimentally with the simultaneous induction of strong electric (E) and magnetic (H) dipolar moments. Nevertheless, this condition supposes significant technological limitations connected to the accessibility of materials operating at frequencies of technological interest.

Such artificial sheet materials, usually consisting of metallic patches or dielectric etchings in planar or multilayer structures of subwavelength thickness, have the benefits of a light weight, ease of processing, and the ability to control the wave propagation on the substratum and in the neighboring free space. This chapter introduces several important applications based on MSs. MSs are subwavelength thickness two-dimensional (2D) or planar versions of metamaterials (hereafter represented as MMs). Due to their light weight and straightforward manufacturing, they are widely investigated and implemented in EM implementations. MSs have the distinctive capacity to block, absorb, concentrate, disperse, or guide waves from the MWv through visible (VIS) frequencies, both on the surface at grazing incidence and in space at normal and oblique incidence. The importance of this field can be understood by the number of publications in the last 12 years (2008–2019). The data is collected from Scopus indexed database as shown in Figure 5.1.

DOI: 10.1201/9781003439165-5

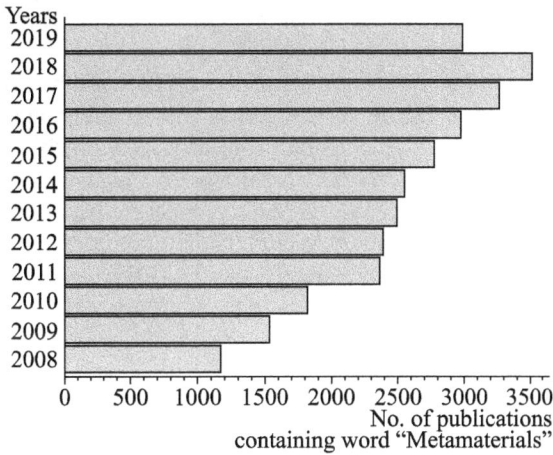

FIGURE 5.1 Number of publications containing the word "metamaterials" over the years 2008–2019. The data is taken from the Scopus database.

5.2 WHAT ARE METASURFACES?

The prefix "meta" suggests that the material's features go beyond what is apparent in nature. MMs are structural materials that are deliberately constructed and receive their characteristics from the internal microstructure instead of the chemical conformation present in normal materials. The core model of MMs is to create materials by means of artificially designed and engineered structural units to realize the preferred characteristics and features. Such structural units can be customized in shapes and size, the lattice constant and interatomic interface can be synthetically adjusted, and defects can be designed and positioned at the anticipated positions. By integrating these nanoscales (n-scales) unit cells into the preferred construction or geometry, the refractive index (hereafter represented as RI) of the MM can be adjusted to positive, near-zero, or negative values.

As a result, MMs may be endowed with characteristics and functions that those normal materials are not. For example, for invisibility cloak technology to conceal an item or, on the other hand, to avoid refraction by a perfect lens and permit direct viewing of an individual protein in a light microscope, the material must be able to regulate the direction of light similarly. MMs provide this potential. While MMs have already revolutionized optics, their execution has been hampered by their inability to operate over large light bandwidths (BWs). Modeling an MM that functions throughout the whole VIS spectrum remains a big issue. Such a material must be able to interact with both E- and H-light fields.

At VIS and infrared (IR) wavelengths, some natural materials are immune to the H-field of light. Previous attempts at MM have resulted in the creation of artificial atoms comprised of two components: one that interacts with the E-field and the other with the H-field. The advantage of this method is that the separate components interact with different colors of light, making it difficult to overlap them throughout a large range of wavelengths.

Three-dimensional (3D) MMs, which are primarily periodic artificial materials consisting of metal and dielectric, have received a lot of attention in the last decade because of their extraordinary interaction with EM waves, which outperform natural or homogeneous materials. Their exceptional wave management ability stems from their high interaction with E- and/or H-fields, which are generally given via resonant effects regulated by unit cell geometry. Such properties enable a wide range of implementations, including antenna efficacy increase, perfect absorbers (PAs), superlenses, cloaking and energy harvesting, among others. On the other hand, significant losses, and problems in 3D manufacturing, particularly at micro- and n-scales, exclude many MM uses.

5.3 NEGATIVE REFRACTION

The negative RI of light and other radiation is one of the most demanding properties of MMs. Negative refraction is founded on the equations established by the Scottish physicist James Maxwell in 1861. All identified natural materials have a positive RI so that light flowing from one medium to another becomes slightly bent in the transmission direction. For example, air has the lowest RI in nature at standard conditions, which is just above 1. Water's RI is 1.33. That's about 2.4 of the diamond. The higher the RI of the material, the more it distorts the original direction of light. Negative refraction occurs in some MMs, though, such that light and other radiations bends backwards as it approaches the system, as shown in Figure 5.2.

As early as the middle of the 20th century, the existence of substances with a negative RI was anticipated. In 1976, Soviet physicist V.G. Veselago published a scientific paper that described their properties, theoretically, as comprising unusual light refraction. In 1999, Roger Walser suggested the term MMs for such substances. But researchers were only able to determine how materials of any sort could be created to achieve negative refraction in the early 2000s. The first MM specimens were made from thin wire formations and operated only with MWv radiation. In electronics, lithography (litho), biomedicine, isolation coatings, heat transfer, space

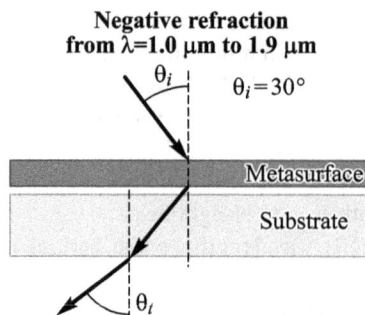

Negative refraction from $\lambda=1.0$ μm to 1.9 μm

θ_i $\theta_i = 30°$

Metasurface

Substrate

θ_t

FIGURE 5.2 Schematic of negative refraction of light. In negative RI materials, the light gets bent the opposite way.

implementations, and perhaps even novel methods to optical computation and energy harvesting are possible with such negative refraction materials.

5.4 METASURFACES

Traditional optical components use refractive, diffractive, and reflective principles to direct light. These devices work by using transmission inside the component's material to change the characteristics of an incident wavefront (e.g. phase, amplitude, and polarization). On the other hand, others work by creating interference patterns that result in Huygens–Fresnel sources that create far-field distributions. The propagation characteristics of a component are determined by differences in component shape and material properties. The volumetric management of the RI on a subwavelength scale allows for extraordinary control of light transmission in MMs, allowing for unique functionalities like negative refraction.

Metasurfaces (MSs, MTSs) are thin artificial layers comprising regular placements of tiny inclusions in a dielectric host medium. They can be customized to attain unusual reflection/transmission of space waves and/or to change the dispersion properties of surface/guided waves. The scientific community has given significant attention to MTS technology and has been employed in several implementations. MTSs are usually formed in MW antenna implementations by a standard texture of small units that are printed on a grounded slab with or without shorting vias. Additionally, MTSs can be made up of a concentrated formation of metallic pins on a ground plane at sub-millimeter wavelengths.

Surface waves (SWs) can be controlled by the design and management of the phase or groups velocity of MS unit cells with the correct impedance. These are designed to guide or separate waves in certain directions and are employed in dispersal control utilizations. Different effective surface refractive indices can be obtained by controlling the sizes and shapes of the MS unit cells and designing the surface to deliver numerous purposes. These can be utilized for the development of 2D MWvs/ optical lenses such as Luneburg and fish-eye lenses, which are employed in surface WGs for antenna systems and planar MWv sources.

When the SW interacts with the modulated, locally periodic boundary conditions (BCs) imposed by the MTS, it helps to replicate the desired aperture field that radiates through a leaky-wave (LW) effect. The conventional MTS configuration has a low profile, is light in weight and is simple to put together using traditional circuit board techniques. Furthermore, the feeding unit is incorporated in the MTS plane, allowing for the avoidance of external protrusion and (sub-) reflectors. Because of these attractive qualities, as well as the possibility of precise control and shape matching of the aperture field, various novel antenna configurations have recently been presented in the literature.

Challenging configurations were also conceivable from the perspective of beam scanning through the evolution of MTSs modeling. Huygens MTSs were suggested to acquire extra degrees of freedom in radiation field control. The prospect of receiving multi-beam radiation from a single aperture has also been shown by the coupling of the MTS and Luneburg lenses. MTSs use units whose geometries gradually change from cell to cell. At the operating frequency, the components that enforce the impedance BCs are small in terms of wavelength (usually between $\lambda/5$ and $\lambda/10$, with λ

being the free-space wavelength). Therefore, the interacting SW sees the impedance BC interface as a continuum. The constitutive units of MTS act like pixels in a printed black and white image, the grayscale of which is implemented by modifying the proportions of the units within a regular lattice. In a regular lattice, the pixels are arranged with elementary cell size in the range $\lambda/10$–$\lambda/5$. The lattice is typical of the Cartesian type, but hexagonal lattices could also be manipulated.

5.4.1 MS DESIGN PROCESS

While EM, or effective medium theory, is superior at describing the scattering properties of individual subwavelength units, Fourier optics can explain the combined activity of all the units and their spatial transfer function. Both phase and amplitude modulation can be utilized in MS configuration.

Let us consider a plane wave $I = Ae^{-j(k_x x + k_y y + k_z z)}$ is incident upon a phase mask in the $z = 0$ plane described by a transfer function $H = e^{j\Phi(x,y)}$, as indicated in Figure 5.3. This system provides an MS model with unit magnitude and a phase distribution $\Phi(x, y)$. If the phase distribution about the point $r_0 = (x_0, y_0)$ is certified as a Taylor expansion then to first order it can be expressed as follows:

$$\Phi(x,y) \approx \Phi(x_o, y_o) + (x - x_o)\frac{\partial \Phi}{\partial x}\bigg|_{r=r_o} + (y - y_o)\frac{\partial \Phi}{\partial y}\bigg|_{r=r_o}. \tag{5.1}$$

If the transfer function H is then applied to the plane wave I, the coefficients of the linear terms in the expansion are added along with the x and y wave vector components of the incident wave to produce the result where all the constant phase terms are merged into θ:

$$T = IH \approx Ae^{-j\left[\left(\frac{\partial \Phi}{\partial x}\big|_{r=r_o} + k_x\right)x + \left(\frac{\partial \Phi}{\partial y}\big|_{r=r_o} + k_y\right)y + k_z^z z + \theta\right]}. \tag{5.2}$$

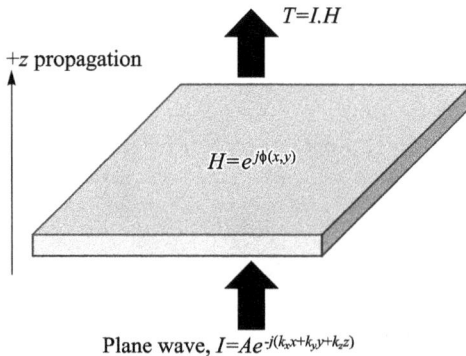

FIGURE 5.3 Schematic representation portraying the functioning of a MS. An incident plane wave I is acted upon by the spatial transfer function of the MS to produce a transmitted wave T.

The gradient of the phase distribution has the effect of altering the wave vector of the incident wave, as indicated in (5.2). This change in wave vector caused by a mask's spatial phase function gives a fundamental understanding of how MSs may alter wavefront unpredictably. To perform an optical transformation, it is necessary to first determine how the incident wave vector must change, and then utilize this knowledge to calculate the appropriate spatial phase function.

The first step in the MS configuration process is to establish the necessary spatial phase function. This step provides a continuous function of position with the ideal transfer characteristics, yet such a phase contour cannot be manufactured in terms of implementation because the infinitesimal resolution would be necessary. MSs phase contours are divided into a grid of finite unit cells representing a spatially discrete function, comparable to signals in discrete-time systems with specific sampling periods. The MS must be designed to meet the criterion of the Nyquist–Shannon sampling theorem for the spatial sampling period. A suitable scattering unit is positioned in each unit cell to accomplish the required phase shift at that point. To physically implement the discrete phase function on the MS, it is necessary to achieve arbitrary phase functions by scattering units that can induce phase shifts from 0 to 2π. Due to configuration constraints, appropriate selection of scattering units is often a challenge. The primary restriction is the requirement to be subwavelength for the period of the scattering units. This ensures that only the zero-order diffraction mode is transmitted by the MS. A second constraint is that the patterned units are of the same height across the surface, allowing the MS to be manufactured with a single litho stage. Due to the uniform height of the scattering structures, researchers are left with limited options to alter the phase. By changing the geometry or orientation of the scattering formations, it is usually possible to achieve phase shifts ranging from 0 to 2π with uniform magnitude. For instance, most MSs use Pancharatnam–Berry (PB) phase shift, which is proportional to the orientation angle of the unit within its cell. Other phase tuning methods can also be utilized, including modifying the widths, diameters, or periodicity of the scattering units.

5.4.2 State of the Art Metasurfaces and Their Limitations

In the last few years, MSs have experienced several major developments. Many of the initial MS-based optics have been using surface plasmon based-scattering units that permit the MS to be made of metal. These systems are limited to working in reflection mode, with metals being so lossy in the optical regime. MSs based on silicon helped to address this setback, allowing MSs that can work in the near-IR mode with high efficacy in both reflection and transmission. More recent developments have revealed that low RI materials have been employed to operate efficiently in the VIS system [1, 2].

Various generic optical components were replicated with the assistance of MS implementations and were fruitfully tested. Many of these replicated generic optical components in their MS-based implementations have been considerably miniaturized. Further advanced features were also accomplished. Polarization-multiplexed formations were introduced allowing various optical makeovers to be performed depending on the polarization state of the input light; on the other hand, these systems were principally limited to operating at the IR region. MSs also enables novel functionalities as well as imitating functionalities of traditional refractive and diffractive optics.

Although MSs offer an impressive range of capabilities, these devices remain subject to substantial limits that remain open to research. While some researchers suggested some devices working in the VIS region, most of them are limited to functioning in the IR region. Since the materials utilized are in general too lossy, with high band gaps or opaque at VIS wavelengths. The scope of the implementations for which MSs can work is significantly reduced without the capability to operate at VIS wavelengths. Such formations often face several aberrations, perhaps most notably chromatic aberrations. Like any system that works based on diffractive principles, as the wavelength varies, the functioning of MS-based devices deviates greatly from their intended functioning. Attempts to accomplish achromatic MSs have been made, but these methods have only validated multi-wavelength functioning instead of delivering a truly broadband achromatic apparatus.

Besides the aberrations and material constraints of MSs that limit the VIS regime functioning, one of the major problems in MS research is the development of a method for achieving phase units that can be dynamically changed to attain reconfigurable phase contours. Attaining this capability would permit the optical equivalent of a field-programmable gate array (FPGA) to be implemented and reprogrammed as preferred. Many approaches to achieving reconfigurable MSs consist of local tuning of the RI of the phase-shifting units. Regrettably, it is difficult to adjust the RI to achieve arbitrary changes from 0 to 2π phase shift. The range over the RI can be modulated is limited, and MS n-scale size constraints restrict the transmission distance and change in phase accumulation that could result from a shift in RI. Some methods are available for a substantial change of the RI, for example, germanium-antimony tellurium ($Ge_2Sb_2Te_5$, GST). However, depending on the operating wavelength, these materials can show a substantial loss. A stable and reconfigurable MS system would need phase units that can easily be modified without significant loss in performance.

5.5 APPLICATIONS OF METASURFACES

Although MM-based optical devices allow distinctive functionality, they are restricted by their reliance on light having to transmit through a bulk medium. Signal loss in the bulk medium is a significant restriction on execution, and the resolution of 3D manufacturing techniques limits the range of viable devices from a practical point of view. Because of these limitations, the 2D analogue of bulk MMs has attracted considerable interest in recent years. Instead of modulating the RI over a volume, the MSs are ultra-thin formations patterned on a subwavelength scale with quasi-periodic arrangements of scattering components. MSs implement spatial transfer functions that transform the input wavefronts over a wavelength-scale range by correctly designing and patterning the scattering components. The designed transfer functions spatially modulate the phase and or amplitude. This practice has facilitated MSs antennas, lenses, PAs, sensors, WG, cloaking, modulators, frequency selective surfaces, polarizers, etc. This section presents several implementations based on

MSs discussed one by one along with examples taken from the recent publications in this field.

5.5.1 METASURFACE ANTENNAS

MSs, apart from their wide implementations in surface and free-space wave management, have exclusive ability and distinctive advantages as media for radiating EM waves into free space. For many implementations, antenna systems require a large gain to meet the requirements of communications links. Because of their large aperture size and narrow beam width, large planar formations are implemented in several antenna systems. Nevertheless, feeding systems are typically complex to diminish loss and deliver phase control for each antenna component to determine the radiation pattern.

MSs can be employed in antenna implementations to boost antenna gain, improve BW and execution, minimize backward radiation, transform polarization, and steer antenna beams. They can be utilized to create a Fabry–Perot cavity to stimulate the LW mode of the formation for antenna gain enhancement. Moreover, MSs can also be tailored to act as a lens to produce an exceedingly focused beam and acts as a spatial filter to transform and transfer wave vectors parallel to the transmission direction. BW enhancement of antennas by MSs is typically credited to the increased impedance matching or parasitic resonance generated by the MS superstrate, and the lessening of the back radiation is generally due to the stopband of the EM band-gap structure formed by the MS. LW antennas allow for the removal of feeding networks for certain implementations, by using a traveling wave feed technique. When an LW antenna is fed by a point source, the field of the surface can be characterized as follows:

$$\psi(\rho, z) = \psi_o \exp(-j\gamma\rho - jk_z z), \tag{5.3}$$

where $\gamma = \alpha + j\beta$ represents the propagation constant on the surface of the LW antenna and $k_z = \sqrt{k_o^2 - \beta^2}$ is the propagation constant normal to the antenna surface. If $\beta < k_o$, k_z is strictly real in the fast wave regime, it means that the energy is coupled as radiation into free-space. It is possible to configure the path of radiation by manipulating the value β. The unit cells of the MS are specifically calculated to match the phase delay on the surface with that of the chosen direction of the beam in the free-space. These also have frequency-dependent beam scanning properties (beam squint), which, depending on the implementation, can be an advantage or a shortcoming.

In MWv antenna implementations, MSs are typically designed by a regular texture of small units printed on a grounded slab with or without shorting vias. Conversely, MSs can be made up of a dense formation of metallic pins on a ground plane at submillimeter wavelengths. When the SW interacts with an MS imposed modulated locally periodic boundary conditions (BCs), it permits one to replicate the anticipated aperture field that radiates through an LW effect. The traditional MS configuration has a low profile, weight, and complexity. Furthermore, it can be fabricated utilizing standard printed circuit board techniques. In addition, the feeding unit is fixed in the MS plane, so that an external protruding or backfeeding arrangement or (sub-) reflector can be avoided.

These appealing characteristics, together with the potential of fine control and shape adjustment of the aperture field, result in a few interesting antenna devices. It was suggested that Huygens MSs gain extra degrees of freedom in the control of the radiating field. In contrast, the probability of receiving multi-beam radiation from a single aperture was shown by combining MSs and Luneburg lenses. The MSs use units whose geometries progressively transform from cell to cell. At the operating frequency, the units implementing the impedance BCs are small in terms of the wavelength (typically, between $\lambda/5$ and $\lambda/10$, with λ being the free-space wavelength). As a result, the interacting SW considers the interface impedance BC to be a continuum. The MS constituent units function similarly to pixels in a black and white printed picture, with grayscale achieved by varying the dimension of the components inside a regular lattice. The pixels are organized in a consistent lattice with primary cell sizes ranging from $\lambda/10$–$\lambda/5$. Characteristically, the lattice is of Cartesian type, but hexagonal lattices could also be utilized.

High gain holographic antennas are successfully designed and produced with MSs [3, 4]. The SW is the primary incident wave for holographic antennas. Radiation occurs for such antennas when a phase matches an LW to a reverse one. Forward LW emerges from the grid, which has a greater periodicity, while the shorter period of grating corresponds to a backward leaky wave. The source with elliptical grating is surrounded by both periodicities. LW modes for such a grating are matched, leading to directive radiation [5]. Circularly polarized leaky wave antennas with a gain of 26 dB are suggested [3] based on this strategy. The grating effect was achieved through the modulation of the surface impedance of the MS. The major benefit of this method is that the MS modulation, i.e. surface impedance, is optimized instead of modifying the antenna shape to model a particular response. An MS antenna was presented [6] that offers the opportunity for both right and left circular polarization. Both TE and TM SWs were released and for their independent control, the polarization decoupling of the two modes was employed. Such modes are phase-matched, and the modulated MS was rotationally symmetric. A circular WG fed the antenna, and a corrugated hat was located on top of the WG to block the space-wave radiation. The phase-matching is important for LW MS antennas, and in case of mismatching, the gain of the structure decreases. Moreover, the BW of such antennas is restricted by the dispersion of the MS. For these antennas, a BW gain expression can be found in [7].

Marco Faenzi et al. [8] suggested novel MS (MTS) antenna models, implementation, and experiments made up of subwavelength components reproduced on a grounded dielectric slab. To generate an essentially arbitrary aperture field, these antennas employ the interaction of a cylindrical SW wavefront and an anisotropic impedance BC. They are very thin and may be stimulated by a simple in-plane monopole. Such antennas may be substantially modified in terms of beam shape, BW, and polarization by constructing the printed components. For the first time, testing results demonstrate that these antennas can have an aperture efficacy of up to 70 percent, a BW of up to 30 percent, create two distinct high-gain direction beams and identical beams at two distinct frequencies, and display outcomes not ever attained before.

MS antenna model for beam steering implementations is described here. The MS superstrate is made up of identical square rings (SRs) that are evenly spaced. The anticipation of SRs in series with varying phase delays is the basic mechanism of

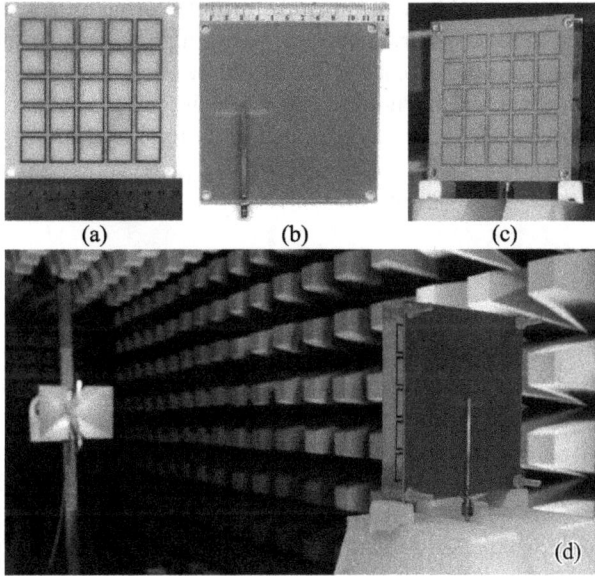

FIGURE 5.4 The manufactured single-slit MS antenna: (a) top view of the MS, (b) bottom view of the slit radiator for $P = 30$, (c) standpoint view when the slit radiator and MS are gathered, (d) photo of the MS setup [9].

radiation. The entire emitted beam may be directed in distinct directions based on the primary excitation of SR components. To attain beam steering, either a single-slit radiator with mechanical drive or a double-slit model with fixed slit locations but variable feeding phase differences between two slits can be utilized as the excitation source. A beam-steering angle of −35 to 35 degrees can be reached with a single-slit arrangement, and −30 to 30 degrees can be achieved with a double-slit configuration. The constructed single-slit MS antenna is presented in Figure 5.4.

5.5.2 METASURFACE LENSES

MS has fascinated major research interests in the field of MMs as originally presented to generalize Snell's law, which is, in fact, a distinct case of the grating equation. It allows management of the transmission of EM waves with a geometric gradient phase contour. The 2D artificial structure consisting of subwavelength scatter formations may create anomalous phenomena of reflection and refraction, which is technically more attractive in comparison with the 3D phenomena. The MS lens is mostly studied with the hope that the outdated bulky lens will be replaced by a piece of structured surface layer, which may optimistically lead to noteworthy advances in modern optical systems. The MS lens can accommodate both transmission magnitude and phase as well as polarization in subwavelength scale resolutions compared to the present thin-film based Fresnel's lens which is mostly to produce a specified phase contour.

The increased freedoms would expand the lens functions to include, for example, the removal of unwanted diffractions or reflections and further enhance focusing/imaging capabilities. A technological benefit is that the manufacturing process can be simplified by single-step litho. Dielectric materials with a high RI have been developed to reduce losses to improve the efficacy of the MS lenses. By putting all these developments into viewpoint, a bright future could be envisaged for MS lenses. However, some practical problems still need to be managed for implementations, which mostly involve device strength and BW. A MS should have a phase contour given by Eq. (5.4) to focus the EM energy at a distance d:

$$\Phi_L(x,y) = \frac{2\pi}{\lambda}\left(\sqrt{x^2 + y^2 + d^2} - d\right). \tag{5.4}$$

This contour also transforms the wavefront's shape from planar to spherical, which is required for focusing. A high numerical aperture (NA) efficacy may be attained if the EM wave reaches the surface at a normal angle. When the angle of incidence (AOI) is not perpendicular, however, a phenomenon known as "coma" develops, which can significantly reduce the NA's effectiveness. It is possible to reduce the effect of "coma" by employing the surface on a curved dielectric unit [10]. It has been shown that the MS composed of V-shaped antennas focuses power at telecommunication frequencies [11]. Another reflected array-based MS lens is introduced [12].

Previous solutions to the development of miniature optical components included gradient RI materials, such as graded-index (GRIN) lenses, traditional Fresnel-type lenses, and binary gratings. These are presumably still bulky due to the wavelength of the light for which they are constructed and depend on classical optical phenomena, i.e. refractive optics, involving the bulk RI to shape the light's wavefront. However, MSs are real miniature optical units, where the thickness is typically less than one wavelength of light. These MSs are typically an area made up of periodic scatterers (either metallic or dielectric) that are themselves subwavelength in the in-plane direction. Nonetheless, to avoid diffraction effects, MSs must have their periodicity less than the wavelength of interest.

The development of lensing MSs, namely metalenses or ultra-thin lenses, has been a major focus in recent years. The early execution of these lenses takes advantage of plasmonic resonance along with the well-known Pancharatnam–Berry phase (Geometric phase) [13]. However, a problem with these lenses is the very low efficacy of less than 15% – generally inherent in plasmonic devices – along with the small lens areas in the VIS to near-IR regimes. Also, the plasmonic structures that often rely on the use of gold (Au), prevent these lenses from operating below the wavelength of 550 nm and thus could not be employed to operate across the whole VIS spectrum. The problem of wavelength can be addressed by selecting other plasmonic metals such as silver (Ag) or aluminum (Al), but the problem of low efficacy remains. Here are some basics of MS-based lenses.

5.5.2.1 Control of the Phase Profile

Refractive lenses are commonly utilized in a variety of optical systems, for instance telescopes and microscopes. Despite having upright characteristics in phase control and polarization, traditional high-numerical-aperture refractive lenses are frequently bulky and pricy. Moreover, the complicated process of macro or mesoscale production also depends on conventional methods of optics manufacturing established over 100 years ago. Refractive lenses are typically intended with distinct shapes to satisfy optical specifications. A metalens, however, offers novel prospects to overcome these constraints. For example, by varying the meta building blocks (MBBs), the phase contour can be changed. The hyperbolic phase contour mandatory to focus a normal incident beam inside the substratum that remains collimated can be articulated as:

$$\varphi(r) = -\frac{2\pi}{\lambda}\left(\sqrt{r^2 + f^2} - f\right), \tag{5.5}$$

where f is the focal length of the incident light and r is the radial coordinate. The MS is designed to create a phase contour to transform the incident planar wavefront into a spherical shape at focal length f from the lenses.

5.5.2.2 Plasmonic Metasurface-Based Lenses

Typically, lenses established on MSs employ MBBs to adjust their optical features. One of the most common methods is to generate plasmonic effects on the surface. By means of advanced electron beam litho (EBL), and a straightforward lift-off process, a plasmonic antenna can be easily micro-machined. It is possible to transform the concentrated incident light into a smaller region that matches its wavelength and produces oscillations. The metalens has drawn considerable interest in the field of optics by getting the plasmonic effect on their MS. For instance, the experimental results stated in [14] showed that the micro-structures have a plasmonic effect on the surface of Ag thin film and effectively formed a focal point at the focusing plane. Alternative research establishes that the focusing could be accomplished and tuned assuming a variety of n-antenna shapes for example elliptical and circular blocks [15].

5.5.2.3 Lenses Rely on All-Dielectric Metasurfaces

The energy absorption loss of the incident light could possibly be reduced meaningfully by utilizing dielectric phase shifters MBBs. Researchers have benefited from this and involved dielectric phase shifters in several novel optical configurations. A polarization-insensitive lens with dielectric building blocks in the form of circular silicon formations has been suggested, as shown in Figure 5.5(a–f). With the incident light at a wavelength of 850 nm, a high transmission efficacy of 70% was achieved [16].

It has been shown that a single dielectric n-antenna could be built as an effective building block that could offer maximum phase coverage. The spatial image resolution, based on engineered n-structures, has excellent quality and relatively high transmission efficacy [17]. Moreover, dielectric metalens has greater execution in VIS range spectral implementations. μ-fabricated titanium dioxide (TiO_2) n-pillars have been employed to develop polarization-insensitive metalens. For incident wavelengths of 532 nm and 660 nm, the metalens could accomplish a comparatively high NA of 0.85 with an efficacy of more than 60% [18].

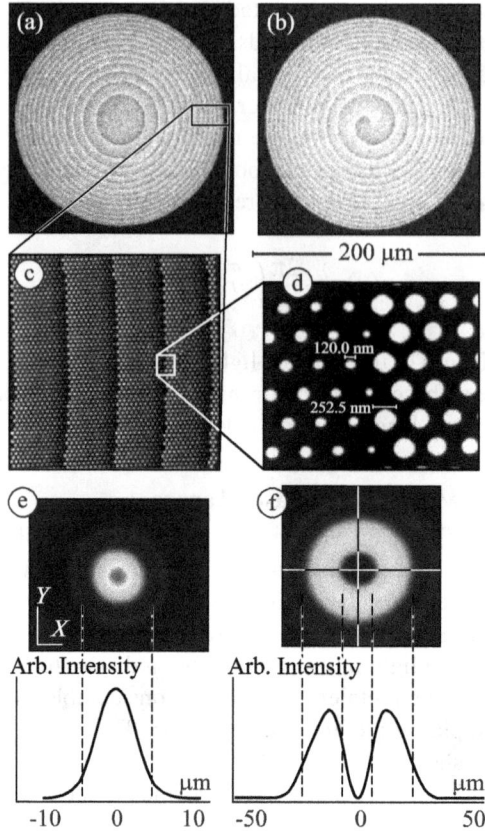

FIGURE 5.5 High-contrast gratings (HCG) lenses. (a) $l = 0$ focusing lens. (b) Vortex HCG lens. (c) and (d) SEM images show an expanded view of the fabricated lens. (e) Measured image intensity contour from the $l = 0$ focusing lens. (f) Measured image intensity contour from the $l = 1$ vortex lens [16].

5.5.3 Metasurface Perfect Absorbers

MMs have made it possible to demonstrate several exotic EM phenomena and have stimulated some interesting potential utilizations. Although bulk MMs present severe manufacturing challenges, mainly in the optical regime, MSs offer alternative ways to achieve desirable functionalities, including wavefront management, the transformation of polarization, and absorption/emission engineering. In various implementations, for example sensing [19], compressive imaging [20], and thermal management [21], MS PAs with a thickness much less than the operating wavelength are desirable.

Also of great interest are broadband PAs covering the whole solar spectrum in the harvesting of solar energy. Using dense n-rods and n-tube films, multilayer planar photonic structures, and photonic crystals, several examples of material structures as high-execution solar PAs have been created. The complex multi-resonator unit cell MS has evolved into a powerful and versatile platform for achieving flawless multiband and broadband absorption, particularly in the MWvs, terahertz, and IR regimes.

Here, we are presenting an example of MS broadband solar PA demonstrated by [22] based on a metallic MS architecture, which achieves an absorptance of more than 90% in the VIS and near-IR range of the solar spectrum and shows low absorptivity at mid and far-IR wavelengths. The complex unit cell of the MS solar PA is made up of eight pairs of Au n-resonators isolated by a thin silicon dioxide spacer from a Au ground plane. The absorbed power by the 2D sheet (resonator formation) is given by:

$$P_{abs}^{sheet} = \frac{1}{2} \text{Re} \int_A K.E^* \, dA, \tag{5.6}$$

where $K = \sigma E$ is the surface current density, E is the electric field at the sheet, and A is the area. The absorptance of the sheet is then given by:

$$A_{sheet} = \frac{\left(\dfrac{P_{abs}^{sheet}}{A} \right)}{I_o} = \frac{\text{Re}(\sigma)|1+r|^2}{c\varepsilon_o}, \tag{5.7}$$

where I_o is the incident intensity, r is the reflection coefficient of the multilayered system, c is the speed of light in a vacuum, and ε_o is the vacuum permittivity. The power degenerated within the ground plane is expressed as:

$$P_{abs}^{gp} = \frac{1}{2} \text{Re} \int_A J.E^* \, dA = \frac{\varepsilon_o \omega}{2} \text{Im} \left(n_{Au}^2 \right) \int_{V_{gp}} |E|^2 \, dV, \tag{5.8}$$

where n_{Au} is the complex RI of Au, and the integration extends across the entire volume of the ground plane. However, since the ground plane acts as an almost perfect mirror, we can approximate the field by its value at the interface spacer-ground plane and restrict the integral to the volume $A\delta/2$, where $\delta/2 = c/[2\omega \, \text{Im}(n_{Au})]$ is the energy penetration depth. The absorptance of the ground plane is then expressed as:

$$A_{gp} = \frac{\left(\dfrac{P_{abs}^{gp}}{A} \right)}{I_o} = \frac{\text{Im} \left(n_{Au}^2 \right) |t|^2}{2 \, \text{Im} \left(n_{Au} \right)}, \tag{5.9}$$

where t is the field transmission into the ground plane.

Figure 5.6(a) depicts the schematic diagram of the MS PA, which is based on the metal-insulator-metal architecture. It is made up of a formation of 50-nm-thick Au n-resonators parted by a 60-nm-thick SiO_2 and a 200-nm-thick ground plane. To achieve broadband absorption, a supercell having 16 resonators of various sizes and shapes was employed, and they all had four-fold symmetry to give a polarization-insensitive response.

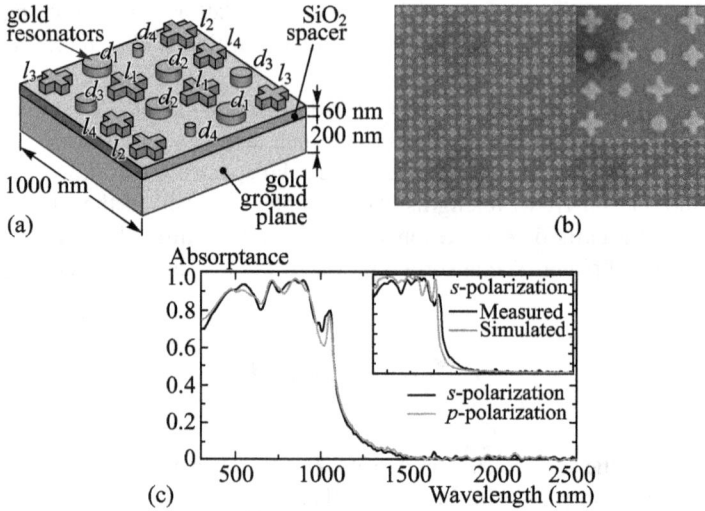

FIGURE 5.6 (a) Graphic depiction of the supercell of the broadband MS PA consisting of 16 resonant units forming a square formation. (b) SEM image of a portion of the fabricated PA. (c) Experimentally measured absorptance for *s*- and *p*-polarizations at 20° AOI [22].

Figure 5.6(b) depicts a SEM image of the sample, with an inset displaying a magnified view of the supercell. Figure 5.6(c) depicts the observed absorptance at 20° AOI for both s- and p-polarizations, demonstrating the polarization-insensitive strong absorption throughout roughly the full solar spectrum. MS PA achieves absorptance of more than 90% in the wavelength range 450–920 nm.

5.5.4 METASURFACES FOR SENSOR APPLICATIONS

Currently, MMs and MSs are increasingly being considered, due to their anomalous and highly tunable light dispersion characteristics. Their execution is reliant on their subwavelength spacing and geometric variables of each meta-atom (MA). MAs are made up of one or more subwavelength n-structures made of noble metals or high RI dielectrics. They are intended to display desired effective local optical responses in terms of amplitude and phase, as well as E- and H- polarization. Various possible implementations with unusual characteristics, such as negative RI, perfect absorption, and transmissivity, have been identified. The most practical and typical implementation of many possible applications, such as superlens, slow light, and cloaking technologies, is RI biosensing. Biomolecular interactions in analyte layers induce RI alterations. The RI sensor offers distinguishing capabilities for sensitive and label-free biochemical tests, allowing it to play a vital role in a wide range of chemical and biological implementations.

Resonant EM spectra, which are dominated by the ambient environment, can be largely tuned by designing individual MAs and their arrangements. The RI of the neighboring bimolecular analytes can be checked with a variety of scattered output spectra due to such resonant properties. The configuration of suitable

MAs at the target wavelengths and certain configurations is therefore essential. Furthermore, sensing systems based on MM and MS offer numerous benefits over traditional SPP-based biosensors in terms of manufacturing tolerance and read-out signal stability. Second, periodic MA configurations enable reduced radiative damping and greater quality factor through intriguing physical phenomena such as plasmonically generated transparency and Fano resonances. MS might eventually be added to the capability of a single n-photonic RI sensor. Several resonances and broadband slow-light phenomena can be produced by delicate and intricate compositions of multiple distinct MAs in a unit-cell or supercell, which are difficult to permit in SPP-based sensors.

Sensing light through light–matter interactions, on the other hand, has been a well-studied issue in optical society. Fundamental characteristics of any incident EM waves may be detected and studied via interference between multiple beams, aniso-tropic scattering, and dispersive responses from specific EM devices. Polarization and spectral composition are essential aspects of interest, and they are examined using polarimeters and spectrometers, which are types of industrial bulk-optic equipment. MSs are projected to be a viable platform for implementing these characteristics with small volume and light weight, as integration and downsizing requirements for these light-sensing components have grown considerably for electronic devices.

5.5.4.1 Sensors Based on Surface-Enhanced Raman Scattering

The inelastic scattering of photons by molecules driven to higher vibrational or rotational energy levels is known as Raman scattering. Because it is based on molecular states, the Raman scattered light spectrum is widely utilized to categorize and study materials. It appears to have worth in that it can be utilized to identify and evaluate not just fundamental things like solids, liquids, and gases but also more complicated materials, including human tissues, biological species, and chemical compounds. Raman scattering has been an essential way of assessing the structural fingerprint by which molecules may be recognized in many chemical processes since the discovery and development of lasers in the 1960s.

Despite its flexibility, applying Raman scattering directly to the sensor requires a lengthy time due to its tiny scattering cross section. When matter and light scatter, most photons are elastically dispersed, which is known as Rayleigh scattering, and the scattered photons have the same frequency as the original photons. Only approximately 1 in 10 million scattered photons are dispersed due to the expectation of scattered photons with a frequency differing from that of input photons. In low sample concentration settings, such as thin film covered with diluted solutions or monolayers, sufficient Raman spectra are difficult to obtain due to the low chance of molecule Raman scattering. Another significant difficulty is that, unlike pure or colorless samples, materials having color or fluorescence are difficult to study using Raman spectra. Also, when a laser in a VIS area is lit, reliable findings are extremely difficult to acquire in the event of a sample that emits greater background-fluorescent signals than Raman scattering of the sample itself.

In 1974, Fleischmann et al. found that electrochemically roughed Ag improved the pyridine-adsorbed Raman signals. A series of experiments in 1977 verified that Raman signals could be significantly enhanced when molecules are adjacent to rough metal surfaces or n-structures. It was referred to as surface-enhanced Raman scattering (SERS) or surface-enhanced Raman spectroscopy [23]. Scientists have been essentially employing noble metals such as Au or Ag and distinct n-structures such as plasmonic antennas for several decades to obtain more amplified signals. This is due to the ability of metal n-antennas to support nonpropagating excitations of the conduction electrons tied to the EM field, known as localized surface plasmons (LSPs).

MMs and MSs are ideal candidates for SERS substratum because they can excite surface plasmon polaritons (SPPs) which helps to reinforce Raman signals, where EM responses can be regulated by self-engineered MAs in the whole range of VIS to near-IR spectrum to detect biomolecules and proteins. Recently, researchers have developed SERS substratum by utilizing split ring resonators MMs and attained chemical sensing [24]. The tunable VIS to near-IR MMs based n-sensor with SERS molecular detection has also been demonstrated [25].

5.5.5 Metasurface Waveguides

Impedance surfaces are metal surfaces that are subwavelength in form and printed on dielectric media. Because of the substratum's small electric thickness, basic 2D surfaces may be modeled. Because it is frequently periodic or pseudo-periodic, even the structures may be precisely and efficiently described using periodic boundary conditions and eigenmode models. The surface impedance and dispersion properties are determined by the size and form of the periodic patterns. If the wave vector along the surface (k) is larger than its free-space counterpart (k_o), the impedance surface supports and binds the propagating wave along the surface, resulting in a planar WG.

For guiding SWs along a constrained route, simple WG structures entailing a thin strip of high impedance surrounded by two low impedance sheets have been proven [26]. The most popular approach for creating materials with variable SW indices is to employ a grid of varying size metallic patches on a dielectric substratum. The use of asymmetric patches yields a tensor SW index. SW WG formation interacts with SWs similarly to how an optical fiber transmission line interacts with light. The physical phenomenon is the same; the wave travels in an ideal high RI zone bordered by a low RI region. In the case of dielectric WG, the high RI and low RI areas are achieved with high- and low-permittivity materials. In the case of SW WG, the high and low RI areas can be achieved by variable size and/or shape metallic patches on a dielectric substratum.

5.5.5.1 Surface Wave Media

Surface wave medium (SWM) refers to any structure that may sustain a SW. They are a subsection of the larger category of MMs identified as artificial impedance surfaces and may serve SWs polarized in either transverse electric (TE) or transverse magnetic (TM) modes. The SW index (n_{SW}) or the SW impedance (Z_{TE} and Z_{TM}) is used to describe the SWM. The SW index is equivalent to an optical RI, and it is associated to SW impedance by:

$$n_{SW} = \left(1 + \frac{Z_{TM}^2}{n_o^2}\right)^{\frac{1}{2}} \quad \text{and} \quad n_{SW} = \left(1 + \frac{n_o^2}{Z_{TE}^2}\right)^{\frac{1}{2}},$$

for TM and TE modes, respectively, where,

$$n_o = \left(\frac{\mu_o}{\varepsilon_o} \right)^{1/2}.$$

A grounded dielectric sheet is the simplest form of an SWM. The transverse resonance method can be utilized to analytically calculate its SW RI. The grounded dielectric is not practical at frequencies less than 10 or 20 GHz because it must be very thick or use a substratum with excessively high permittivity to support SWs effectively. The most common way to prevail over this limitation is to distribute a grid of metallic patches (in any possible shape and distributes in any pattern) on a dielectric substratum. Nevertheless, the distribution on a uniform, periodic formation makes simulations, calculations, and development easier. A uniform periodic formation of square patches is the simplest grid. A frequency-independent capacitance defines a grid of squares or any convex shape. The SW RI of SWMs generated with these forms is insufficient due to the substratum's RI. Grids constructed out of filamentary forms, such as the Jerusalem cross, have equivalent capacitance, inductance, and resonance [27]. SWMs with these forms have a greater SW RI than square-grid SWMs and are not subject to the same SW RI restrictions. They have lower BW, though, since the grid's resonant structure decreases the SW band-gap frequency. Generally, SWM constructed of metallic patches of any shape and pattern obey the following rules: SW RI rises monotonically with SW frequency, formation period, patch size relative to the formation period, substratum permittivity, and thickness. SWMs are classified as either scalar or tensor. In a scalar SWM, the SW RI is independent of the transmission direction, but in a tensor SWM, the RI is direction dependent, and tensor quantities represent the RI and the impedance.

5.5.5.2 Surface Waves Waveguide Theory

Here three methods for manipulating the transmission of SWs are presented: (1) use abrupt variations in SW RI as confinement boundaries, (2) guide the phase fronts along curved paths deprived of alteration by grading the SW RI across the subwavelength gratings (SWG) width, and (3) align the lower-RI principle axis of a tensor SWM with transmission direction. The easiest SWG consists of a strip of constant SW RI encircled by a low-RI SWM, which is a 2D counterpart of a 3D dielectric WG.

It can be viewed as a high RI 2D fiber encircled by a low RI medium. The simulation results for straight and curved SWGs are presented in [26]. The SW is excited by a dipole feed. The high-impedance channel is one free-space wavelength wide. The corresponding SW indices are $n_1 = 1.44$, and $n_2 = 2.24$. Simulation is done using FastScat, a high-order integral-equation solver. Higher field values are represented as brighter colors. Power outside the SWG is more than 20 dB below the peak of the SWG. The phase front is approximately perpendicular to the transmission path. Some phase front falsification arises as the SWG width is 2.24 times the SW wavelength, and more than one transmission mode is existing. A curved SWG guides the SW around a curvature radius of 6.7 λ_{sw}, where λ_{sw} is the wavelength of the SW. The SW is confined to the SWG, but some power is leaked as it negotiates the curve.

The leaked power is 15–20 dB below the peak levels at the SWG center. The leakage of power is due to the path length, which differs across the width of the curve, which triggers the wavefront to diverge from being perpendicular to the WG axis. By classifying the SW RI across the width of the SWG, this problem can be solved so that the electrical path length is constant across the width. This results in the progressive rotation of the phase front to match the curvature of the transmission path.

5.5.6 CLOAKING

One of the most fascinating and well-known characteristics of MS is the potential to create an invisibility cloak. It has been theoretically demonstrated and experimentally confirmed in several scenarios that the anomalous wave interaction of artificial materials and MMs can be tailored to significantly reduce an object's overall visibility in distinct frequency regions spanning radio frequencies (RF), IR, and VIS light [28].

The presence of anomalous localized resonance near superlenses has been demonstrated to contribute to cloaking phenomena. This happens when a polarizable line or point dipole-generated resonant field acts back on the polarizable line or point dipole, cancelling out the field operating on it from outside sources. Cloaking is proven in the quasi-static limit for finite collections of polarizable line dipoles that are all within a certain distance of a coated cylinder with a shell permittivity $\varepsilon_s \approx -\varepsilon_m \approx \varepsilon_c$, where ε_m is the permittivity of the neighboring matrix, and ε_c is the core permittivity [29].

The first suggested models employed 3D anisotropic MMs and were based on the notion of transformation optics [30]. Such cloaks, on the other hand, were overly heavy and narrowband, resulting in enormous losses. An alternate technique called "mantle cloaking" was suggested, which involves placing ultra-thin MS screens at a distance over the item to be cloaked.

Although the theoretical basis of all these MMs-related cloaking methods is well established, the practical implementation of these systems, based on existing MM technologies, is far from optimal. At the moment, experimental demonstrations on cloaking technology have resulted in very modest scattering reduction, which falls well short of theoretical projections. This is largely owing to the intrinsic difficulties of establishing bulk MMs as collections of tiny inclusions with the necessary unusual bulk characteristics that may be handled as a continuum. These constraints are especially apparent for transformation-based cloaks, which need complicated and accurate inhomogeneity contours that might be difficult to implement with finite granularity formations. Furthermore, some of these cloaking approaches need an average thickness equal to the size of the cloaked zone, which may result in BW limits and greater sensitivity to realistic material loss.

5.5.6.1 The Challenges Faced by the First Cloaking Device

On October 19, 2006, a tiny cloaking device was utilized to illustrate the idea of cloaking at frequencies in the MWv radiation range for the first time. It was smaller than 13 mm in height and 125 mm in diameter, and it effectively directed MWvs around it. The object to be hidden, a tiny cylinder, has been put in the device's center.

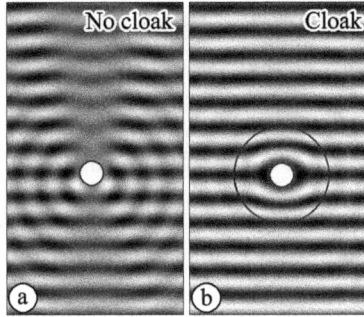

FIGURE 5.7 (a) The cross section of a PEC cylinder subject to a plane wave (only the E-field component of the wave is shown). The field is dispersed. (b) A circular cloak, designed using transformation optics methods, is employed to cloak the cylinder. In this case, the field remains unaffected outside the cloak and the cylinder is unseen electromagnetically.

The invisibility cloak deflected MWv beams, causing them to flow around the cylinder inside with only a tiny alteration, almost as if nothing was there at all. Typically, such a device entails enclosing the item to be cloaked with a shell that alters the passage of light close to it. The object's reflection of EM waves was decreased. Unlike a homogeneous natural substance with the same material characteristics everywhere, the cloak's material properties vary from point to point, with each point being tailored for specific EM interactions (inhomogeneity) and varying in various directions (anisotropy). This achieves a gradient in the properties of the material. This can be implemented in an MM in which split-ring resonators known to deliver an H-response that can be custom-made are located with their axes along the radial direction [31]. The interaction of plane wave around the cloaked and uncloaked cylinder is revealed in Figure 5.7 (a and b). Three notable limitations can be shown, despite a successful demonstration.

1. Since the experimental demonstration was successful only in the MWv range, the small object is somewhat transparent only at MWv frequencies. It implies that invisibility had not been achieved for the VIS spectrum since the VIS spectrum's wavelengths are significantly shorter than MWvs. However, this was the first step toward a cloaking device for VIS light, although more advanced nanotechnology-related techniques would be needed due to the short wavelengths of light.
2. According to the experiment conducted in 2006, only small objects can be made to appear as the neighboring air.
3. For any given demonstration, cloaking can only take place over a narrow frequency band. This means that there is no broadband cloak, on hand currently, that works across the EM spectrum, from radio frequencies to MWv to VIS spectrum to X-ray. This is because of the dispersive nature of today's MMs. The transformation optics requires exceptional material variables that are merely accessible using resonant units that are essentially narrowband, and dispersive in resonance.

5.5.7 All-dielectric Metasurfaces Visible Light Modulator

Although the field of MSs has started to evolve based on metallic materials, the poor execution of these devices for short wavelengths, mainly due to ohmic dissipation and dispersion losses, has prompted the scientific community to shift its attention to all-dielectric alternatives. In the case of MSs utilizing high RI dielectric n-antennas, they benefit not only from the lower ohmic losses but also from the naturally rich phenomenology of optical modes, including both E- and H-multipolar Mie resonances, which permit a great deal of flexibility in adapting their scattering directivity.

Most of the MSs that have been shown to date are static. Nevertheless, implementations in the display and other optoelectronic device industries require compact and electrically tuned optical components. Several strategies have been explored to dynamically control light modulation by an MS. By using diodes, one can easily achieve that at MW frequencies. For optical frequencies where this strategy does not work, mechanically deformable MAs or substratum may be introduced and MEMS can be used to tune their resonances or alter the dielectric MS environment using birefringence crystals, graphene, gated semiconductors, phase change materials, or electro-optical polymers.

Nonetheless, due to the small change in permittivity, mainly at high frequencies, the tuning range for the cases of charge injection or electro-optical effects is inadequate, while phase change materials appear to be lossy in the VIS spectrum. At room temperature, the liquid crystal has a large birefringence and is almost lossless in the VIS range. Together with the innovative liquid crystal technology, it makes them the ideal platform for active devices for the VIS spectral range, which can be addressed via electrical, optical, H-, and thermal approaches.

Several works have recently been reported based on tunable devices through the combination of liquid crystals with plasmonic and dielectric MSs. To date, most of the liquid crystal–based high RI dielectric MSs have been documented to have used silicon as MAs and show broad tuning of the supported E- and H-dipole resonances under external and electrical control in the near-IR frequency range. Nevertheless, given the low band-gap of silicon, which is approximately equal to 1.1 eV, significant losses are likely for such MSs within the VIS spectral range. Meanwhile, the methodical discussion with the liquid crystal birefringence on the evolution of EM resonance within MS n-antennas is still inadequate.

As shown in Figure 5.8(a and b), a novel EO-polymer-based metallic MS modulator operating at 1.55-μm wavelength that employs bimodal resonance is suggested [32]. EO polymer has several advantages among various active materials, such as it exhibits a large EO coefficient above 200 pm/V, ultrahigh-speed modulation capability over 40 GHz, simple and low-cost spin-coating-based manufacturing that allows flexible device configuration, and high material reliability proven in commercial products. The device configuration is modeled to excite two metal-insulator-metal (MIM) resonant modes inside the EO polymer. A sharp dip arises in the reflected spectrum by employing the strong coupling between these two modes, with a meaningfully higher Q-factor than that attained when using a single-mode Fabry–Perot (FP) resonance, as demonstrated in Figure 5.8(c). A transmission-type, liquid crystal-based, tunable MS that efficiently operates in the VIS spectral range

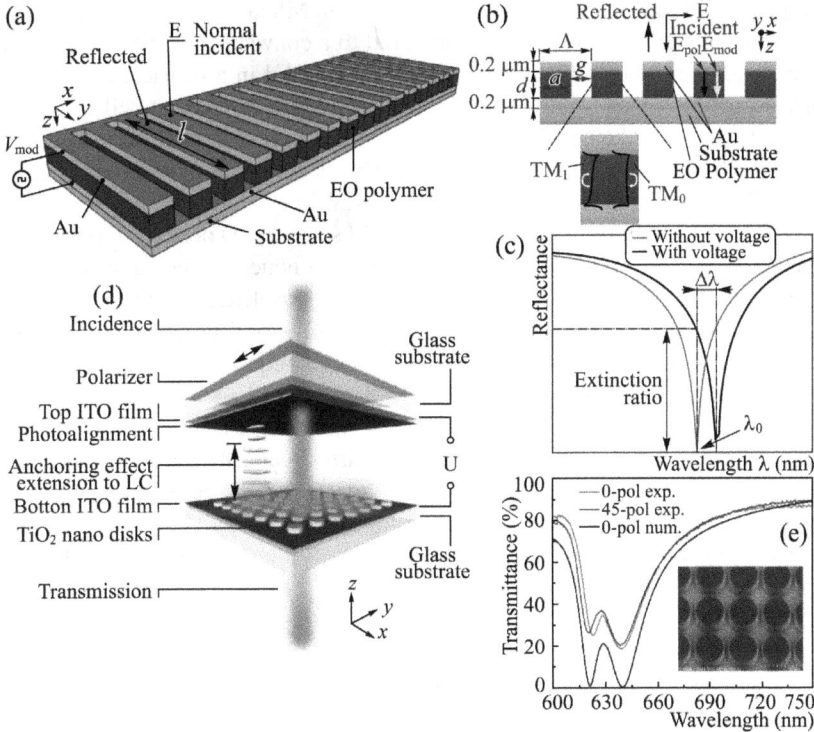

FIGURE 5.8 (a) Diagram of an active MS modulator with electro-optic polymer. (b) Cross-sectional view. (c) Modulation principle of the surface normal MS modulator [32]. (d) Plan of an electrically tunable transmission type TiO$_2$ MS with liquid crystal infiltration sandwiched by ITO-coated glass substratum. (e) Assessment of calculated and measured transmittance spectra of the TiO$_2$ n-discs MS. The inset gives SEM image of TiO$_2$ MS [33].

is demonstrated where titanium dioxide (TiO$_2$) n-antennas are used as demonstrated in Figure 5.8(d). TiO$_2$ has a wide band-gap and is lossless throughout the VIS spectrum. The TiO$_2$ n-discs are mounted in a thin liquid crystal cell to achieve resonance tuning. The E- and H-resonances are dynamically controlled by switching the liquid crystal alignment under electrical voltages [33]. When modulation voltage is put on by means of the same electrodes, it is possible to change the RI of the EO polymer using the Pockels effect and shift the resonant wavelength (λ_0) so that the reflected light intensity is modulated at (λ_0), as demonstrated in Figure 5.8(e).

5.5.8 FREQUENCY SELECTIVE SURFACES BASED ON METASURFACES

A frequency selective surface (FSS) is a periodic arrangement with each component resonating at the resonance frequency. Usually, the FSSs have a periodicity equal to half the wavelength of the resonant frequency. These have often been used to deliver spectral filtering for signal communications. They are also utilized as diplexers, resonant beam splitters, and antenna radomes. Due to the resonant nature of MAs, MSs can replace FSS. Because the building blocks of an MS are subwavelength in nature, they often bring additional benefits compared to a conventional FSS. The subwavelength periodicity allows many unit cells to be packed in a restricted space that is highly useful for space-limited radomes. The tiny size of the unit cell also enables the MSs to have a balanced response to variations in the AOI of the incoming EM wave. FSSs have now been used in dichroic subreflectors, radio frequency identification (RFID), lenses antennas, and shield from EM interference.

FSS is a vigorously researched subject of EM science, 2D periodic structures with planar metal formation units (apertures) on a nonconductor substratum that display transmission and reflection at a particular resonant wavelength. Subject to the layout of the formation component, the inbound plane wave will be either transmitted or reflected, in whole or in part. This happens when the frequency of the plane wave corresponds to the resonant frequency of the FSS units. An FSS can, therefore, transfer or block EM waves with certain frequencies, so they are best known as space filters.

According to the circuit theory, FSS structures (capacitive and inductive), also known as spatial filters, are like MWv filters. It is possible to categorize FSS filtering characteristics into four groups, including high pass, low pass, stopband and bandpass, as demonstrated in Figure 5.9. High-pass FSS filters allow a higher frequency range to pass through the structure while bypassing the lower-frequency range. Using the Babinet principle, low-pass FSS filter functioning is a counterpart of the high-pass filter function. Likewise, the stopband FSS filter blocks unwanted frequencies while the band-pass FSS filter only allows frequency ranges. FSSs are devised by periodic arrangement of metal patches and/or slits etched on a dielectric material for a desired resonant functioning. The highly crucial part of the configuration development is the correct choice of FSSs formation units, shape, size, and substratum material. Munk has clarified the functional concept of FSS based structures in detail [34].

The functionality of FSS is attained by a corresponding self-resonating network. Electric currents are stimulated in the formation units when EM waves strike the FSS structure. The amplitude of the currents created is determined by the amount of energy coupling. Induced currents, on the other hand, frequently operate as EM sources, generating further dispersed fields. The resultant field in the neighboring FSS is formed

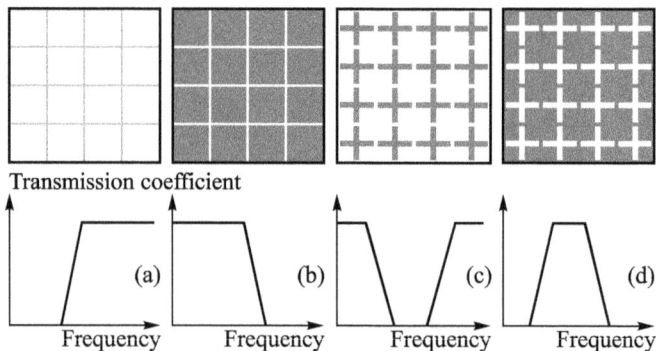

FIGURE 5.9   The frequency response characteristics of the FSS: (a) high pass, (b) low pass, (c) stopband and (d) band-pass.

by incident EM fields coupled with these dispersed fields. As a result, accurately con-structed components may get the required currents and field characteristics, and a filter response can be created. The patch components (dipole FSSs) are meant to function as stopband filters for inbound plane waves and to operate as completely reflecting sur-faces in a narrow band of frequencies. The aperture units, on the other hand, have band-pass properties, which means they operate as transparent surfaces for the incidence of EM waves within the operating frequency range. Traditional FSSs, on the other hand, are limited in their usage to fulfill the practical requirements for a range of EM utiliza-tions due to their poor filter response, low angular stability, and small BW.

FSS patches produce resistance (R) and inductance (L), whereas capacitance (C) is generated by gaps between the FSS units. The basic electrostatic assumption holds the physical meaning of these passive values for various FSS components, e.g. L of two parallel wires and C produced by a parallel plate capacitor. The essential filter response is therefore created by combining these capacitive and inductive compo-nents. Any adjustment in the dimensional variables of FSS, however, results in an analogous difference in the values of L and C. Physically, when the unit cell of the FSS is lit by the EM pulse, it can be transformed into an analogous resonance circuit. The resonance frequency (f_{res}) can be expressed in which L and C indicate the cor-responding inductance and the capacitance:

$$f_{res} = \frac{1}{2\pi\sqrt{LC}}. \tag{5.10}$$

The choice of a suitable formation component is very critical when designing FSS. Although distinct unit cell geometries have been introduced, some of them can be easily controlled and are therefore more common in the FSS community. For instance, a nonresonant unit (patch, wire grid) can be modeled by a capacitance, while a single resonant component (loop, cross, dipole) can be characterized by a series of capacitors and inductor combinations. It should be noted that there is a direct relation between the number of resonances and the number of lumped units.

5.5.9 Polarizers Based on Metasurfaces

Polarization (POL) is an inherent characteristic of EM waves, and transition between POL states is very often suitable for several modern EM and photonic utilizations. For example, the transformation of linear POL to circular POL in advanced com-munication and sensing renders a beam immune to environmental variation, scat-tering, and diffraction in the atmosphere. Thanks to the configuration stability and compactness, POL transformation using MSs has drawn increased interest in recent years. The subsequent ability to tune a phase delay covering the whole 2π spec-trum across a broad BW and with a deep subwavelength resolution may theoretically resolve some critical issues in the development of flat optics.

Simple, substantially symmetrical MAs can be beneficial for maintaining POL states. However, violating the symmetry can give extra flexibility for gaining cus-tomized usage, allowing POL states to be utilized. It is possible to have an identi-cal transmission magnitude but a relative phase delay ($\Delta\varphi$) at a given frequency by

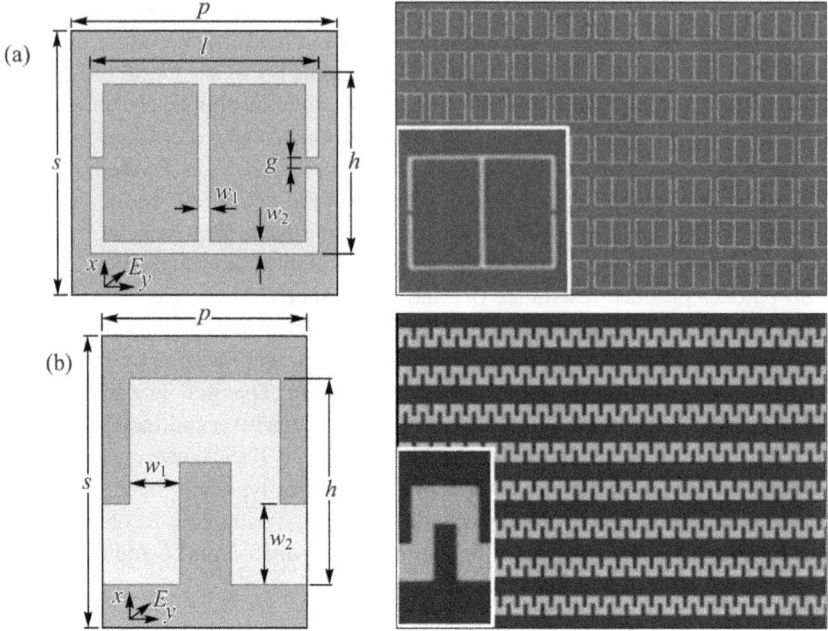

FIGURE 5.10 The individual unit cells for (a) the electric split-ring resonator quarter-wave plate and optical microscope picture (left to right), and (b) the meander line quarter-wave plate and optical microscope picture (left to right) [35].

customizing the two eigenmodes leading to orthogonal linear POLs. Single-layer MSs [35] or multilayer cascading MSs operating from MWv to optical frequencies have achieved narrowband POL transformation between linear and circular POL states. While the conventional meander line has a better BW and transmission magnitude, the electric split-ring resonator is simpler to make since it has only one active Au layer, a more constant phase shift, and a higher peak POL percentage. Figure 5.10 depicts the configuration of an electric split-ring resonator quarter-wave plate and a meander line quarter-wave plate.

Dielectric MSs containing distinct half-waveplate-like hydrogenated amorphous silicon n-posts and are shown to control the wavefront of transmitted VIS light with controllable linear POL angles. The cross-polarized scheme shows a high-quality anomalous beam deflector configuration with an absolute deflection efficacy of 82%, a POL conversion efficacy of 96%, and an extinction ratio of 37 dB. The anomalously deflected light could also hold a high degree of linear POL (>0.96), which can be continuously rotated by adjusting the POL angle of the incident [36].

The low level of POL conversion efficacy can be resolved by implementing MSs with few layers. Using the definition suggested by Jones matrix linearly polarized incident fields (E_x, E_y) through an MS can be described as:

$$\begin{pmatrix} E_x^t \\ E_y^t \end{pmatrix} = \begin{pmatrix} T_{xx} & T_{xy} \\ T_{yx} & T_{yy} \end{pmatrix} \begin{pmatrix} E_x^i \\ E_y^i \end{pmatrix} = \hat{T}_{\text{linear}} \begin{pmatrix} E_x^i \\ E_y^i \end{pmatrix}. \tag{5.11}$$

For circularly polarized incident fields, it becomes:

$$\begin{pmatrix} E_+^t \\ E_-^t \end{pmatrix} = \begin{pmatrix} T_{++} & T_{+-} \\ T_{-+} & T_{--} \end{pmatrix} = \hat{T}_{\text{circular}} \begin{pmatrix} E_+^i \\ E_-^i \end{pmatrix}, \tag{5.12}$$

where $T_{\pm\pm} = \frac{1}{2}\left(T_{xx} + T_{yy}\right) \pm \frac{i}{2}\left(T_{xy} - T_{yx}\right)$ and $T_{\pm\mp} = \frac{1}{2}\left(T_{xx} - T_{yy}\right) \mp \frac{i}{2}\left(T_{xy} + T_{yx}\right)$.

ACKNOWLEDGEMENT

"The work was supported by the Russian Science Foundation grant No. 21-79-20075"

REFERENCES

1. Zhan, A. et al. 2016. Low-contrast dielectric metasurface optics. *ACS Photonics* 3(2):209–14. https://doi.org/10.1021/acsphotonics.5b00660.
2. Astilean, S., Lalanne, P., Chavel, P., Cambril, E., and Launois, H. H. 1998. High-efficiency subwavelength diffractive element patterned in a high-refractive-index material for 633 nm. *Opt. Lett.* 23(7):552–4. https://doi.org/10.1364/OL.23.000552.
3. Minatti, G., Caminita, F., Casaletti, M., and Maci, S. 2011. Spiral leaky-wave antennas based on modulated surface impedance. *IEEE Trans. Antennas Propag.* 59(12):4436–44. https://doi.org/10.1109/TAP.2011.2165691.
4. Gregoire, D. J. 2013. 3-D conformal metasurfaces. *IEEE Antennas Wirel. Propag. Lett.* 12:233–6. https://doi.org/10.1109/LAWP.2013.2247017.
5. Nannetti, M., Caminita, F., and Maci, S. 2007. Leaky-wave based interpretation of the radiation from holographic surfaces. *Proc. 2007 IEEE Antennas and Propagation Society International Symposium*:5813–6. https://doi.org/10.1109/APS.2007.4396873.
6. Pereda, A. T., Caminita, F., Martini, E., Ederra, I., Iriarte, J. C., Gonzalo, R., and Maci, S. 2016. Dual circularly-polarized broadside beam metasurface antenna. *IEEE Trans. Antennas Propag.* 64(7):2944–53. https://doi.org/10.1109/TAP.2016.2562662.
7. Minatti, G., Faenzi, M., Sabbadini, M., and Maci, S. 2017. Bandwidth of gain in metasurface antennas. *IEEE Trans. Antennas Propag.* 65(6):2836–42. https://doi.org/10.1109/TAP.2017.2694769.
8. Faenzi, M. et al. 2019. Metasurface antennas: New models, applications and realizations. *Sci. Rep.* 9:10178. https://doi.org/10.1038/s41598-019-46522-z.
9. Hongnara, T., Chaimool, S., Akkaraekthalin, P., and Zhao, Y. 2018. Design of compact beam-steering antennas using a metasurface formed by uniform square rings. *IEEE Access* 6:9420–9. https://doi.org/10.1109/ACCESS.2018.2799551.
10. Yu, N., and Capasso, F. 2014. Flat optics with designer metasurfaces. *Nat. Mater.* 13:139–50. https://doi.org/10.1038/nmat3839.
11. Aieta, F., Genevet, P., Kats, M. A., Yu, N., Blanchard, R., Gaburro, Z., and Capasso, F. 2012. Aberration-free ultrathin flat lenses and axicons at telecom wavelengths based on plasmonic metasurfaces. *Nano Lett.* 12(9):4932–6. https://doi.org/10.1021/nl302516v.
12. Pors, A., Nielsen, M. G., Eriksen, R. L., and Bozhevolnyi, S. I. 2013. Broadband focusing flat mirrors based on plasmonic gradient metasurfaces. *Nano Lett.* 13(2):829–34. https://doi.org/10.1021/nl304761m.
13. Chen, X., Huang, L., Mühlenbernd, H., Li, G., Bai, B., Tan, Q., Jin, G., Qiu, C.-W., Zentgraf, T., and Zhang, S. 2013. Reversible three-dimensional focusing of visible light with ultrathin plasmonic flat lens. *Adv. Opt. Mater.* 1(7):517–21. https://doi.org/10.1002/ADOM.201300102.

14. Yin, L., Vlasko-Vlasov, V. K., Pearson, J., Hiller, J. M., Hua, J., Welp, U., Brown, D. E., and Kimball, C. W. 2005. Subwavelength focusing and guiding of surface plasmons. *Nano Lett.* 5(7):1399–402. https://doi.org/10.1021/nl050723m.
15. Liu, Z., Steele, J. M., Srituravanich, W., Pikus, Y., Sun, C., and Zhang, X. 2005. Focusing surface plasmons with a plasmonic lens. *Nano Lett.* 5:1726–9.
16. Vo, S., Fattal, D., and Wayne, V. et al. 2014. Sub-wavelength grating lenses with a twist. *IEEE Photonics Technol. Lett.* 26(13):1375–8. https://doi.org/10.1109/LPT.2014.2325947.
17. Arbabi, A., Horie, Y., Ball, A. J., Bagheri, M., and Faraon, A. 2015. Subwavelength-thick lenses with high numerical apertures and large efficiency based on high-contrast transmit arrays. *Nat. Commun.* 6:7069. https://doi.org/10.1038/ncomms8069.
18. Khorasaninejad, M., Zhu, A. Y., Roques-Carmes, C., Chen, W. T., Oh, J., Mishra, I., Devlin, R. C., and Capasso, F. 2016. Polarization-insensitive metalenses at visible wavelengths. *Nano Lett.* 16:7229–34. https://doi.org/10.1021/acs.nanolett.6b03626.
19. Liu, N., Mesch, M., Weiss, T., Hentschel, M., and Giessen, H. 2010. Infrared perfect absorber and its application as plasmonic sensor. *Nano Lett.* 10(7):2342–8. https://doi.org/10.1021/nl9041033.
20. Watts, C. M., Shrekenhamer, D., Montoya, J., Lipworth, G., Hunt, J., Sleasman, T., Krishna, S., Smith, D. R., and Padilla, W. J. 2014. Terahertz compressive imaging with metamaterial spatial light modulators. *Nat. Photon.* 8:605–9. https://doi.org/10.1038/nphoton.2014.139.
21. Shi, N. N., Tsai, C.-C., Camino, F., Bernard, G. D., Yu, N., and Wehner, R. 2015. Keeping cool: Enhanced optical reflection and radiative heat dissipation in Saharan silver ants. *Science* 349(6245):298–301. https://doi.org/10.1126/science.aab3564.
22. Azad, A. K. et al. 2016. Metasurface broadband solar absorber. *Sci. Rep.* 6:20347. https://doi.org/10.1038/srep20347.
23. Goul, R., Das, S., Liu, Q., Xin, M., Lu, R., Hui, R., and Wu, J. Z. 2017. Quantitative analysis of surface enhanced raman spectroscopy of rhodamine 6G using a composite graphene and plasmonic au nanoparticle substrate. *Carbon* 111:386–92. https://doi.org/10.1016/j.carbon.2016.10.019.
24. Xu, X., Peng, B., Li, D., Zhang, J., Wong, L. M., Zhang, Q., Wang, S., and Xiong, Q. 2011. Flexible visible-infrared metamaterials and their applications in highly sensitive chemical and biological sensing. *Nano Lett.* 11(8):3232–8. https://doi.org/10.1021/nl2014982.
25. Cao, C., Zhang, J., Wen, X., Dodson, S. L., Dao, N. T., Wong, L. M., Wang, S., Li, S., Phan, A. T., and Xiong, Q. 2013. Metamaterials-based label-free nanosensor for conformation and affinity biosensing. *ACS Nano* 7(9):7583–91. https://doi.org/10.1021/nn401645t.
26. Gregoire, D. J., and Kabakian, A. V. 2011. Surface-wave waveguides. *IEEE Antennas Wireless Propag. Lett.* 10:1512–5. https://doi.org/10.1109/LAWP.2011.2181476.
27. Simovski, C., de Maagt, P., and Melchakova, I. V. 2005. High-impedance surfaces having stable resonance with respect to polarization and incidence angle. *IEEE Trans. Antennas Propag.* 53(3):908–14. https://doi.org/10.1109/TAP.2004.842598.
28. Chen, X., Luo, Y., Zhang, J., Jiang, K., Pendry, J. B., and Zhang, S. 2011. Macroscopic invisibility cloaking of visible light. *Nat. Commun.* 2:176. https://doi.org/10.1038/ncomms1176.
29. Milton, G. W., and Nicorovici, N.-A. P. 2006. On the cloaking effects associated with anomalous localized resonance. *Proc. Math. Phys. Eng. Sci.* 462(2074):3027–59. https://doi.org/10.1098/rspa.2006.1715.
30. Fleury, R., Monticone, F., and Alù, A. Invisibility and cloaking: Origins, present, and future perspectives. *Phys. Rev. Appl.* 2015; 4(3):037001. https://doi.org/10.1103/PhysRevApplied.4.037001.

31. Schurig, D. et al. 2006. Metamaterial electromagnetic cloak at microwave frequencies. *Science* 314(5801):977–80. https://doi.org/10.1126/science.1133628.
32. Zhang, J., Kosugi, Y., Otomo, A., Nakano, Y., and Tanemura, T. 2017. Active metasurface modulator with electro-optic polymer using bimodal plasmonic resonance. *Opt. Express* 25(24):30304–11. https://doi.org/10.1364/OE.25.030304.
33. Sun, M. et al. 2019. Efficient visible light modulation based on electrically tunable all dielectric metasurfaces embedded in thin-layer nematic liquid crystals. *Sci. Rep.* 9:8673. https://doi.org/10.1038/s41598-019-45091-5.
34. Munk, B. A. 2000. *Frequency selective surfaces: Theory and design.* Hoboken, NJ: Wiley Online Library. ISBN: 978-0-471-37047-5.
35. Strikwerda, A. C., Fan, K., Tao, H., Pilon, D. V., Zhang, X., and Averitt, R. D. 2008. Comparison of birefringent electric split-ring resonator and meanderline structures as quarter-wave plates at terahertz frequencies. *Opt. Express* 17(1):136–49. https://doi.org/10.1364/OE.17.000136.
36. Gao, S., Park, C.-S., Lee, S.-S., and Choi, D.-Y. 2019. All-dielectric metasurfaces for simultaneously realizing polarization rotation and wavefront shaping of visible light. *Nanoscale* 11(9):4083–90. https://doi.org/10.1039/C9NR00187E.

6 Optical Fiber Sensors Based on Diffractive and Fiber Periodic Microstructures

Sergey Vladimirovich Karpeev[1,2] and
Svetlana Nikolaevna Khonina[1,2]
[1]Image Processing Systems Institute –
Branch of the Federal Scientific Research Centre,
"Crystallography and Photonics" of Russian
Academy of Sciences, Samara, Russia
[2]Samara National Research University,
Samara, Russia

6.1 INTRODUCTION

Optical fiber sensors are now widely used as parts of measuring systems for monitoring various technical and natural systems. Their main advantages are compatibility with existing fiber-optic data transmission networks, high resistance to external chemical and radiation influences and insensitivity to electromagnetic fields, high sensitivity to selected parameters, and ease of integration into measuring systems. Depending on the type of optical fiber sensor, the output signal can be either the amplitude of the transmitted light, or the phase or the frequency shift. In the case of amplitude modulation, multimode optical fibers can be used, and in this case it is possible to increase the sensitivity and adjust the measurement ranges using the diffractive optical elements (DOEs) incorporated into the measuring system. However, amplitude-type sensors are susceptible to noise when using coherent radiation and can be used with small lengths of coupling fibers, especially in the case of filtering individual modes. Optical fiber sensors that implement the frequency shift of the transmitted radiation are much more resistant to interference due to the fundamental nature of the wavelength. They are also easily combined into measuring networks with the identification of each individual sensor and allow maintaining the measurement mode at a sufficiently large distance from the object of study. These sensors are based on variously microstructured fibers for creating Bragg fiber gratings. Structuring methods are now actively developing and have therefore found their place in this chapter.

DOI: 10.1201/9781003439165-6

6.2 AMPLITUDE OPTICAL FIBER SENSORS BASED ON DIFFRACTIVE OPTICAL ELEMENTS MATCHED WITH TRANSVERSE MODES

6.2.1 DESIGN PRINCIPLES OF OPTICAL FIBER CONVERTERS BASED ON MODE FILTERING

Depending on the underlying physical processes, optical fiber converters can be categorized as interference or amplitude [1]. Interference-aided converters commonly utilize a pair of single-mode optical fibers, with one fiber exposed to an external physical influence and the other used as a reference. By deciphering the resulting interference pattern, the magnitude of the physical influence is evaluated. Amplitude-aided converters utilize a single multimode optical fiber, measuring variations in the amplitude of light in the fiber caused by the external influence. The amplitude variations occur because the energy of a guided mode is partly transferred to unguided modes owing to the physical influence. With different converters utilizing different types of influence, irrespective of the type, the energy is transferred into the unguided modes differently for different modes.

For higher-order modes, which are nearer to the cutoff frequency, the light energy is confined closer to the fiber core boundaries, which means that the transfer to the unguided modes takes place at smaller external influences, compared to the lower-order modes. For the amplitude optical fiber converters, it is the difference in sensitivity of different modes to the external influence that gives a gain in the converter parameters when an analyzer and a transverse mode shaper are inserted into the beam. In terms of the scalar theory of diffraction, this is equivalent to evaluating variations in the optical properties of a wave medium through variations in the transverse mode spectrum. It is worth noting that thanks to the use of mode filtering, characteristics of the interference optical fiber converters can also be improved. Here, with the interference taking place between the same fiber modes, the design of the converter can be simplified compared with the known designs [1]. The same principle also applies when dealing with an important task of controlling the refractive index profile in the fiber.

Among the optical fiber converters of physical magnitudes, a significant place belongs to micro-displacement sensors that work by measuring power losses on fiber microbendings. A physical principle behind the design of an optical fiber converter is mode coupling due to fiber deformation, with the energy redistributed between the modes and partly transferred to unguided modes [2]. As a result, the energy of individual modes changes and the total energy flux in the fiber is decreased. Let us analyze a converter from [3, 4], which utilizes an optical fiber that supports guided modes with scalar-approximated complex amplitudes $\{\psi_p(\mathbf{x})\}$ and the mode numbers $\mathbf{p} = (p, l)$.

The modes of interest may include both classical (e.g. Laguerre–Gaussian, Hermite–Gaussian, and Bessel) optical fiber modes [5–8] and more general types of laser beams [9–11]. Mode excitation occurs under illumination by a laser beam [12–14] with complex amplitude:

$$F(\mathbf{x}) = \sum_{\mathbf{p}} F_{\mathbf{p}} \psi_{\mathbf{p}}(\mathbf{x}), \quad \mathbf{x} \in G.$$

The propagation of the beam in a fiber can be described using a linear operator $\hat{T}$ acting on the $F(\mathbf{x})$ function. The complex amplitude $F^1(\mathbf{x})$ of the beam at the fiber output is described by the equation:

$$F^1 = \hat{T}F.$$

In the output beam, the mode spectrum $\{F_1\}$:

$$F_1(\mathbf{x}) = \sum_{p'} F^1_{p'} \psi_{p'}(\mathbf{x}) = \sum_{p} F_p \hat{T} \psi_p(\mathbf{x}), \tag{6.1}$$

is defined by the input beam mode content $\{F_p\}$ and the fiber properties described by the operator $\hat{T}$. The form of the operator $\hat{T}$ depends on the fiber state, which depends on the perturbations of fiber physical parameters, characterized by measurable parameters $\mathbf{V} = (V_1, V_2, \ldots, V_n)$. Hence, we can write $\hat{T} = \hat{T}(\mathbf{V})$, $F^1_p = F^1_p(\mathbf{V})$.

For an unperturbed fiber, $V = 0$, $\hat{T} = \hat{T}(0)$, where the operator $\hat{T}(0)$ describes the eigen-medium propagation of the mode. Using the definition of the propagation constant β_p, we can write:

$$\hat{T}(0)\psi_p(\mathbf{x}) = \exp(i\beta_p z)\psi_p(\mathbf{x}), \tag{6.2}$$

where z is the mode travel distance in the fiber. We note that in an unperturbed fiber, the modes are uncoupled:

$$F^1_{p'} = \exp(i\beta_{p'} z)F_{p'}, \left|F^1_{p'}\right|^2 = \left|F_{p'}\right|^2. \tag{6.3}$$

This is because in an unperturbed fiber, modes are the eigen-functions of the propagation operator. Values of $\mathbf{p}$ correspond to the guided modes numbers. If the optical fiber has a non-zero perturbation $\mathbf{V}$, instead of Eq. (6.2), we introduce coefficients T_{pp} and obtain:

$$\hat{T}(\mathbf{V})\psi_p(\mathbf{x}) = \sum_{p'} T_{pp'}(\mathbf{V})\psi_{p'}(\mathbf{x}). \tag{6.4}$$

The coefficients $T_{pp'}(\mathbf{V})$ form a matrix of operator $\hat{T}$ components, being equal to the values of mode overlap integrals. Because the mode basis ψ_p is orthogonal, we obtain:

$$T_{pp'}(\mathbf{V}) = \int_G \psi^*_{p'}(\mathbf{x})\hat{T}\psi_p(\mathbf{x})d^2\mathbf{x}. \tag{6.5}$$

The coefficients $F^1_{p'}$ can be derived from:

$$F^1_{p'}(\mathbf{V}) = \int_G F^1(\mathbf{x})\psi_{p'}(\mathbf{x})d^2\mathbf{x}, \tag{6.6}$$

$$F_{p'}^{1}(\mathbf{V}) = \sum_{p} T_{pp'}(\mathbf{V}) F_{p}. \tag{6.7}$$

Considering that the Parseval equality holds for orthogonal bases, the beam power at the fiber input and output is defined by series:

$$E = \sum_{p} |F_{p}|^{2}, E'(\mathbf{V}) = \sum_{p'} |F_{p'}^{1}(\mathbf{V})|^{2}. \tag{6.8}$$

The majority of familiar fiber-optic sensors operate by measuring the total output beam energy $E'(\mathbf{V})$ as a function of perturbations $\mathbf{V}$. From (6.8), the value of $E'(\mathbf{V})$ is seen to be averaged over the individual-mode power $|F_{p}(\mathbf{V})|^{2}$ for the entire ensemble of the excited modes with numbers $\mathbf{p}'$. A converter proposed in [3, 4] performs selective measurements of the power $\left| F_{p'}^{1}(\mathbf{V}) \right|^{2}$ of individual modes. We note that in a special case of single-mode excitation, the F_{p} coefficient is non-zero only at one value of $\mathbf{p}$. We term the quantity $|T_{p'p}(\mathbf{V})|^{2}$ a mode excitation coefficient because it determines the power of the fiber mode with the number $\mathbf{p}'$ excited by the mode with the number $\mathbf{p}$.

6.2.2 Analysis of Mode Coupling Due to Periodic Microbendings of a Graded-Index Fiber

The physical principle behind the design of an optical fiber converter is mode coupling due to fiber deformations, which leads to the energy being redistributed between the modes and partially transferred to non-propagating modes [15]. This causes changes in the energy of individual modes and a decrease in the total light flux in the fiber. Many well-known converter designs are based on measuring the total light flux. But this effect is secondary and as such is essentially less sensitive (four to five times for a gradient fiber) to fiber deformations compared to when measuring individual modes. The sensitivity of the total light flux can be increased by using converter designs in which the fiber has multiple periodic bendings with a period close to the light wavelength, causing resonance phenomena contributing to the increase of losses.

It was in an optical fiber converter with microbendings that the mode selection was first demonstrated [3]. For this purpose, the device for transverse mode excitation and analysis was complemented with a deformation device, as shown in a schematic diagram in Figure 6.1.

The deformation device was put 0.3 m away from the input end of a 1.5-m long fiber. The periodic microbendings were implemented via pulling them through a pair of profiled plates, with the fiber-facing plates' surface relief forming triangular teeth of period $\Lambda = 1.5$ mm and the number of periods being $k = 5$. The bending radii were varied through applying an external force to the upper plate and controlled by means

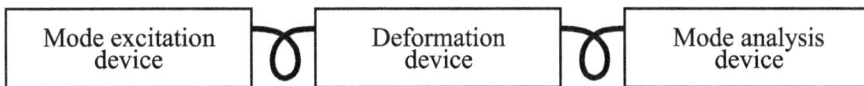

FIGURE 6.1 A schematic diagram of a fiber-optic converter.

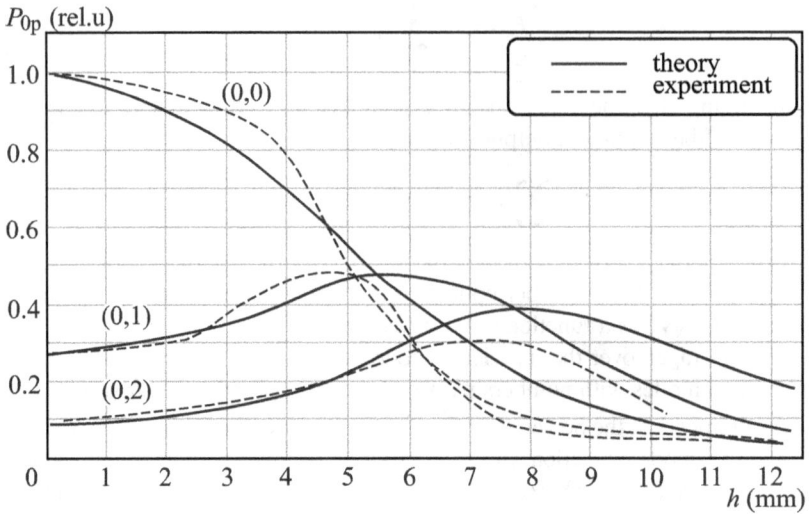

FIGURE 6.2 Mode power versus microbending radii (solid lines – theory, dashed lines – experiment).

of a microscope and an eyepiece micrometer. Experimentally derived curves for the relative mode powers:

$$P_{op}(h) = \frac{\left|\xi_{op}(h)\right|^2}{\left|\xi_{oo}(h)\right|^2_{h=0}}, \qquad (6.13)$$

are shown as dashed lines in Figure 6.2.

From Figure 6.2, the curves are seen to behave in agreement with the theory, but if the deformations exceed 5–6 μm, the rate of mode power variations is somewhat higher (at lower deformations) than theoretical predictions. Arguably, here, reasons are the same as those leading to the variation of mode excitation conditions, because these processes have a similar mechanism.

For comparison with the earlier measured characteristics of the converters under analysis, the integrated light flux was measured as a function of the microbending radius. The corresponding plot is presented in Figure 6.3 in comparison with that for the principal mode power.

From Figure 6.3, the sensitivity for the principal mode is seen to be twice as high as that for the total light flux. Note that although the curve for the total light flux features a better linearity, after the proper correction the advantage will not be essential.

6.2.3   STUDY OF THE OUTPUT MODE POWER FOR A GRADED-INDEX FIBER AGAINST THE MICROBENDING MAGNITUDE

Along with being complex in design, the just-discussed converter with multiple periodic fiber bendings is rather massive, which deteriorates its mechanical frequency

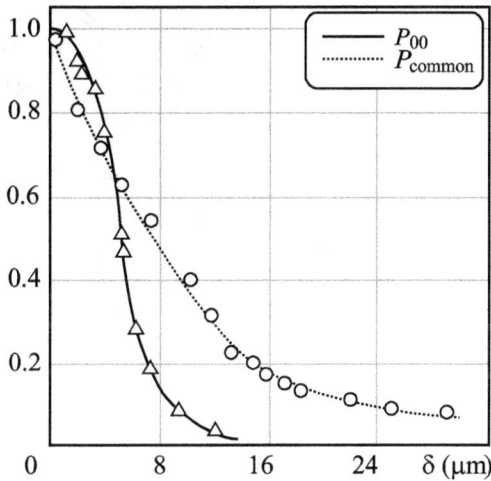

FIGURE 6.3 Variations in the total light flux and principal mode power versus the micro-bending radius.

characteristics. Besides, the entire range of measurable deformations falls within 10–15 μm. On the other hand, mass-produced optical fibers endure 100-μm-deep bendings without being damaged. In [4], a simpler mechanical design of the deformation device was proposed, producing a single 50- to 60-μm-deep microbending. The device has the advantage of lighter weight and smaller dimensions, which also results in better frequency characteristics. At the same time, thanks to mode filtering, the sensitivity of the converter is close to that of "multi-bending" converters measuring the total flux. Moreover, the proposed design [4] enables the converter sensitivity to be readjusted in real time by measuring the energy of different-order modes. Corresponding experiments with multimode gradient optical fibers were described in [4], and a feasibility of the implementation of such types of converters was demonstrated. However, the approach has a drawback, namely, the power of higher-order modes was found to have a fairly large constant component (~0.3–0.4 of maximum) independent of the bending magnitude. At the same time, in the theoretical model of an ideal parabolic profile such a phenomenon is not expected to occur. Hence, we may infer that its explanation should be sought for in the deviation of the refractive index profile from the theoretical model and the large number of higher-order modes in such fibers, leading to noise, in addition to non-ideal DOEs utilized. On the other hand, higher-selective excitation and more accurate analysis of transverse modes can be achieved in a few-mode fiber with stepped refractive index profile [15]. Experimental measurements of the mode power in a stepped-index fiber against the microbending depth have been reported [16, 17].

For a single, specific-depth bending to be realized, a dedicated deformation device was designed and fabricated, as schematically shown in Figure 6.4.

The device comprises a base (1) (Figure 6.4), mounted on which are a lever (2) with a return spring (3), a pressure plate (4), and an adjustable support unit. The lever axis rests on prisms (5). Lever movements are controlled by means of

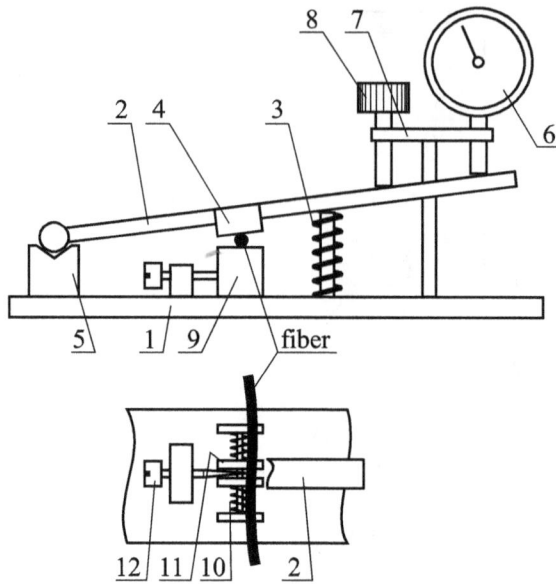

FIGURE 6.4   An outline of the deformation device.

an indicator (6), which sits on the same platform (7) as a pressure bolt (8). The adjustable support unit (Figure 6.4) comprises two supports (9) for springs (10) that thrust supports (11) for the fiber under study. The gap between the supports (11) is adjusted with a taper-shank bolt (12), which passes through them limiting the gap. For a comparison with earlier reported results [3], the inter-support gap was chosen to be ~1 mm. The fiber is being controllably deformed by rotating the pressure bolt, which displaces the lever and, according to a lever arm ratio of 1:7.5, the pressure plate. In the experiment, a commercial 2-m-long single mode fiber (SMF) was used. Some differences in techniques for analyzing the transverse mode powers are due to the use in the experiment of a DOE for laser mode generation (MODAN), intended for a larger number of modes. These phase elements are described in detail in [15].

The mode content of laser light at the fiber output can be analyzed using multi-order and multi-channel diffractive optical elements (DOEs) [18–20]. The analysis is easiest to conduct with the focal length of microlens M2 set equal to 13.9 mm and the focal length of the Fourier-transform lens at 300 mm. With the principal mode radius being about five times larger compared to other modes, for a desired scale picture to be obtained in the CCD camera plane, the phase element should be put into a diverging beam, ensuring an ~1-m distance between the Fourier-transform objective and the CCD camera. In the CCD camera field of view, about nine different modes are observed, including all modes of the fiber under analysis. As we show later in the chapter, such a scale provides an acceptable (noise level) systematic error of measurements. A distinguishing feature of the filter that is utilized is the absence of the principal mode. However, the adjustment criterion of the minimal intensity at the centers of all diffraction orders still holds. Figure 6.5 shows an intensity pattern

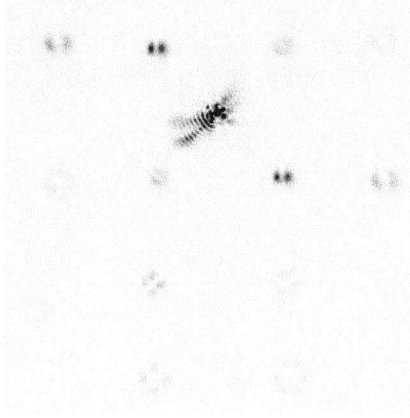

FIGURE 6.5 Intensity pattern at the output plane of a Fourier stage when coupling the principal mode to the phase filter.

at the output plane of a Fourier stage with a spatial filter (correlation field) when coupling the principal mode to the fiber.

The pattern in Figure 6.5 needs to be explained. The Fourier transform of the binary phase element employed in the scheme is centrally symmetric relative to the zero order, which is clearly seen in Figure 6.5. The frame in the pattern is a bit displaced to show a larger number of modes on one side. We note that the image in Figure 6.5 can be processed using software described in [17]. The result of processing an experimental image is depicted in Figure 6.6.

Table 6.1 connects the diffraction order numbers with particular modes.

The next step was measuring the mode power against the fiber deformation magnitude. Figure 6.7 depicts a correlation field upon the excitation of the mode LP_{11}.

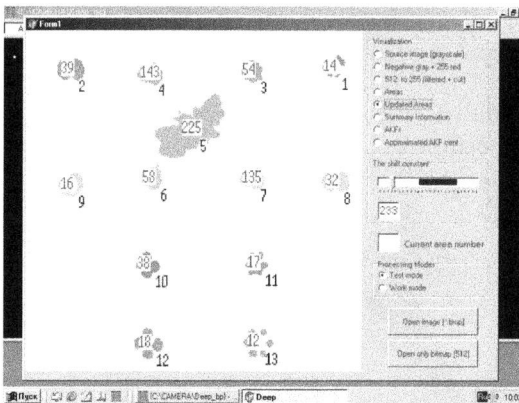

FIGURE 6.6 The result of processing the intensity distribution aimed at finding the diffraction order centers.

TABLE 6.1

Diffraction Order Numbers versus Mode Indices

Order Numbers	1	2	3	4	6	7	8	9	10	11
Modes	LP_{03}	LP_{12}	LP_{02}	LP_{11}	LP_{02}	LP_{11}	LP_{12}	LP_{03}	LP_{21}	LP_{31}

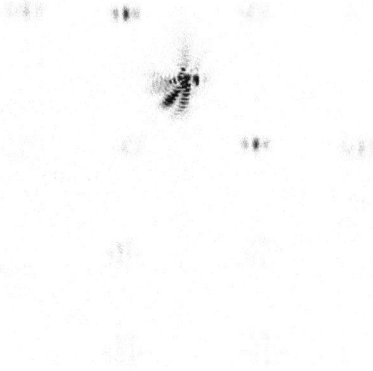

FIGURE 6.7 Intensity distribution at the output plane of the Fourier stage when coupling the mode LP_{11} to the phase filter.

A corresponding correlation peak is clearly seen, with no intensity peaks being observed in other diffraction orders. From Figure 6.8, the correlation peak is seen to occupy 10–15 pixels on either axis, allowing it to be measured with an error not higher than the noise level.

When the fiber is being deformed, the energy is redistributed between the modes, leading to changes in the correlation peak. Simultaneously, the intensity distribution between diffraction orders ceases to be symmetric. Figure 6.9 depicts a correlation field for a bending depth of 60 μm, with the rest parameters of the optical setup being the same as in Figure 6.7.

Shown in Figure 6.10 is an experimental curve of the LP_{11} mode power against the fiber microbending depth, which is seen to be nonlinear. The permanent component is seen to be as low as 0.17, unlike results previously derived for a GRIN fiber [3, 4] where it was found to be 0.3–0.4, with the power stabilized at the same bending

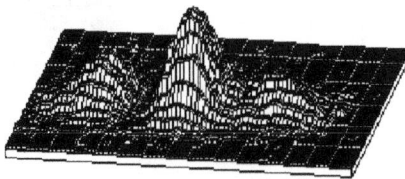

FIGURE 6.8 Intensity distribution in the correlation peak region.

FIGURE 6.9   Diffraction pattern for a 60-μm deep bending.

depth of ~75 μm. The latter result agrees better with theory, enabling the dynamic range to be extended and the accuracy of the bending-based converter enhanced.

In a similar way to Figure 6.3, Figure 6.11 plots the power of the principal mode and, for comparison, the total light flux against the microbending depth.

The curves in Figure 6.11 are seen to be somewhat different from those for a multi-bending GRIN-fiber converter of Figure 6.3. The curves in Figure 6.11 have similar-type nonlinearity, with a twofold increase in sensitivity being observed over the entire range of the analyzed deformations. The two curves have a larger proportion of the permanent component when compared with the multi-bending converter of Figure 6.3, which is apparently due to the fiber's single bending. Besides, the sensitivity over the total light flux is more than twice as high as that for a single-bending GRIN-fiber converter [4] (with the total flux decreasing to a level of 0.8 at half as large fiber deformation),

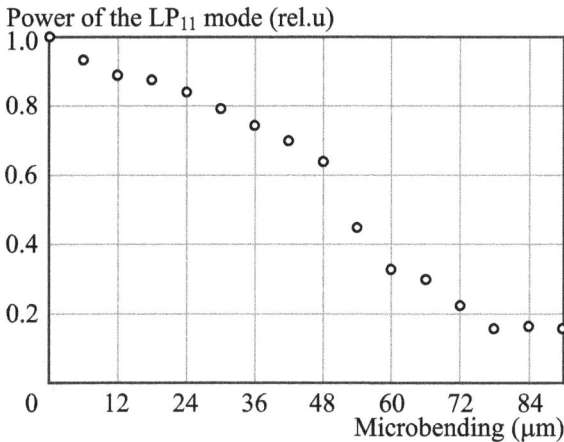

FIGURE 6.10   Power of the LP_{11} mode against the fiber bending.

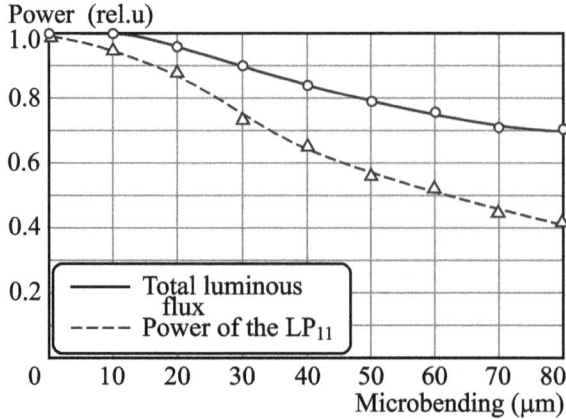

FIGURE 6.11 Variations in the total light flux and the power of the principal mode as a function of the microbending magnitude in a stepped-index optical fiber.

whereas the sensitivity for the principal mode increases considerably less, by ~20%. The explanation can be found in different core diameters, with the stepped-index fiber core being significantly thinner, meaning that as a result of a smaller bending, light will leave the stepped-index fiber sooner than it will the multimode GRIN fiber. Hence, in such converters it is preferable to use higher-order modes for which the proportion of the permanent component is essentially less (see Figure 6.10).

Thus, we have derived experimental curves for the dependence of the LP-modes power and the total light flux on the magnitude of a stepped-index fiber bending. A twofold decrease in the mode power permanent component has been attained, enabling a dynamic range of the converter to be extended.

6.3 SPECTRAL OPTICAL FIBER SENSORS BASED ON PERIODICALLY MICROSTRUCTURED FIBERS

6.3.1 APPROACHES TO MEASURING PHYSICAL QUANTITIES

External physical influences introduce perturbations in the surrounding medium of periodic fiber-optic microstructures, leading to a shift in the Bragg resonance position in the optical reflection and transmission spectra. This effect underlies the use of fiber Bragg gratings (FBG) as optical fiber sensors. A variety of methods for measuring the Bragg wavelength shift have been proposed.

6.3.1.1 Measuring the Intensity

One method for measuring the physical quantities of interest utilizes a narrowband laser and a photodetector, which provides higher processing speed. In this case, the Bragg wavelength shift in the grating is transformed into intensity variations of the optical signal on the photodetector.

The magnitude of the external force can be evaluated from a magnitude of the frequency shift for the reflection spectrum or a shift of the intensity curve dip for the

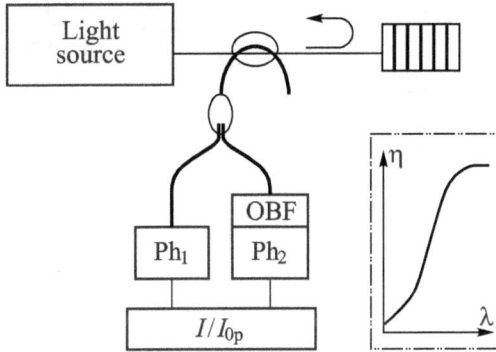

FIGURE 6.12 Spectrum analyzer for the reflection spectrum of a Bragg grating.

transmission spectrum. Accordingly, spectral devices need to be used for the spectral analysis. A schematic diagram of a bandpass-filter-aided spectrum analyzer for the reflection spectrum of a Bragg grating is depicted in Figure 6.12.

In this setup, one part of light reflected from the Bragg grating serves as a reference signal, being fed to the photodetector. The other part of light comes to the second photodetector through a bandpass filter, with the filter transmission depending on the wavelength. With a varying external force exerted on the grating structure, the reference signal intensity at the first photodetector remains unchanged; meanwhile the intensity of the light coming to the second photodetector varies due to changing conditions of the Bragg reflection in the fiber grating. A signal processing unit calculates the ratio of the intensities coming to the photodetectors, producing the information on the force magnitude in a desired form. Disadvantages of this scheme stem from the mechanical instability of the optical setup and instability of characteristics of the wideband light source.

The bandpass filters may be in the form of scanned Fabry–Perot interferometers, acoustic-optical filters, and filters based on tunable Bragg gratings.

In the scheme with a scanned interferometer in Figure 6.13, the reflected light comes to the input of a filter that has the near-unit transmission on resonance

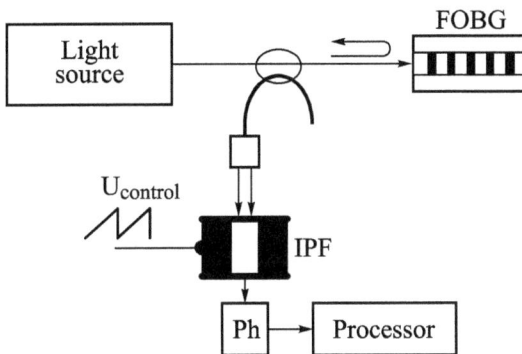

FIGURE 6.13 Schematic of a scanned Fabry–Perot interferometer.

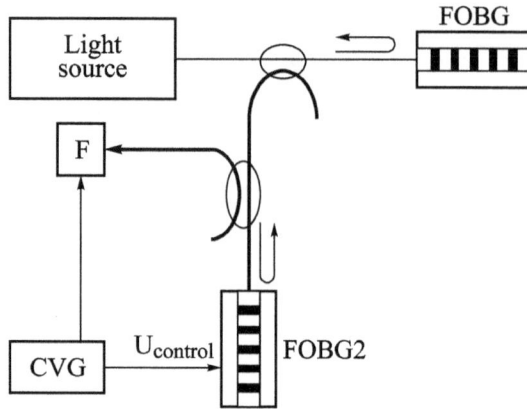

FIGURE 6.14   Schematic of a tunable Bragg grating filter.

wavelengths. In the FP interferometer, the position of the maximum transmission depends on the inter-mirror distance. With one of the mirrors being mounted on a piezoelectric substrate, its position is adjusted by varying the command voltage. With a sawtooth command voltage, the interferometer resonance frequency is chirped during the forward stroke, enabling the reflected signal spectrum to be obtained within a single sawtooth period.

The fiber-optic Bragg grating may also be used as a tunable spectral filter (Figure 6.14). In this case, the wideband radiation is fed to a measuring Bragg grating, after which the reflected radiation comes to the second grating. The optical fiber and the second grating are located on a piezoceramic substrate.

In this grating, the resonance wavelength is specified by the command voltage applied across the piezoelectric element electrodes from a command voltage generator (CFG). If both gratings have the same resonance wavelengths, after reflection at the second grating the signal comes to the photodetector. In this way, it is also possible to analyze spectral characteristics of the radiation reflected at the first grating.

6.3.1.2   Reflectometry Analysis

In the practical use of fiber-optic Bragg gratings (FOBG), major shortcomings are linked with their complex design and costly equipment for spectral measurements of the wavelength shift. The majority of spectral systems provide a redundant accuracy when measuring physical influences. Because of this, simplified measurement setups can be utilized, where some physical parameters are measured with lower accuracy. A simplified scheme can be implemented on the principle of detecting the reflected signal amplitude at a fiber grating.

An optical time-domain reflectometry (OTDR) method enables distributed measurements to be performed over the entire length of the fiber-optic path. Equipment for optical reflectometry comprises a pulse generator, a sensitive photodiode, and a signal processing unit. The parameter under measurement is the power of the backscattered light due to effects of Rayleigh, Mandelstam–Brillouin, or Raman scattering. An optical reflectometer measures the power of backscattered light against distance.

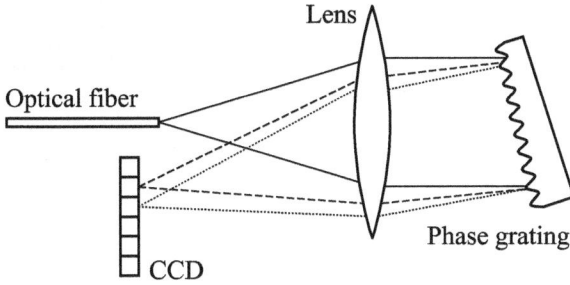

FIGURE 6.15 Optical setup of a CCD spectrometer.

6.3.1.3 Spectral Analysis

Measurements of reflection/transmission spectra in fiber-optic structures using a wideband light source and a spectrum analyzer are most comprehensive. The method is insensitive to possible losses in an optical channel. Disadvantages include the high cost of equipment and its limited performance.

With this method, it is possible to determine in more detail spectral characteristics linked with mode interference, energy exchange, polarization effects, and absorption.

The scheme of an optical spectrum analyzer comprises a monochromator, a diffraction grating, and a CCD array (Figure 6.15). After being reflected at the Bragg grating, the light goes to the diffraction grating, which sends the reflected light to the CCD array. At the diffraction grating, the angle of diffraction depends on the incident wavelength, meaning that different wavelengths come to different CCD array areas. Considering these properties, CCD spectrometers can be utilized for detecting a large number of fiber-grating-based sensors. With the sensitivity threshold of CCD arrays being essentially higher compared to photodiodes, the former can detect lower-energy light.

For several fiber-optic channels to be sensed, a spectral arrangement with a curved diffraction grating can be used. Radiation from different fiber-optic channels hits different CCD array columns, thus enabling a two-dimensional analysis (Figure 6.16).

6.3.2 Optical Fiber Sensors

Modes traveling in an optical fiber cladding are used in a wide class of sensors for measuring physical quantities. For a cladding mode to be excited as a result of matching with the principal mode, the fiber needs to have an inhomogeneity, which may be introduced by a fiber Bragg grating, a long-period fiber grating, or coupling fibers with unmatched refractive index profiles.

Sensors based on in-fiber Bragg gratings have been widely used in both quasi-distributed and point sensing [21–26]. Spectral characteristics of Bragg gratings, such as the dynamic range and sensitivity to physical influences, essentially depend on the fiber material, doping agent concentration, and techniques of fabricating the grating structure and processing the fiber core/cladding material.

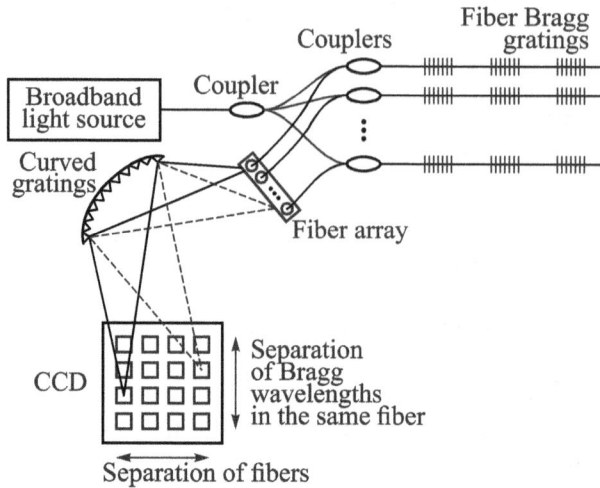

FIGURE 6.16   Scheme of a two-dimensional spectral analysis.

A sensor on the basis of a sapphire Bragg grating implemented using a Talbot interferometer was described in [21] and utilized for high temperature measurements. The sensor benefits an extended dynamic range of measured temperatures up to 1,900°C.

The said structures can be synthesized in a standard single-mode fiber using femtosecond laser pulses [22] and remain stable at temperatures of up to 1,000°C.

In the standard gratings, the temperature sensitivity is 8 pm/°C. In [23], the temperature sensitivity was increased by placing the fiber grating into a thermoplastic cylinder with high thermal expansion coefficient, which was fabricated using 3D printing. The exposure to high temperatures leads to a small decrease in the reflectivity and a redshift of the resonance wavelength owing to the material deformation. The temperature sensitivity was reported to be 139 pm/°C.

The sensitivity of standard fiber Bragg gratings to the on-axis deformation is about 0.44 pm/με and can be improved by modifying the design of the optic fiber Bragg grating (OFBG)-aided sensors. For instance, by etching the grating cladding, the deformation sensitivity can be increased to 5.5 pm/με [24].

Unlike in-fiber Bragg gratings, OFBG-based superstructures [27] have higher sensitivity to mechanical and temperature influences. The fiber optic superstructures are long-period gratings, which are formed by alternating Bragg grating sections [27–36]. However, such long-period gratings have a limited sensitivity because changes in the cladding properties due to external influences are compensated for by similar changes in the fiber core. The deformation sensitivity of photo-induced superstructures is about 1.7 and 0.44 pm/με, whereas the temperature sensitivity is 9 and 64 pm/°C [29]. A structure that has the axial-strain sensitivity of 2.8 and 1.06 pm/με and temperature sensitivity of 80 and 10 pm/°C has also been described [30].

The effect of microbending strain on the Bragg grating may be used for multiparameter measurements of external physical factors like pressure and temperature.

In this case, a shift in the Bragg-grating-aided resonance peaks compensates for thermal effects that occur when applying pressure to an optical fiber.

In a superstructure formed by acoustically induced microbendings in a homogeneous Bragg grating [37, 38], the reflection spectrum has more resonance side-lobes with an increased distance between them. Longitudinal acoustic vibrations produce radial fiber deformations in the form of periodic diameter widening and narrowing. This acoustic-optical effect causes changes in the effective refractive index for the principal and the opposite modes, also changing the grating period [37–39]. For low-frequency vibrations, bending acoustic modes are excited, with longitudinal modes prevailing in the high-frequency range. A noticeable decrease in the reflectivity for the main peak and an increase in the amplitude of the side-lobes are the result of coupling between the fiber principal mode and higher-order cladding modes [40].

When classical Bragg gratings are exposed to high temperatures, short-period fiber structures get erased. Because of this, prior to generating thermo-induced superstructures [41–44], segments of an optical fiber are first preheated, after which Bragg gratings are written. Besides, phase discontinuity in the Bragg grating can be generated by means of an electric arc discharge, which will partly erase the recorded periodic structure [41].

A technique for producing a helical superstructure by means of an electric arc and UV light has been reported [42]. Thanks to an increased twist period of the long-period grating section at the unchanged period of the Bragg grating, the inter-peak distance in the reflection spectrum is decreased. If the Bragg grating period is increased but the helical structure period is fixed, the reflection spectrum peaks are redshifted with the inter-peak intervals remaining unchanged. The shift resonance wavelength of the helical superstructure was measured with an accuracy of 64.6 $nm \cdot mm \cdot rad^{-1}$ in the interval from -0.075 to $+0.075$ $rad \cdot mm^{-1}$ and was found to be 9.36 nm. Unlike the shift of a wideband dip, FBG resonances remain practically unchanged upon twisting. The fundamental mode of the fiber is practically insensitive to the helical structure twisting. The temperature sensitivity is 60.51 pm/°C for a broadband dip generated by long-period sections, being 10.12 pm/°C for a narrowband peak that the Brag grating generates for temperatures ranging from 26°C to 75°C. In this case, the FBG spectral response allows obtaining information on temperature-related effects.

A superstructure has been reported to be fabricated by heating the Bragg grating surface with an IR laser light [44]. In a two-stage procedure, a Bragg grating is first written with 800-nm 120-fs laser pulses passing through a phase mask, before using a 0.5-W CO_2 laser for the pointwise heating of the surface of a short-period structure. In the process, one fiber end remains rigid and the other has a 5-g weight attached to it to exclude the appearance of microbendings in the course of heating. When increasing the length of the long-period grating section, intervals between the reflection spectrum peaks also increase. When increasing the duration of the Bragg grating heating, the intensity of reflection peaks decreases.

Another type of fiber-optic superstructure is realized by etching the Bragg grating cladding. The loss bands are highly sensitive to twisting because the accompanying strain is higher on the etched sections compared to the non-etched sections. Stretching a corrugated fiber grating leads to a higher attenuation in the transmission spectrum.

However, a lower mechanical strength of etched structures imposes limits on the deformation measurements. If the fiber Bragg grating has a corrugated surface [45], changes in the propagation and coupling of the cladding mode will be insignificant, weakly affecting the reflection spectrum of the fiber Bragg grating as the principal mode is mainly confined to the fiber core. When applying an axial strain to the corrugated superstructure, the resonance wavelength in the reflection spectrum is redshifted. With increasing tension stress, the side-lobes in the reflection spectrum increase, which can be explained by the interaction of the principal mode with the cladding modes due to a greater stretching of the etched segments compared to the non-etched ones.

However, alongside fiber-optic superstructures, sensing elements with a simpler design may also be utilized. For instance, we may recall a fiber Mach–Zehnder interferometer built by connecting conventional single-mode fibers with nonmatching profiles. The insert may be in the form of a multimode fiber segment, a W-shaped fiber, or a photonic crystal fiber segment.

In [46], a fiber interferometer with an insert of a multimode stepped-index fiber with periodic twisted inhomogeneities on either side of the input and output ends was proposed. In the experimental bending tests in the range from 0 to 1.739 m^{-1} the sensitivity was found to be 2.42 nm/m^{-1}. In the temperature tests in the range from 21 to 120°C, the sensitivity was shown to be 10 pm/°C.

In [47], an interferometer with a two-layered optical fiber segment was proposed. Inter-mode coupling between the fundamental mode and cladding modes at the interface of fibers SM630 and SMF-28 with different refractive indices was analyzed. Such structures are formed by welding together standard single-mode fibers SMF-28. After passing through the first welded joint, the energy from the core is partly redistributed to the insert-fiber cladding, leading to the excitation of cladding modes. When the energy is coupled into the core of the third fiber at the second joint, there is interference between the fundamental mode and cladding modes. For such a structure, the sensitivity of the transmission spectrum to stretching and temperature variations is, respectively, 0.3 pm/µε and 10 pm/°C.

Fiber-optic interferometers on the basis of photonic crystal fiber segments are also feasible. Such sensors are fabricated by welding the end of a photonic crystal fiber with standard single-mode optical fibers. Another approach utilizes the collapse of photonic crystal holes at two sections. Such a sensor with an insert of the dual-core photonic crystal fiber has been described [48]. In the transmission spectrum of a composite fiber structure, multiple interference intensity peaks were observed. The temperature and deformation sensitivity was 25 pm/°C and 1.26 pm/µε.

REFERENCES

1. Fang, Z., Chin, K., Qu, R., and Cai, H. 2012. *Fundamentals of optical fiber sensors.* Hoboken, NJ: John Wiley and Sons Ltd. ISBN: 978-0-470-57540-6.
2. Soifer, V. A., and Golub, M. A. 1994. *Laser beam mode selection by computer generated holograms.* Boca Raton, FL: CRC Press. ISBN: 978-0-8493-2476-5.
3. Garitchev, V. P., Golub, M. A., Karpeev, S. V., Krivoshlykov, S. G., Petrov, N. I., Sissakian, I. N., Soifer, V. A., Haubenreisser, W., Jahn, J.-U., and Willsch, R. 1985. Experimental investigation of mode coupling in a multimode graded-index fiber, caused by periodic microbends using computer-generated spatial filters. *Opt. Commun.* 55(6):403–5.

4. Golub, M. A., Sisakyan, I. N., Soifer, V. A., and Uvarov, G. V. 1991. Mode-selective fiber sensor operating with computer generated optical elements. *Proc. SPIE* 1572:101. https://doi.org/10.1117/12.50022.

5. Kotlyar, V. V., Soifer, V. A., and Khonina, S. N. 1998. Rotation of multimodal Gauss-Laguerre light beams in free space and in a fiber. *Opt. Lasers Eng.* 29(4–5):343–50. https://doi.org/10.1016/S0143-8166(97)00121-8.

6. Khonina, S. N., Kotlyar, V. V., and Soifer, V. A. 1998. Diffraction optical elements matched to the Gauss-Laguerre modes. *Opt. Spectrosc.* 85(4):636–44.

7. Khonina, S. N., Kotlyar, V. V., and Soifer, V. A. 1999. Self-reproduction of multimode Hermite-Gaussian beams. *Tech. Phys. Lett.* 25(6):489–91. https://doi.org/10.1134/1.1262525.

8. Kharitonov, S. I., and Khonina, S. N. 2018. Conversion of a conical wave with circular polarization into a vortex cylindrically polarized beam in a metal waveguide. *Comput. Opt.* 42(2):197–211. https://doi.org/10.18287/2412-6179-2018-42-2-197-211.

9. Khonina, S. N., Striletz, A. S., Kovalev, A. A., and Kotlyar, V. V. 2010. Propagation of laser vortex beams in a parabolic optical fiber. *Proc. SPIE* 7523:75230B. https://doi.org/10.1117/12.854883.

10. Kirilenko, M. S., Mossoulina, O. A., and Khonina, S. N. 2016. Propagation of vortex eigenfunctions of bounded Hankel transform in a parabolic fiber. *IEEE Proc. Int. Conf. on Laser Optics*:R4–32. https://doi.org/10.1109/LO.2016.7549787.

11. Monin, E. O., Ustinov, A. V., and Khonina, S. N. 2018. Propagation modeling of vortex generalized Airy beams in parabolic fiber. *Proc. Progress in Electromagnetics Research Symposium (PIERS)* F134321:583–9. https://doi.org/10.1109/PIERS.2017.8261809.

12. Khonina, S. N., and Karpeev, S. V. 2004. Excitation and detection of angular harmonics in an optical fiber using DOE. *Comput. Opt.* 26:16–26.

13. Karpeev, S. V., and Khonina, S. N. 2007. Experimental excitation and detection of angular harmonics in a step-index optical fiber. *Opt. Mem. Neural Netw.* 16(4):295–300. https://doi.org/10.3103/S1060992X07040133.

14. Savelyev, D. A., and Khonina, S. N. 2017. Investigation of vortex laser beam injection into an optical fiber. *J. Phys. Conf. Ser.* 917:062035. https://doi.org/10.1088/1742-6596/917/6/062035.

15. Karpeev, S. V., Pavelyev, V. S., and Kazanskiy, N. L. 2005. Steplike fiber modes excitement with binary phase DOEs. *Opt. Mem. Neural Netw.* 14(4):223–8.

16. Karpeev, S. V., Pavelyev, V. S., and Khonina, S. N. 2005. High-effective fiber sensors based on transversal mode selection. *Proc. SPIE* 5854:163–9. https://doi.org/10.1117/12.634603.

17. Karpeev, S. V., Pavelyev, V. S., Khonina, S. N., Kazanskiy, N. L., Gavrilov, A. V., and Eropolov, V. A. 2007. Fibre sensors based on transverse mode selection. *J. Mod. Opt.* 54(6):833–44. https://doi.org/10.1080/09500340601066125.

18. Khonina, S. N., Skidanov, R. V., Kotlyar, V. V., Jefimovs, K., and Turunen, J. 2003. Phase diffractive filter to analyze an output step-index fiber beam. *Opt. Mem. Neural Netw.* 12(4):317–24. https://doi.org/10.1117/12.509468.

19. Lyubopytov, V. S., Tlyavlin, A. Z., Sultanov, A. K., Bagmanov, V. K., Khonina, S. N., Karpeev, S. V., and Kazanskiy, N. L. 2013. Mathematical model of completely optical system for detection of mode propagation parameters in an optical fiber with few-mode operation for adaptive compensation of mode coupling. *Comput. Opt.* 37(3):352–9.

20. Khonina, S. N., Kazanskiy, N. L., and Soifer, V. A. 2012. Optical vortices in a fiber: Mode division multiplexing and multimode self-imaging, Ch. 15. In *Recent progress in optical fiber research*, eds. M. Yasin, S. W. Harun, and H. Arof. Croatia: INTECH Publisher. https://doi.org/10.5772/28067.

21. Anania, S., Unnikrishnan, A., Aparna, A., Parvathi, G. R., Baby Sreeja, S. D., and Mohan, P. 2018. Analytical study of FBG spectrum for temperature sensing applications. *Proc. 2nd Int. Conf. on Inventive Communication and Computational Technologies (ICICCT 2018)*:1109–13. https://doi.org/10.1109/ICICCT.2018.8473283.

22. Habisreuther, T., Elsmann, T., Pan, Z., Graf, A., Willsch, R., and Schmidt, M. 2015. Sapphire fiber Bragg gratings for high temperature and dynamic temperature diagnostics. *Appl. Therm. Eng.* 91:860–5. https://doi.org/10.1016/j.applthermaleng.2015.08.096.

23. Leal-Junior, A., Casas, J., Marques, C., José Pontes, M., and Frizera, A. 2018. Application of additive layer manufacturing technique on the development of high sensitive fiber Bragg grating temperature sensors. *Sensors.* 18(12):4120. https://doi.org/10.3390/s18124120.

24. Sridevi, S., Vasu, K. S., Asokan, S., and Sood, A. K. 2016. Enhanced strain and temperature sensing by reduced graphene oxide coated etched fiber Bragg gratings. *Opt. Lett.* 41(11):2604–07. https://doi.org/10.1364/OL.41.002604.

25. Yang, S., Hu, D., and Wang, A. 2017. Point-by-point fabrication and characterization of sapphire fiber Bragg gratings. *Opt. Lett.* 42(20):4219–22. https://doi.org/10.1364/OL.42.004219.

26. Zhang, C., Yang, Y., Wang, C., Liao, C., and Wang, Y. 2016. Femtosecond-laser-inscribed sampled fiber Bragg grating with ultrahigh thermal stability. *Opt. Express.* 24(4):3981–8. https://doi.org/10.1364/OE.24.003981.

27. Tan, R. X., Yap, K., Tan, Y. C., Tjin, S. C., Ibsen, M., Yong, K. T., and Lai, W. J. 2018. Functionalized fiber end superstructure fiber Bragg grating refractive index sensor for heavy metal ion detection. *Sensors.* 18(6):1821. https://doi.org/10.3390/s18061821.

28. Chi, H., Tao, X.-M., and Yang, D. 2001. Simultaneous measurement of axial strain, temperature, and transverse load by a superstructure fiber grating. *Opt. Lett.* 26(24):1949–51. https://doi.org/10.1364/OL.26.001949.

29. Sengupta, S., Ghorai, S. K., and Biswas, P. 2016. Design of superstructure fiber Bragg grating with efficient mode coupling for simultaneous strain and temperature measurement with low cross-sensitivity. *IEEE Sens. J.* 16(22):7941–9. https://doi.org/10.1109/JSEN.2016.2611002.

30. Guan, B.-O., Tam, H.-Y., Tao, X.-M., and Dong, X.-Y. 2000. Simultaneous strain and temperature measurement using superstructure fiber Bragg grating. *IEEE Photonics Technol. Lett.* 12(6):675–7. https://doi.org/10.1109/68.849081.

31. Zhang, A.-P., Guan, B.-O., Tao, X.-M., and Tam, H.-Y. 2002. Mode couplings in superstructure fiber Bragg gratings. *IEEE Photonics Technol. Lett.* 14(4):489–91. https://doi.org/10.1109/68.992587.

32. Zeng, X., and Liang, K. 2011. Analytic solutions for spectral properties of superstructure, Gaussian-apodized and phase shift gratings with short- or long-period. *Opt. Express* 19(23):22797–808. https://doi.org/10.1364/OE.19.022797.

33. He, Y.-J., Hung, W.-C., and Lai, Z.-P. 2016. Using finite element and eigenmode expansion methods to investigate the periodic and spectral characteristic of superstructure fiber Bragg gratings. *Sensors.* 16(192):192. https://doi.org/10.3390/s16020192.

34. Rodriguez-Cobo, L., Cobo, A., and Lopez-Higuera, J. M. 2012. Sampled fiber Bragg grating spectral synthesis. *Opt. Express.* 20(20):22429–41. https://doi.org/10.1364/OE.20.022429.

35. Luo, Z., Ye, C., Cai, Z., Dai, X., Kang, Y., and Xu, H. 2007. Numerical analysis and optimization of optical spectral characteristics of fiber Bragg gratings modulated by a transverse acoustic wave. *Appl. Opt.* 46(28):6959–65. https://doi.org/10.1364/ao.46.006959.

36. Idrisov, R. F. 2020. A review on fiber-optical superstructures based on in-fiber bragg grids. *Appl. Phys. Math.* 3:15–27.

37. Pohl, A., Oliveira, R., Da Silva, R., Marques, C., de Tarso, P., Cook, K., Canning, J., and Nogueira, R. 2013. Advances and new applications using the acousto-optic effect in optical fibers. *Photonic Sens.* 3(1):1–25. https://doi.org/10.1007/s13320-013-0100-0.

38. Oliveira, R. A., Marques, C. A. F., Mayer, C. E. N., Pereira, J. T., Nogueira, R. N., and Pohl, A. A. P. 2009. Single device for excitation of both flexural and longitudinal acousto-optic effects in fiber Bragg gratings. *SBMO/IEEE MTT-S International Microwave & Optoelectronics Conference* (IMOC):546–9. https://doi.org/10.1109/IMOC.2009.5427526.

39. Silva, R., Tiess, T., Becker, M., Eschrich, T., Rothhardt, M., Jäger, M., Pohl, A., and Bartelt, H. 2015. All-fiber 10 MHz acousto-optic modulator of a fiber Bragg grating at 1060 nm wavelength. *Opt. Express.* 23(20):25972–8. https://doi.org/10.1364/OE.23.025972.

40. Gao, Z., Chang, P., Huang, L., Gao, F., Mao, D., Zhang, W., and Mei, T. 2019. All-fiber frequency shifter consists of fiber Bragg grating modulated via acoustic flexural wave for optical heterodyne measurement. *Opt. Lett.* 44(15):3725–8. https://doi.org/10.1364/OL.44.003725.

41. Ouyang, Y., Liu, J., Xu, X., Zhao, Y., and Zhou, A. 2018. Phase-shifted eccentric core fiber Bragg grating fabricated by electric arc discharge for directional bending measurement. *Sensors* 18(6):1168. https://doi.org/10.3390/s18041168.

42. Li, J., Chen, G., Ma, P., Sun, L.-P., Wu, C., and Guan, B. 2018. Sampled Bragg gratings formed in helically twisted fibers and their potential application for the simultaneous measurement of mechanical torsion and temperature. *Opt. Express.* 26(10):12903–11. https://doi.org/10.1364/OE.26.012903.

43. Guo, G. 2019. Superstructure fiber Bragg gratings for simultaneous temperature and strain measurement. *Optik.* 182:331–40. https://doi.org/10.1016/j.ijleo.2019.01.011.

44. Fang, X., He, X. Y., Liao, C. R., Yang, M., Wang, D. N., and Wang, Y. 2010. A new method for sampled fiber Bragg grating fabrication by use of both femtosecond laser and CO_2 laser. *Opt. Express.* 18(3):2646–54. https://doi.org/10.1364/OE.18.002646.

45. Lin, C., Chern, G., and Wang, L. 2001. Periodical corrugated structure for forming sampled fiber Bragg grating and long-period fiber grating with tunable coupling strength. *J. Lightw. Technol.* 19(8):1212–20. https://doi.org/10.1109/50.939803.

46. Tian, K., Xin, Y., Yang, W., Geng, T., Ren, J., Fan, Y.-X., and Farrell, G. 2017. A curvature sensor based on twisted single-mode–multimode–single-mode hybrid optical fiber structure. *J. Lightw. Technol.* 35(9):1725–31. https://doi.org/10.1109/JLT.2017.2650941.

47. Zlodeev, I. V., and Ivanov, O. V. 2013. Transmission spectra of a double-clad fibre structure under bending. *Quantum Electron.* 43(6):535–41. https://doi.org/10.1070/QE2013v043n06ABEH014971.

48. Chen, W., Lou, S., and Wang, L. 2011. In-fiber modal interferometer based on dual-concentric-core photonic crystal fiber and its strain, temperature and refractive index characteristics. *Opt. Commun.* 284(12):2829–34. https://doi.org/10.1016/j.optcom.2011.02.019.

7 Wave Front Aberration Sensors Based on Optical Expansion by the Zernike Basis

Svetlana Nikolaevna Khonina[1,2], Pavel Alexeyevich Khorin[2], and Alexey Petrovich Porfirev[1]
[1]Image Processing Systems Institute – Branch of the Federal Scientific Research Centre "Crystallography and Photonics" of Russian Academy of Sciences, Samara, Russia
[2]Samara National Research University, Samara, Russia

7.1 INTRODUCTION

Studying the deviation of a wave front from the desired shape is one of the most significant problems in optics. There are many well-known methods for solving this problem, and new techniques are being constantly developed. The most common and versatile among them is interferometry [1, 2], which has unsurpassed accuracy and allows one to directly obtain a pattern of wave front deviations at very large apertures. The accuracy of interferometers, especially heterodyne interferometers, exceeds $\lambda/100$. The disadvantages of interferometry are well known and include complexity of decoding interferograms, sensitivity of measuring equipment to vibrations, and a need for the physical presence of a reference wave front. At the initial stages of the development of optical production, the schlieren (or shadow) method was used to control spherical surfaces; [3] however, shadow patterns are difficult to quantify, and schlieren systems, like interferometers, must have high rigidity and be vibration-proof.

The Hartmann method [4], which appeared later, differs from the previous techniques in the fact that the wave front deviations are calculated from a set of sub-apertures, with some step covering the full size of the region to be studied. Wave front deviations are calculated using ray tracing data, with the rays passing through sub-apertures. A further development of the Hartmann method was a Shack–Hartmann wave front sensor [5]. In this version of the sensor, the data on the wave front deviations are transferred to the photodetector plane by installing a lens raster. Each lens forms a sub-aperture for which an average wave front deviation is calculated. Information about the wave front phase within the sub-aperture is contained in the coordinates of the focused light spot. The main advantage of both the Hartmann method and the Shack–Hartmann wave front sensor consists in the fact that there is no need to use a reference wave front in calculations. However, these

DOI: 10.1201/9781003439165-7

techniques are not exempt from disadvantages, such as sensitivity to vibrations and initially discrete nature of the measurements. This disadvantage arises if too much distorted wave front incidents on a sensor matrix. If aberrations are too much, a local light spot leaves its own sub-area and jumps into the neighboring section of the sensor matrix, in the Shack–Hartmann scheme. In this case, data on a part of the surface of the wave front are inevitably lost. In addition, a false spot can be wrongly detected. In order to eliminate errors due to cross talk, fairly complex algorithms are used. Many Shack–Hartmann sensors currently being manufactured have no more than 10^3 sub-apertures, which does not satisfy the requirements of many tasks. A new approach to the formation of microlens arrays based on mesoscale square cubic dielectric particles [6] was proposed, which will significantly increase the dimension of the lens raster.

In the methods just described, to visualize the phase, some additional digital operations are required to process the obtained measurements. Direct methods of phase visualization are also known. The Zernike phase contrast method [7] is a powerful tool for transforming the spatial phase information of an optical beam into a spatial intensity distribution without absorbing light. The basic principle is to split a light beam into its Fourier components using a lens and a filter. The introduced phase shift creates an intensity distribution in accordance with the phase information carried by the higher spatial frequencies. This method has been successfully applied to analyze aberrations and improve resolution in telescopes, in decoding phase-coded information, as well as in biological tissue microscopy.

Another approach to phase visualization is the optical knife method, as well as the radial Hilbert transform, which is implemented on the basis of a spiral phase plate. The radial Hilbert transform is used both for the selection of the edges of the image and for phase contrast [8].

One of the tools for describing wave fronts, in addition to the deviation patterns, is the decomposition of aberrations over various bases. The most famous decomposition bases include Zernike polynomials [9], as well as Seidel aberrations. Note that the generally accepted representation of wave front aberrations [10–12], including in the individual optical system of the human eye, is a series of Zernike polynomials [7, 13–16]. The human eye can be described as a system of lenses consisting of three basic components: the cornea, pupil and lens. The cornea (including the tear film) structure is the dominant eye optical power (it provides about 70% of the optical power of the eye). Accordingly, it is a major source of aberrations in the eye. The front surface of the cornea has an elongated profile; that is, the central region is steeper than at the periphery. This shape helps reduce the amount of spherical aberration in the whole eye. However, the cornea shape may be significantly different for different people, and this results in asymmetric astigmatism and high-order aberration (for example, coma). Optical elements of the human eye optical systems work concertedly to create an image on the retina. However, the image in the real system of a human eye is never perfect. The emergence of additional optical aberrations associated with aging or disease degrades the image quality significantly. To compensate the wave front distortions, it is necessary to determine exactly which aberration led to the distortions. Wave front analysis based on an expansion in the Zernike basis facilitates diagnosis in clinical ophthalmology and allows the creation of the most advanced lenses, presently providing the highest quality vision. If the wave front is represented by a linear combination of Zernike polynomials, it has a number of useful properties. First, Zernike polynomials are easy to relate to classical aberrations. Second, polynomials are usually defined by procession of values at the points using ordinary least squares. Consequently, since Zernike polynomials are orthogonal on the unit circle, any of the terms of the expansion are also the best approximations of the ordinary least squares. Thus, to prevent the shift of focus or tilt of the wave front, it is necessary that the corresponding coefficients are equal to zero. The average value of the aberration is determined by the value of each respective member, so it is not necessary to perform a new approximation using ordinary least squares.

Aberrational representations are more efficient [17] in terms of data volumes and also allow one to make use of the wave front features that are important for solving specific problems. Direct measurement of aberration coefficients is possible only for some types of aberrations. Calculation of the Zernike aberration coefficients [18–21] on the basis of a two-dimensional (2D) array of measured values of the wave front deviations in each of the sub-apertures is provided in the data processing programs supplied with Shack–Hartmann sensors, as well as with ophthalmic aberrometers. However, it should be noted that due to the rather rough discretization of wave front data, the calculation of high-order aberrations is difficult.

Active employment of Zernike polynomials for representation of wave aberrations stimulates the development of new sensors, including for direct measurements of expansion coefficients by the Zernike basis. In this work, we propose a new sensor for measuring aberration coefficients based on a special multichannel diffractive optical element [22–24]. The developed sensor provides a sensitivity to wave front deviations no worse than $\lambda/20$, is resistant to vibrations, and does not require the use of reference optical elements.

Diffractive optical elements (DOEs) for the integral calculation of the expansion coefficients of amplitude–phase distributions of light fields over various bases [25–28], including the basis of Zernike functions [29–31], have been developed and used in fiber-optic sensors [32–34] for measuring the angular momentum of laser beams [35–37], for optical communication using mode and polarization (de)multiplexing [38–44], and in testing problems [45, 46]. These elements make it possible to simultaneously obtain the values of the decomposition coefficients in the given elements of the photodetector matrix. In contrast to the Shack–Hartmann sensor, in which the calculation of aberration coefficients requires mathematical processing of a 2D data array, the values of aberration coefficients in multichannel diffractive optical sensors are proportional to the intensities of diffraction maxima located at the photodetector matrix points with constant coordinates. Thus, the entire area of the tested beam is simultaneously involved in the formation of the values of each coefficient, while in Shack–Hartmann sensors and especially in Hartmann sensors, information about part of the wave front area is not involved in the measurements and remains unknown. It should also be noted that the calibration function of the proposed sensor is substantially nonlinear, which leads to a decrease in the dynamic range of aberrations being measured. However, this is quite enough for most practically significant cases of certification of optical systems; for example, it is believed that the average aberration should not exceed $\lambda/10$ for budget imaging systems and $\lambda/100$ for high-end systems.

7.2 METHOD OF OPTICAL EXPANSION BY THE ZERNIKE BASIS

7.2.1 THEORETICAL AND NUMERICAL INVESTIGATIONS

In the problem of analysis of laser light, it is advisable to use spatial filters separating the analyzed beam diffraction on individual components corresponding to angular harmonics. Angular harmonics of different orders m are the complex harmonic functions with a single module and a linear dependence on the polar angle $\exp(im\phi)$. Such harmonics appear in the optical implementation of the higher-order Hankel transform based on the spiral phase plates [47, 48]. The complex amplitude of monochromatic light wave $E(r, \phi)$ represented in polar coordinates (r, ϕ) can always be expanded in a functional Fourier series of angular harmonics. Such functional series is written as follows:

$$E(r,\phi) = \sum_{m=-\infty}^{\infty} E_m(r)\exp(im\phi), \tag{7.1}$$

$$E_m(r) = \frac{1}{2\pi}\int_0^{2\pi} E(r,\phi)\exp(-im\phi)\,d\phi. \tag{7.2}$$

Similarly, the expansion (7.1) of the angular harmonics of the spatial spectrum of the light field can be performed. Indeed, if we take the Fourier transform of the function $E(r, \phi)$, we find that:

$$FT\left[E(r,\phi)\right] = F(\rho,\theta) = \sum_{m=-\infty}^{\infty} F_m(\rho)\exp(im\theta), \tag{7.3}$$

$$F_m(\rho) = \frac{4\pi^2 i^m}{\lambda f}\int_0^{\infty} E_m(r)J_m\left(\frac{2\pi}{\lambda f}r\rho\right)r\,dr, \tag{7.4}$$

where (ρ, θ) is the polar coordinates in the focal plane, λ is the radiation wavelength, f is a focal length of the optical system, and $J_m(x)$ is the m-th order Bessel function of the first order. It is clearly seen that the expressions (Eq. 7.1) and (7.3) are similar in structure, while the function $F_m(\rho)$ and $E_m(r)$ are associated with the Hankel transform of m-th order (see Eq. (7.4)). Optical expansion in Eq. (7.3) can be performed using tandem "lens+DOE," where the spherical lens with a focal length f performs the Fourier transform, and the DOE provides a decomposition of the incident radiation with a wavelength λ on the set of "vortex" basis.

The expansion in Eq. (7.3) is useful for the invariant to the rotation analysis and detection, since the rotation of the analyzed field $F(\rho, \theta)$ on the angle $\Delta\theta$ will not change the experimentally measured modulus of the coefficients $|F_m(\rho)|$. There is a complete set of orthogonal functions with angular harmonics in a circle of radius R. These are the circular Zernike polynomials corresponding to a complete set of orthogonal functions in polar coordinates (r, ϕ) in a circle of radius r_0 [13]:

$$Z_n^m (r,\phi) = A_n R_n^m (r) \begin{Bmatrix} \cos(m\phi) \\ \sin(m\phi) \end{Bmatrix}, \tag{7.5}$$

where $A_n = \sqrt{(n+1)/\pi}$ and $R_n^m (r)$ are the radial Zernike polynomials:

$$R_n^m (r) = \sum_{p=0}^{(n-m)/2} (-1)^p (n-p)! \times \tag{7.6}$$
$$\times \left[p! \left(\frac{n+m}{2} - p \right)! \left(\frac{n-m}{2} - p \right)! \right]^{-1} \left(\frac{r}{r_0} \right)^{n-2p}.$$

Examples of Zernike polynomials are shown in Figure 7.1.

The expansion of the light field with complex amplitude $E(r, \phi)$ into a series in terms of the functions in Eq. (7.5) is given by:

$$E(r,\phi) = \sum_{n=0}^{\infty} \sum_{m=-n}^{n} C_{nm} Z_n^m (r,\phi). \tag{7.7}$$

The expansion coefficients for the wave front in the Zernike orthogonal functions (Eq. 7.5) allow one to determine the deviations (aberrations) from the ideal wave front [13–16, 18, 45, 46]. Taking into account expression (Eq. 7.4), the spatial spectrum of the field $E(r, \phi)$ can be represented as an expansion in Zernike polynomials:

$$F(\rho,\theta) = -\frac{2i\sqrt{\pi}}{\lambda f R} \sum_{n=0}^{\infty} \sum_{m=-n}^{n} \sqrt{n+1} (-i)^m \times \tag{7.8}$$
$$\times C_{nm} \exp(im\theta) \int_0^R Z_n^m (r) J_m \left(\frac{2\pi}{\lambda f} r\rho \right) r\, dr.$$

The integral in expression (Eq. 7.8) may be taken explicitly [1]:

$$W_{nm}(\rho) = \int_0^R Z_n^m (r) J_m \left(\frac{2\pi}{\lambda f} r\rho \right) r\, dr = (-1)^{(n-m)/2} R^3 \frac{J_{n+1} \left(\frac{2\pi}{\lambda f} R\rho \right)}{\left(\frac{2\pi}{\lambda f} R\rho \right)}. \tag{7.9}$$

Let us consider an aberrated wave front in the form of a field:

$$g(r,\phi) = \exp[i\psi(r,\phi)], \tag{7.10}$$

N	n	m	Trigonometric Representation	Aberration Type	Amplitude and Phase
1	0	0	1	Constant	
2	1	−1	$2r\sin(\theta)$	Tilt	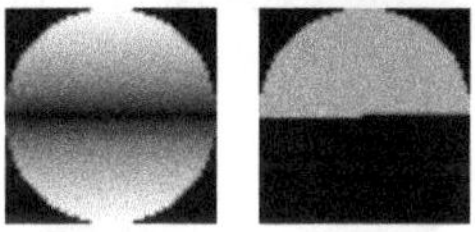
3	1	1	$2r\cos(\theta)$	Tilt	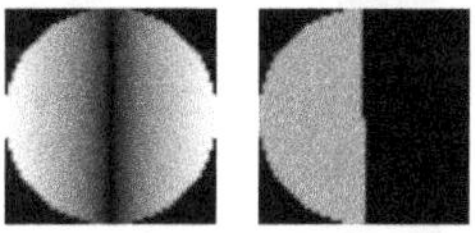
4	2	−2	$\sqrt{6}r^2\sin(2\theta)$	Astigmatism	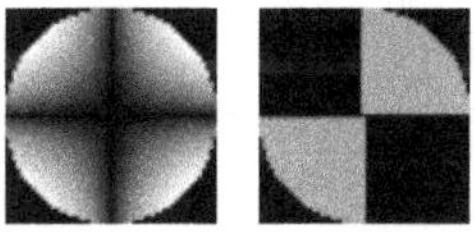
5	2	0	$\sqrt{3}(2r^2-1)$	Defocus	
6	2	2	$\sqrt{6}r^2\cos(2\theta)$	Astigmatism	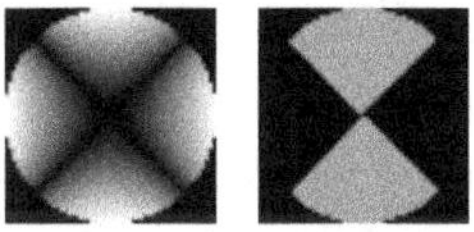
7	3	−3	$2\sqrt{2}r^3\sin(3\theta)$	Zero curvature coma (trefoil)	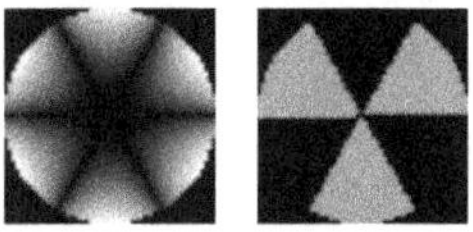
8	3	−1	$2\sqrt{2}(3r^3-2r)\sin(\theta)$	Pure coma	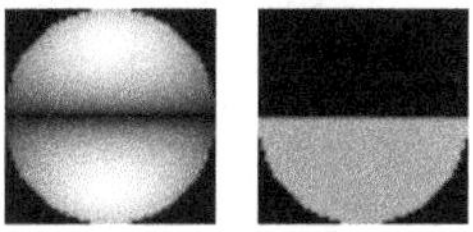
9	3	1	$2\sqrt{2}(3r^3-2r)\cos(\theta)$	Pure coma	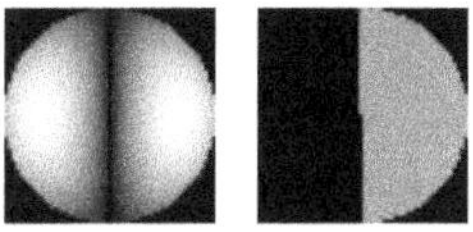
10	3	3	$2\sqrt{2}r^3\cos(3\theta)$	Zero curvature coma (trefoil)	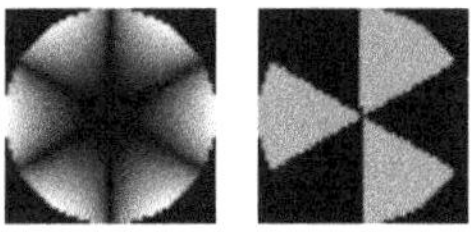

FIGURE 7.1 Examples of Zernike polynomials.

whose phase is a superposition of Zernike functions:

$$\psi(r,\phi) = 2\pi\alpha \sum_{n,m} b_{nm} Z_n^m(r,\phi), \tag{7.11}$$

where b_{nm} are coefficients of the superposition.

The expansion coefficients of field (7.10) in basis (7.5) are calculated as follows:

$$c_{pq} = \int_0^{r_0}\int_0^{2\pi} g(r,\phi) Z_p^q(r,\phi) r\,dr\,d\phi. \tag{7.12}$$

We represent field (7.10) as the following expansion:

$$\begin{aligned}
g(r,\phi) &= \exp\left[i\psi(r,\phi)\right] = \\
&= 1 + i\psi(r,\phi) - \frac{1}{2}\psi^2(r,\phi) - \frac{i}{6}\psi^3(r,\phi) + \ldots = \\
&= 1 + i2\pi\alpha \sum_{n,m} b_{nm} Z_n^m(r,\phi) - 2(\pi\alpha)^2 \left[\sum_{n,m} b_{nm} Z_n^m(r,\phi)\right]^2 + \ldots.
\end{aligned} \tag{7.13}$$

For small aberrations (i.e. the value of α), expression (Eq. 7.13) can be significantly simplified:

$$g(r,\phi) \underset{\alpha\to 0}{\approx} 1 + i2\pi\alpha \sum_{n,m} b_{nm} Z_n^m(r,\phi). \tag{7.14}$$

Thus, if aberrations are small enough to leave only the first two terms in expansion (7.13), then the field can be considered as a superposition of the functions themselves. In this case, the expansion coefficients of field (7.12) will be proportional to the coefficients in superposition (Eq. 7.11):

$$\begin{aligned}
c_{pq} &\underset{\alpha\to 0}{\approx} \int_0^{r_0}\int_0^{2\pi} \left(1 + i2\pi\alpha \sum_{n,m} b_{nm} Z_n^m(r,\phi)\right) Z_p^q(r,\phi) r\,dr\,d\phi = \\
&= A_p + i2\pi\alpha \cdot \delta_{pq,nm} \sum_{n,m} b_{nm} = A_p + i2\pi\alpha b_{pq},
\end{aligned} \tag{7.15}$$

where A_p is the normalization constant, and $\delta_{pq,nm}$ is the Kronecker delta.

In this case, the type and magnitude of aberrations can be detected using a multichannel filter matched with the Zernike functions [29, 31, 49]. Approximation error of decomposition in Eq. (7.14) increases with increasing of meaning of aberrations. It happens because of a finite quantity of Zernike polynomials in the sum (Eq. 7.14). Figure 7.2 shows simulation results of wave front aberrations analysis

Incident Field (Intensity and Phase Distribution)	Spatial Spectrum (Amplitude)	Expansion Coefficients, $\lvert b_{nm}\rvert^2$	Zernike Coefficients Expansion Approximation $\hat{g}(r,\phi)=\displaystyle\sum_{n,m\,\in\,\Omega} b_{nm}, \Psi_{nm}(r,\phi)$
$g(r,\phi)=Z_3^1(r,\phi)+Z_3^{-1}(r,\phi)$			
$g(r,\phi)=\exp\{i\,0.2\pi\times \times(Z_3^1(r,\phi)+Z_3^{-1}(r,\phi))\}$			
$g(r,\phi)=\exp\{i\,0.4\pi\times \times(Z_3^1(r,\phi)+Z_3^{-1}(r,\phi))\}$			
$g(r,\phi)=\exp\{i\,0.6\pi\times \times(Z_3^1(r,\phi)+Z_3^{-1}(r,\phi))\}$			
$g(r,\phi)=\exp\{i\,0.8\pi\times \times(Z_3^1(r,\phi)+Z_3^{-1}(r,\phi))\}$			

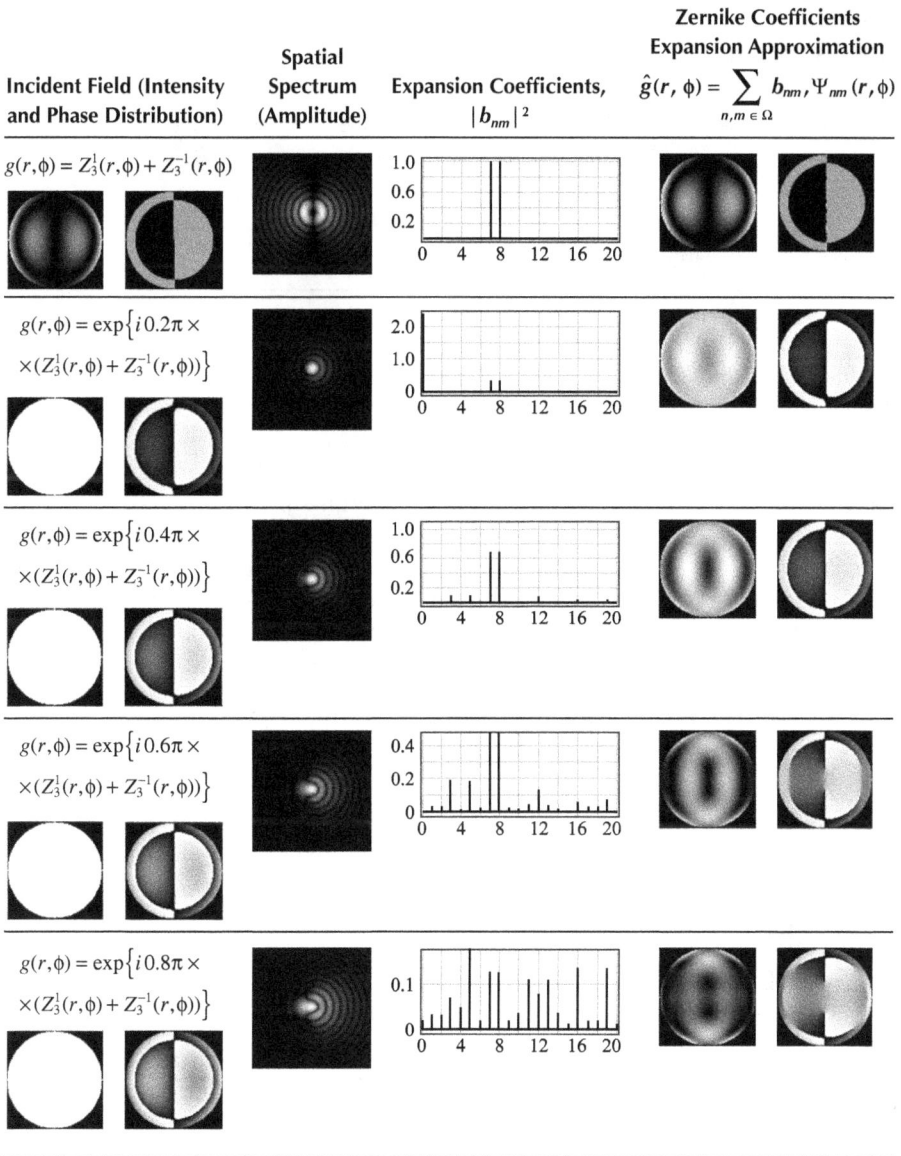

FIGURE 7.2   Wave front aberration analysis with Zernike polynomials expansion.

using a 20-order optical filter that is matched with 20 Zernike polynomials: $n \in \overline{0,5}$ $m \in -n, n$.

Physically, this corresponds to an average path difference between an ideal wave front and a wave front with an aberration coefficient of 0.3λ. For visualization of the expansion coefficients, it is convenient to use a single index, rather than a double one. Table 7.1 shows the correspondence of Zernike functions (7.5) to a single index l.

TABLE 7.1

Correspondence of Zernike Functions (1) to a Single Index *l*

l	0	1	2	3	4
$Z_n^m(r,\phi)$	$R_0^0(r)$	$R_1^1(r)\cos\phi$	$R_1^1(r)\sin\phi$	$R_2^2(r)\cos 2\phi$	$R_2^0(r)$
l	5	6	7	8	9
$Z_n^m(r,\phi)$	$R_2^2(r)\sin 2\phi$	$R_3^3(r)\cos 3\phi$	$R_3^1(r)\cos\phi$	$R_3^1(r)\sin\phi$	$R_3^3(r)\sin 3\phi$
l	10	11	12	13	14
$Z_n^m(r,\phi)$	$R_4^4(r)\cos 4\phi$	$R_4^2(r)\cos 2\phi$	$R_4^0(r)$	$R_4^2(r)\sin 2\phi$	$R_4^4(r)\sin 4\phi$

It should be noted that the task of detecting small aberrations is topical, because the point scattering function (PScF) in this case slightly differs from the Airy pattern (diffraction spot) in the absence of aberrations (see Table 7.2).

Table 7.2 shows the phase distribution (wave front) and PScF patterns in the presence of various aberrations corresponding to Zernike polynomials with different values of α. One can see that at $\alpha = 0.4$ (Table 7.2a2–7.2h2), the intensity distributions in the focal plane of the lens (PScF) look approximately the same regardless of the aberration type and differ little from the diffraction spot. This fact makes it possible to establish the applicability criterion for a multichannel filter, matched with the Zernike functions, during the measurement process. The first column of Table 7.2 (see Table 7.2a1–7.2a4) shows the distribution of the coefficients at $\alpha = 0.4$; the correspondence of the index *l* to decomposition coefficients (Eq. 7.12) is shown in Table 7.1. It follows from Table 7.2 (see Table 7.2a2–7.2h4) that only at $\alpha \leq 0.4$ (which corresponds to the average aberration coefficient of $\leq 0.4\lambda$) can we detect (recognize) with confidence the aberration structure.

Approximations (Eq. 7.14) and (Eq. 7.15) become invalid with increasing α, and in expanding the field expands in the Zernike basis, other coefficients will appear, except for those present in superposition (Eq. 7.11).

Knowing this fact, we can determine to some extent how significant the level of aberration is. However, this can be done by measuring a sufficiently large number of factors. Given the need to do this using a single multichannel diffractive sensor, it is desirable to optimize the number of necessary coefficients.

Note that the basis of the Zernike functions with trigonometric functions on angle (Eq. 7.5) is not invariant to rotation, which is inconvenient in practical applications. Another representation of the Zernike functions is also well known:

$$Z_{n,m}(r,\phi) = B_n R_n^m(r)\exp(im\phi), \qquad (7.16)$$

where $B_n = A_n$ at $m = 0$ and $B_n = A_n/2$ at $m \neq 0$.

Obviously, functions (Eq. 7.5) can be represented via a superposition of functions (Eq. 7.16) and vice versa. Compared to expression (Eq. 7.5), representation (Eq. 7.16) is more convenient due to its invariance to rotation. The Zernike basis in form expression (Eq. 7.16) cannot be used in superposition (Eq. 7.11); however, it can be conveniently used for the expansion of the optical field [31]. Amplitude and phase

TABLE 7.2
The Type of Phase (Wave Front) and PScF in the Presence of Various Aberrations Corresponding to Zernike Polynomials

	Expansion coefficients	Input phase and PScF (intensity)		
		$\alpha=0.4$	$\alpha=0.6$	$\alpha=1$
	1	2	3	4

a — $\psi(r,\phi)=0.8\pi R_2^0(r)$

b — $\psi(r,\phi)=0.8\pi R_2^2(r)\cos2\phi$

c — $\psi(r,\phi)=0.8\pi R_3^1(r)\cos\phi$

d — $\psi(r,\phi)=0.8\pi R_3^3(r)\cos3\phi$

e — $\psi(r,\phi)=0.8\pi R_4^0(r)$

f — $\psi(r,\phi)=0.8\pi R_4^2(r)\cos2\phi$

g — $\psi(r,\phi)=0.8\pi R_4^4(r)\cos4\phi$

h — $\psi(r,\phi)=0.8\pi[R_3^1(r)\cos\phi+R_4^2(r)\cos2\phi]$

Function Indices, (n, m)	Vortex Zernike Functions (5)	Standard Zernike Functions (11)	Function Indices, (n, m)	Vortex Zernike Functions (5)	Standard Zernike Functions (11)
1 (0,0)			7 (3,1)		
2 (1,1)			8 (3,−1)		
3 (1,−1)			9 (3,3)		
4 (2,0)			10 (3,−3)		
5 (2,2)			11 (4,0)		
6 (2,−2)					

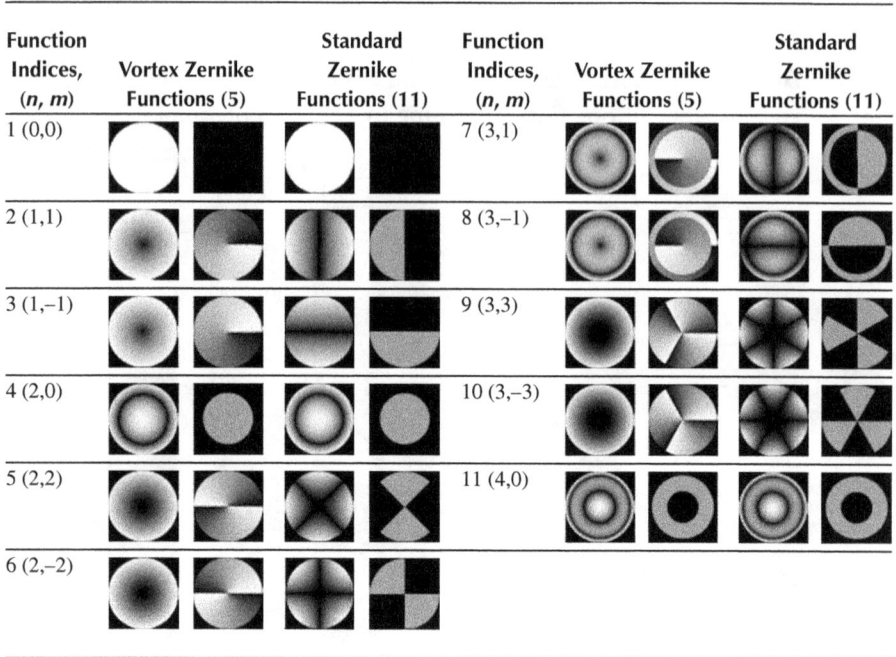

FIGURE 7.3 Amplitude and phase pictures of several first Zernike functions in representation in Eqs. (7.5) and (7.16).

pictures of several first Zernike functions in representation in expression (7.5) and expression (Eq. 7.16) are shown in Figure 7.3. The functions having a zero angular index m coincide in both representations. It should be noted that the Zernike functions in standard representation in expression (Eq. 7.5) are all real-valued. However, when we use for the analysis of wave front multi-order diffractive filters [29, 50–53] matched with Zernike functions, it can be a more convenient vortex representation in expression (Eq. 7.16).

7.2.2 Multi-Order Diffractive Optical Elements for Optical Decomposition

A multichannel diffractive filter matched with Zernike functions has the following representation [53]:

$$\tau(r,\phi) = \sum_{n=0}^{\infty}\sum_{m=-n}^{n} Z_{n,m}^{*}(r,\phi)\exp\left[is_{n,m}(r,\phi)\right], \qquad (7.17)$$

where $s_{n,m}(r,\phi)$ are phase functions corresponding to definite diffractive orders. It is obvious, if they are chosen as follows:

$$s_{n,-m}(r,\phi) = -s_{n,m}(r,\phi), \qquad (7.18)$$

then for vortex functions in expression (7.16) we shall receive the multi-order filter with the real-valued transmission function:

$$\tau_b\left(r,\phi\right)=\sum_{n=0}^{\infty}\sum_{m=0}^{n}\sqrt{\frac{n+1}{\pi R^2}}Z_n^m\left(r\right)\cos\left[s_{n,m}\left(r,\phi\right)+m\phi\right].\qquad(7.19)$$

A phase of the filter in expression (7.19) is binary, which is convenient for manufacturing.

Figure 7.4 shows phases of 4-channel filters matched with functions 2, 3, 5, 6 (see corresponding lines in Figure 7.3) for vortex and standard functions, accordingly. Intensity and phase distributions formed in a focal plane are also shown in Figure 7.4 (we use a lens added to the filter). Apparently from Figure 7.4 the filter's phase of the vortex basis has binary structure (black color corresponds to zero value of a phase, and gray color corresponds to π value of a phase). Distributions in the focal plane show preservation of corresponding vortex or trigonometric structure of basic functions. Note, when we add into a filter Zernike functions with zero angular index m, they should be duplicated to keep binary phase structure of the filter.

Another feature of the vortex basis in expression (Eq. 7.16) compared with the standard one in expression (Eq. 7.5) is: the intensity of expansion coefficients of the analyzed fields are invariant to rotation of the field. This property is illustrated

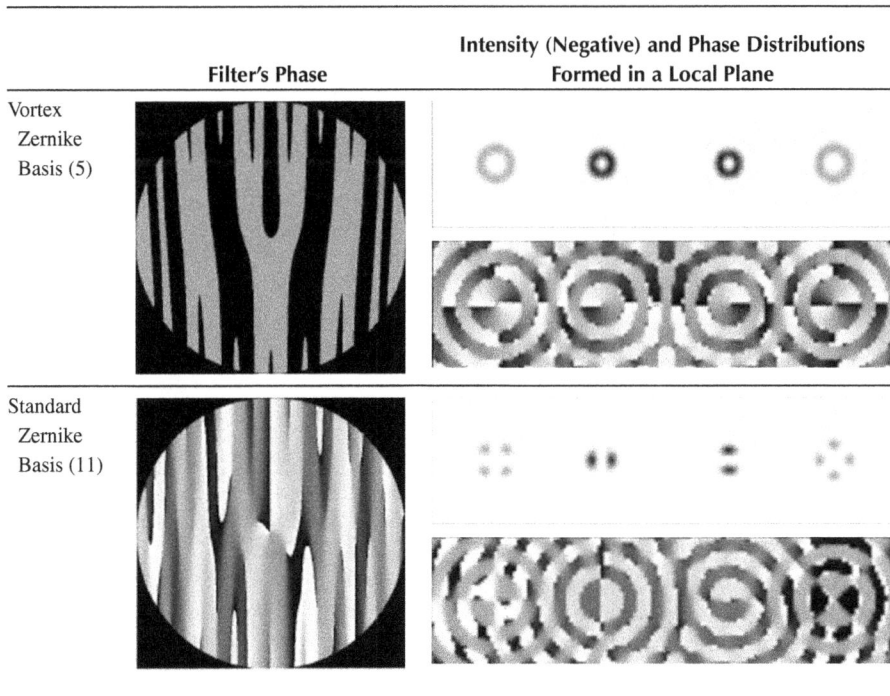

FIGURE 7.4 Type and action of 4-channel filters matched with functions 2, 3, 5, 6 (see corresponding lines in Figure 7.3).

Analyzed Field (Amplitude and Phase)	Intensity Distributions (Negative) Formed in a Local Plane	
	Vortex Zernike Basis (5)	Standard Zernike Basis (5)
$E(r,\phi) = 1,\ r \leq R$		
$E(r,\phi) = r\cos(\phi + \phi_0),\ \phi_0 = 0$		
$E(r,\phi) = r\cos(\phi + \phi_0),\ \phi_0 = -\pi/3$		
$E(r,\phi) = r\cos(\phi + \phi_0),\ \phi_0 = -\pi/2$		
$E(r,\phi) = r^2\cos(r + 2\phi + \phi_0),\ \phi_0 = -\pi/4$		

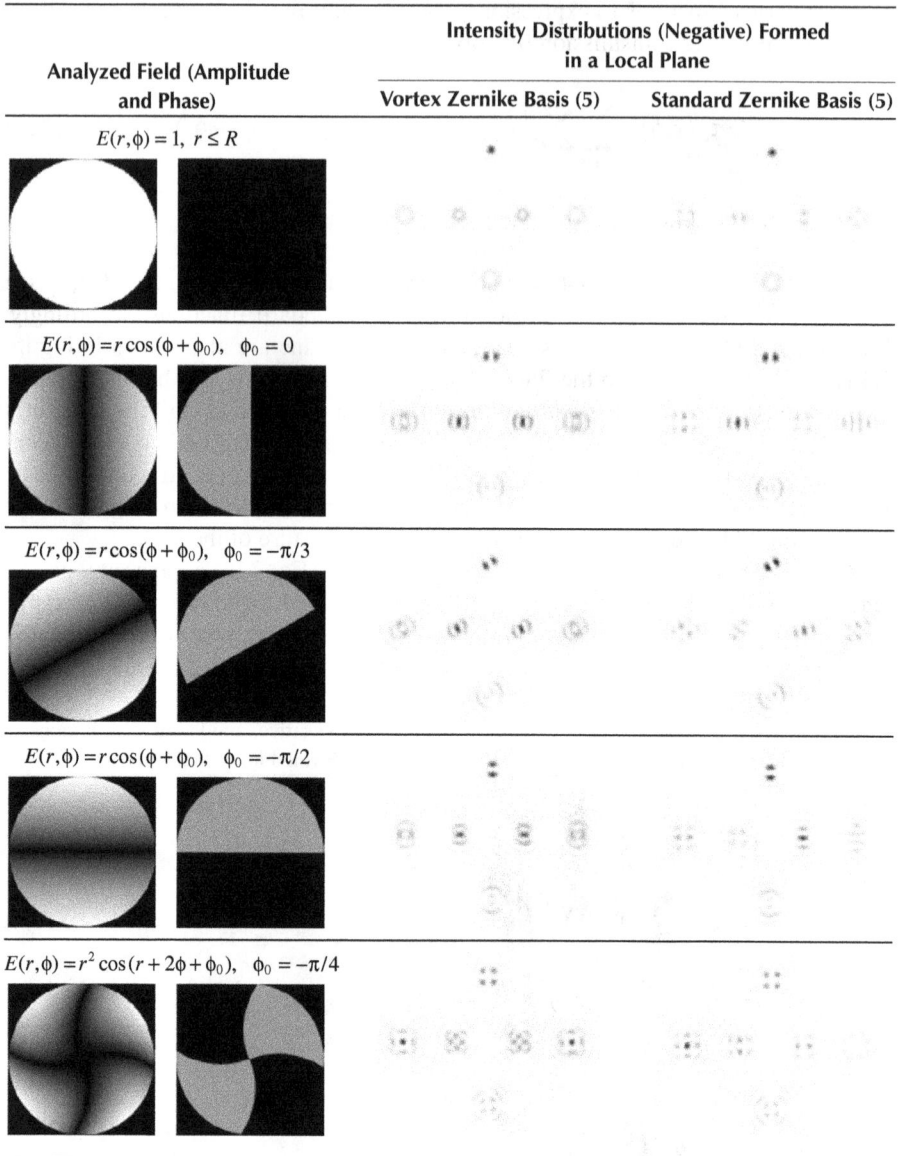

FIGURE 7.5 Results of optical decomposition of light fields by means of 6-channel filters.

in the images in Figure 7.5, which show the results of the expansion of the optical light field $E(r, \phi) = r\cos(\phi + \phi_0)$, where ϕ_0 is the angle of rotation of the field. We use 6-channel filters, consistent with the functions 1–6 of Figure 7.3 for two types of Zernike functions.

Diffraction orders corresponding to the functions (0, 0) and (2, 0) are vertically arranged: (0, 0) is at the top and (2, 0) is at the bottom. Other functions are arranged as shown in Figure 7.4. Note that we can observe the Fourier transform of the analyzed

TABLE 7.3
Correspondence of the Considered Zernike Functions (9) to a Single Index l

l	0	1	2	3	4	5	6	7	8
(n, m)	(0,0)	(1,1)	(1,–1)	(2,2)	(2,0)	(2,–2)	(3,3)	(3,1)	(3,–1)
l	9	10	11	12	13	14	15	16	17
(n, m)	(3,–3)	(4,4)	(4,2)	(4,0)	(4,–2)	(4,–4)	(5,5)	(5,3)	(5,1)
l	18	19	20	21	22	23	24	25	26
(n, m)	(5,–1)	(5,–3)	(5,–5)	(6,6)	(6,4)	(6,2)	(6,0)	(6,–2)	(6,–4)
l	27	28	29	30	31	32	33	34	35
(n, m)	(6,–6)	(7,7)	(7,5)	(7,3)	(7,1)	(7,–1)	(7,–3)	(7,–5)	(7,–7)
l	36	37	38	39	40	41	42	43	44
(n, m)	(8,8)	(8,6)	(8,4)	(8,2)	(8,0)	(8,–2)	(8,–4)	(8,–6)	(8,–8)

field in the diffraction order of the function (0, 0). The presence of non-zero intensity in the center of a diffraction order indicates the presence of the corresponding basis function in the analyzed beam with a factor that is proportional to the intensity. The first row of Figure 7.5 shows results of the expansion of the uniform field $E(r, \phi) = 1$, $r \le R$. In this case, only the single function from six Zernike functions written in these filters will have a non-zero coefficient, namely, the center of the top diffraction order (0,0) has a correlation peak. The second row of Figure 7.5 shows results of expansion of the cosine $E(r, \phi) = r\cos(\phi + \phi_0)$, $\phi_0 = 0$. In this case, the vortex basis gives correlation peaks in the left (1,1) and the right (1,–1) diffraction orders, while standard basis gives the response only in the cosine order. In the third and fourth rows of Figure 7.5, results of decomposition of $E(r, \phi) = r\cos(\phi + \phi_0)$ with $\phi_0 = \pi/6$ and $\phi_0 = \pi/2$ are shown, correspondingly. Obviously, the vortex basis shows invariance to rotation, in difference to the standard basis. However, the advantage of the standard basis is the ability to unequivocally respond to standard aberration distributions as it is apparently seen, from rows 2, 4, and 5 of Figure 7.5. Therefore we are planning to use these two types of Zernike functions together. Besides, the additional information can be used for measurements of not only intensity of expansion coefficients but also their relative phases [29, 52, 53].

Basis (15) implies positive and negative values of the index m. The correspondence of a single index l to the pair indices (n, m) is shown in Table 7.3.

7.2.3 ANALYSIS OF THE INFLUENCE OF THE LEVEL OF ABERRATION ON THE DETECTION ACCURACY

Next we will consider some examples. For the convenience of further analysis, we write out explicit expressions for several Zernike polynomials in Table 7.4.

1. *Defocusing* $Z_2^0(r,\phi)$.

$$g(r,\phi) = \exp\left[i2\pi\alpha Z_2^0(r,\phi)\right] = \exp\left[i2\pi\alpha A_2\left(2r^2 - 1\right)\right]. \quad (7.20)$$

TABLE 7.4

Explicit Expressions for Some Zernike Polynomials

(n, m)	$R_n^m(r)$	(n, m)	$R_n^m(r)$
(0,0)	1	(4,0)	$6r^4 - 6r^2 + 1$
(1,1)	r	(4,2)	$4r^4 - 3r^2$
(2,0)	$2r^2 - 1$	(4,4)	r^4
(2,2)	r^2	(5,1)	$10r^5 - 12r^3 + 3r$
(3,1)	$3r^3 - 2r$	(5,3)	$5r^5 - 4r^3$
(3,3)	r^3	(5,5)	r^5

Taking into account expansion (7.13), field (7.17) can be represented as:

$$\exp\left[i2\pi\alpha Z_2^0(r,\phi)\right] =$$
$$= 1 + i\pi\alpha Z_{2,0}(r,\phi) - (\pi\alpha A_2)^2\left(4r^4 - 4r^2 + 1\right) + ... = \quad (7.21)$$
$$= D_0 + i\alpha D_1 Z_{2,0}(r,\phi) - \alpha^2 D_2 Z_{4,0}(r,\phi) + ... ,$$

where D_j are the reduced constants.

It can be seen from expression (Eq. 7.21) that for large values of the parameter α, when field (7.20) is expanded in the Zernike basis, in addition to coefficients with indices (0,0) ($l = 0$) and (2,0) ($l = 4$) there will appear a coefficient with the index (4,0) ($l = 12$), as well as higher-order coefficients (see Figure 7.6).

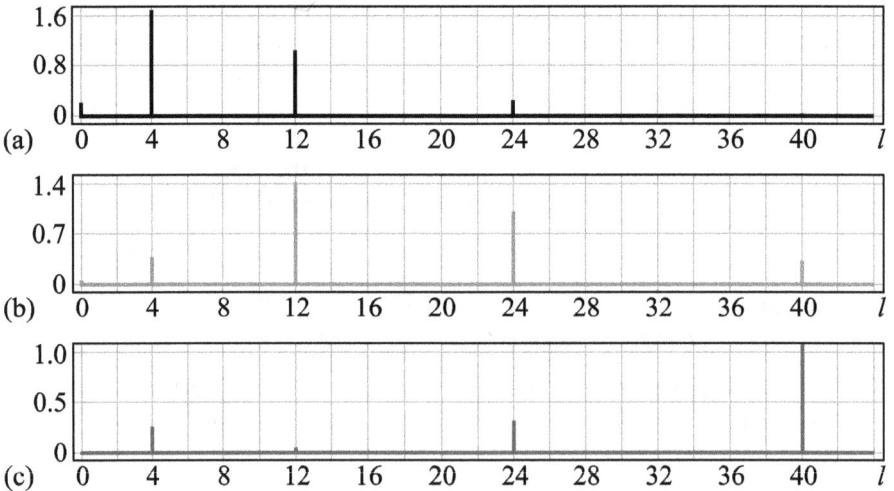

FIGURE 7.6 Wave front expansion coefficients with defocusing $Z_2^0(r,\phi)$ of various levels: (a) $\alpha = 0.4$, (b) $\alpha = 0.6$, and (c) $\alpha = 1$.

Figure 7.6 shows the results of calculating the expansion coefficients of field (7.20) in basis (7.16) at different values of the parameter α. As can be seen, at small values of α, the coefficient (2,0) ($l = 4$) is the largest [Figure 7.6(a)]. With increasing α, the weight of the coefficient (4,0) ($l = 12$), as well as that of the coefficient (6,0) ($l = 24$), increases [Figure 7.6(b)]. At a high level of defocusing, the weight of the coefficient (8,0) ($l = 40$) is significantly enhanced [Figure 7.6(c)]. Thus, the appearance of energy in high-order aberrations corresponds to a large level of available low-order aberration, and this effect can be detected by optical expansion of the analyzed wave front in the Zernike basis.

2. *Astigmatism* $Z_2^2(r,\phi)$.

$$g(r,\phi) = \exp\left[i2\pi\alpha Z_2^2(r,\phi)\right] = \exp\left[i2\pi\alpha A_2 r^2 \cos 2\phi\right], \qquad (7.22)$$

Taking into account expansion (7.13), field (7.19) can be represented as:

$$\exp\left[i2\pi\alpha Z_2^2(r,\phi)\right] =$$
$$1 + i\pi\alpha Z_{2,\pm2}(r,\phi) - 2\left(\pi\alpha A_2\right)^2 r^4 \left(1 + \cos 4\phi\right) + \dots =$$
$$= D_0 + i\alpha D_1 Z_{2,\pm2}(r,\phi) -$$
$$- \alpha^2 \left[D_2 Z_{2,0}(r,\phi) + D_3 Z_{4,0}(r,\phi) + D_4 Z_{4,\pm4}(r,\phi)\right] + \dots . \qquad (7.23)$$

The presence of $r^4 \cos 4\phi$ in (Eq. 7.23) leads to the appearance of $Z_{4,\pm4}(r, \phi)$, with r^4 resulting in the defocusing of various orders, in particular, $r^4 = \left(R_4^0(r) + 3R_2^0(r) + 2\right)/6$.

Thus, expression (Eq. 7.23) shows that for large values of the parameter α, in addition to coefficients with indices (0,0) ($l = 0$) and (2,±2) ($l = 3,5$), there will appear coefficients with indices (2,0) ($l = 4$), (4,0) ($l = 12$), and (4,±4) ($l = 10,14$), as well as higher-order coefficients (see Figure 7.7).

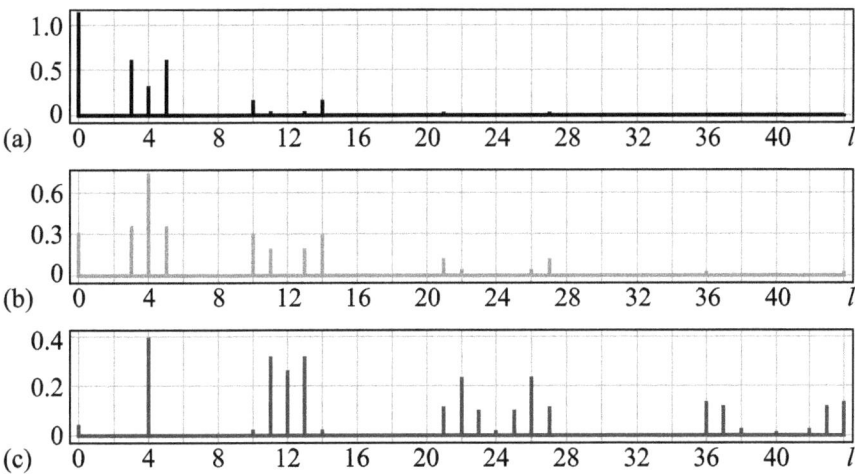

FIGURE 7.7 Wave front expansion coefficients with astigmatism $Z_2^2(r,\phi)$ of various levels: (a) $\alpha = 0.4$, (b) $\alpha = 0.6$, and (c) $\alpha = 1$.

As can be seen from the simulation, for small values of α, the coefficients (0,0) ($l = 0$) and (2,±2) ($l = 3,5$) predicted in (19) are the largest [Figure 7.7(a)]. With increasing α, the coefficients (2,0) ($l = 4$) and (4,±4) ($l = 10,14$) become more significant [Figure 7.7(b)]. It should be noted that with a high level of astigmatism, the field energy is distributed over a large number of coefficients, with defocusing (2,0) ($l = 4$) being the most noticeable [Figure 7.7(c)].

3. *Coma* $Z_3^1(r,\phi)$.

$$g(r,\phi) = \exp\left[i2\pi\alpha Z_3^1(r,\phi)\right] = \exp\left[i2\pi\alpha A_3\left(3r^3 - 2r\right)\cos\phi\right]. \qquad (7.24)$$

Field (7.21) can be represented as:

$$\exp\left[i2\pi\alpha Z_3^1(r,\phi)\right] =$$
$$= 1 + i\pi\alpha Z_{3,\pm1}(r,\phi) - \left(\pi\alpha A_3\right)^2\left(9r^6 - 6r^4 + 4r^2\right)\left(1 + \cos 2\phi\right) + ... = \qquad (7.25)$$
$$= D_0 + i\alpha D_1 Z_{3,\pm1}(r,\phi) -$$
$$- \alpha^2\left[D_2 Z_{6,0}(r,\phi) + D_3 Z_{6,\pm2}(r,\phi) + D_4 Z_{2,\pm2}(r,\phi) + D_5 Z_{4,\pm2}(r,\phi)\right] + ... \ .$$

Expression (Eq. 7.25) is quite complex, but one can clearly see that aberrations with even $n = 2,4,6$ and $m = \pm2$ appear additionally. Obviously, if we also take the cubic term into account, then additional odd aberrations with $m = \pm3$ should appear.

As can be seen from the simulation, for small values of α, the coefficients (0,0) ($l = 0$) and (3,±1) ($l = 7,8$) predicted in (24) are the largest [Figure 7.8(a)]. With increasing α, the coefficients (6,0) ($l = 24$), (6,±2) ($l = 23,25$), and (2,±2) ($l = 3,5$), also predicted in (24), become significant [Figure 7.8(b)]. At a high

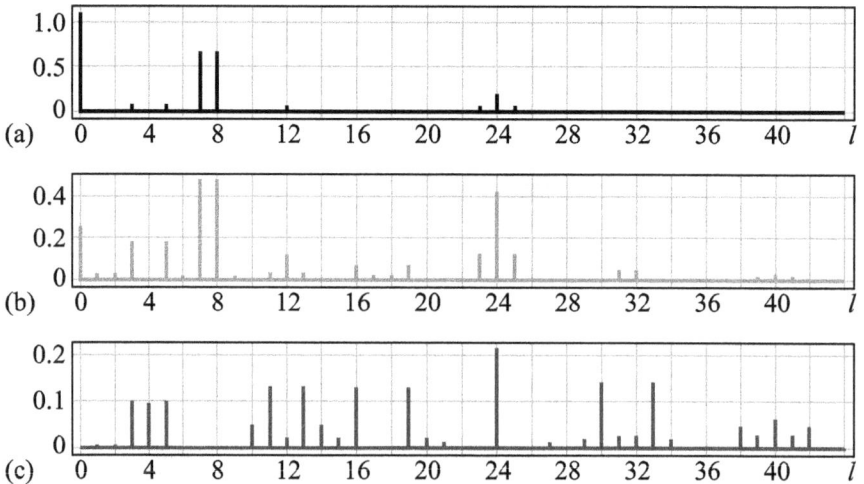

FIGURE 7.8 Wave front expansion coefficients with coma $Z_3^1(r,\phi)$ of various levels: (a) $\alpha = 0.4$, (b) $\alpha = 0.6$, and (c) $\alpha = 1$.

coma level, the field energy is distributed over a large number of coefficients [Figure 7.8(c)]. As expected, aberrations with higher angular multiplicity also appear in this case, namely, (5,±3) (l = 16,19) and (7,±3) (l = 30,33). However, the coefficient (6,0) (l = 24), corresponding to high-order defocusing, becomes the most noticeable [Figure 7.8(c)].

4. *Coma (trefoil)* $Z_3^3(r,\phi)$.

$$g(r,\phi) = \exp\left[i2\pi\alpha Z_3^3(r,\phi)\right] = \exp\left[i2\pi\alpha A_3 r^3 \cos 3\phi\right]. \qquad (7.26)$$

Field (7.23) can be represented as:

$$\exp\left[i2\pi\alpha Z_3^3(r,\phi)\right] =$$
$$= 1 + i\pi\alpha Z_{3,\pm3}(r,\phi) - (\pi\alpha A_3)^2 r^6 (1 + \cos 6\phi) + \ldots = \qquad (7.27)$$
$$= D_0 + i\alpha D_1 Z_{3,\pm3}(r,\phi) -$$
$$- \alpha^2 \left[D_2 Z_{2,0}(r,\phi) + D_3 Z_{4,0}(r,\phi) + D_4 Z_{6,0}(r,\phi) + D_5 Z_{6,\pm6}(r,\phi)\right] + \ldots .$$

The presence of $r^6 \cos 6\phi$ in Eq. (7.27) leads to the appearance of $Z_{6,\pm6}(r,\phi)$, i.e. a multiple increase in the angular dependence. The dependence r^6 can be described by superposition $R_6^0(r)$, $R_4^0(r)$, $R_2^0(r)$ (i.e. defocusing of various orders). In addition, aberrations with the same angular dependence $m = 3$, but with a higher degree of radial polynomials ($n > 3$), may appear. The simulation confirms the theoretical analysis: for small values of α, the coefficients (0,0) ($l = 0$) and (3,±3) ($l = 6,9$) are the largest [Figure 7.9(a)]. With increasing α, the coefficients (5,±3) ($l = 16,19$) and (6,±6) ($l = 21,27$) increase, and the weight of the coefficients associated with defocusing (2,0) ($l = 4$) also

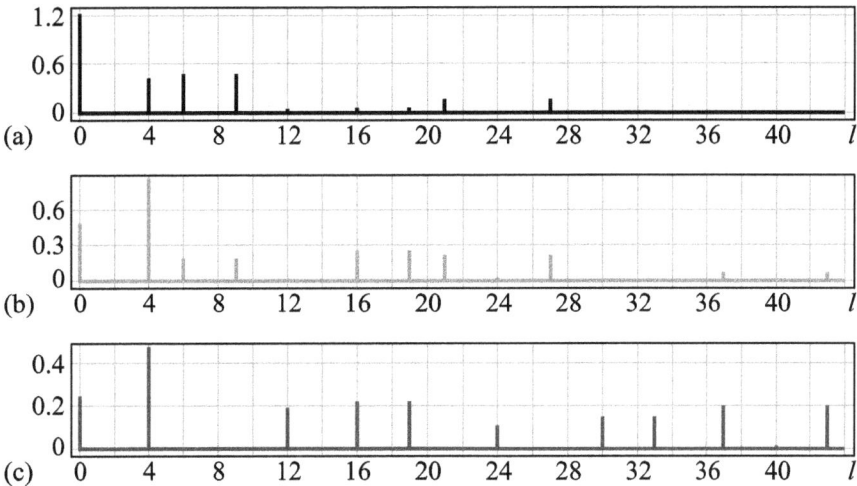

FIGURE 7.9   Wave front expansion coefficients with coma (trefoil) $Z_3^3(r,\phi)$ of various levels: (a) $\alpha = 0.4$, (b) $\alpha = 0.6$, and (c) $\alpha = 1$.

grows [Figure 7.9(b)]. At a high level of aberrations, there also appear aberrations with a higher multiplicity [Figure 7.9(c)].

5. *Defocusing (quatrefoil)* $Z_4^4(r,\phi)$.

$$g(r,\phi) = \exp\left[i2\pi\alpha Z_4^4(r,\phi)\right] = \exp\left[i2\pi\alpha A_4 r^4 \cos 4\phi\right]. \tag{7.28}$$

Field (7.23) can be represented as:

$$\exp\left[i2\pi\alpha Z_4^4(r,\phi)\right] =$$
$$= 1 + i\pi\alpha Z_{4,\pm4}(r,\phi) - (\pi\alpha A_4)^2 r^8 (1 + \cos 8\phi) + ... =$$
$$= D_0 + i\alpha D_1 Z_{4,\pm4}(r,\phi) - \alpha^2 \left[D_2 Z_{2,0}(r,\phi) + D_3 Z_{4,0}(r,\phi) + \right.$$
$$\left. + D_4 Z_{6,0}(r,\phi) + D_5 Z_{8,0}(r,\phi) + D_6 Z_{8,\pm8}(r,\phi)\right] + ... \tag{7.29}$$

Expression (Eq. 7.29) is obtained similarly to the previous example.

The simulation results are shown in Figure 7.10: for small values of α, the coefficients (0,0) ($l = 0$), (2,0) ($l = 4$), and (4,±4) ($l = 10,14$) are the largest [Figure 7.10(a)]. With increasing α, the coefficients (6,±4) ($l = 22,26$) and (8,±8) ($l = 36,44$) increase, and the weight of the coefficients (2,0) ($l = 4$) associated with defocusing also grows [Figure 7.10(b)]. At a large aberration level, aberrations with the same angular dependence but a higher degree of radial polynomials (6,±4) ($l = 22,26$) and (8,±4) ($l = 38,42$) become more significant and the effect of defocusing (2,0) ($l = 4$), (4,0) ($l = 12$), (6,0) ($l = 24$) is enhanced [Figure 7.10(c)].

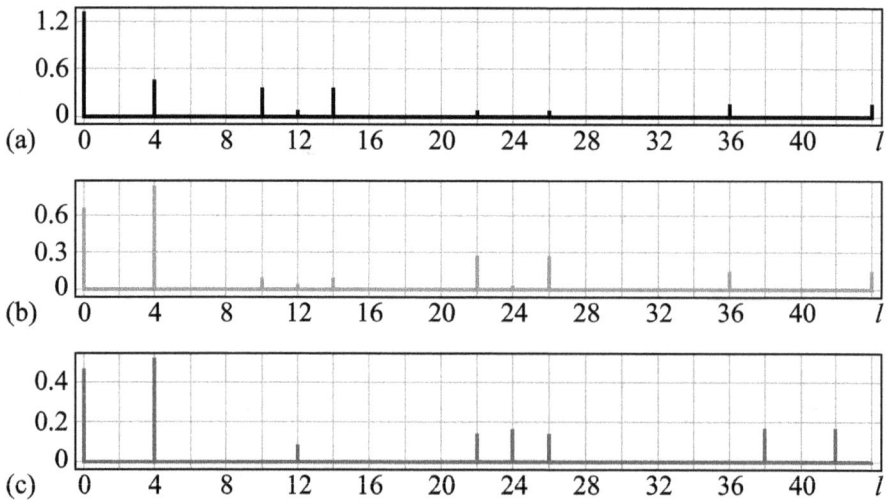

FIGURE 7.10 Wave front expansion coefficients with defocusing (quatrefoil) $Z_4^4(r,\phi)$ of various levels: (a) $\alpha = 0.4$, (b) $\alpha = 0.6$, and (c) $\alpha = 1$.

The preceding examples allow us to identify the main trends associated with an increase in the level of aberrations. If the initial aberration has an angular dependence on the order of m, then aberrations appear with a multiple angular dependence of $2m$, $3m$. In addition, as a rule, there appears defocusing of various orders. The detection of defocusing is associated with an increase in the PScF area, which is always observed with an increase in the aberration level (see Table 7.2).

6. *Superposition* $Z_3^1(r,\phi) + Z_4^2(r,\phi)$.

$$g(r,\phi) = \exp\left\{i2\pi\alpha\left[A_3\left(3r^3 - 2r\right)\cos\phi + A_4\left(4r^4 - 3r^2\right)\cos 2\phi\right]\right\}. \quad (7.30)$$

A theoretical analysis of expression (Eq. 7.30) is rather difficult; therefore, we consider only the results of numerical simulation shown in Figure 7.11.

For small values of α, the expected coefficients (0,0) ($l = 0$), (3,±1) ($l = 7,8$), and (4,±2) ($l = 11,13$), as well as additional defocusing (6,0) ($l = 24$), are the largest [Figure 7.11(a)]. With an increase in α, the weight of the coefficients (2,0) ($l = 4$) and (6,0) ($l = 24$) corresponding to defocusing increases [Figure 7.11(b)], which is explained by an increase in the PScF area.

Further enhancement of aberration leads to an almost uniform distribution of field energy over all coefficients, which should serve as a signal of a high level of wave front distortion in measurements. In this case, other methods, including neural networks, need to be used to recognize and compensate for aberrations [54–56].

Listing of software codes written in the C++ programming language for the Visual Studio development environment for calculating the sensor matched with the Zernike basis functions is provided in Appendix A.

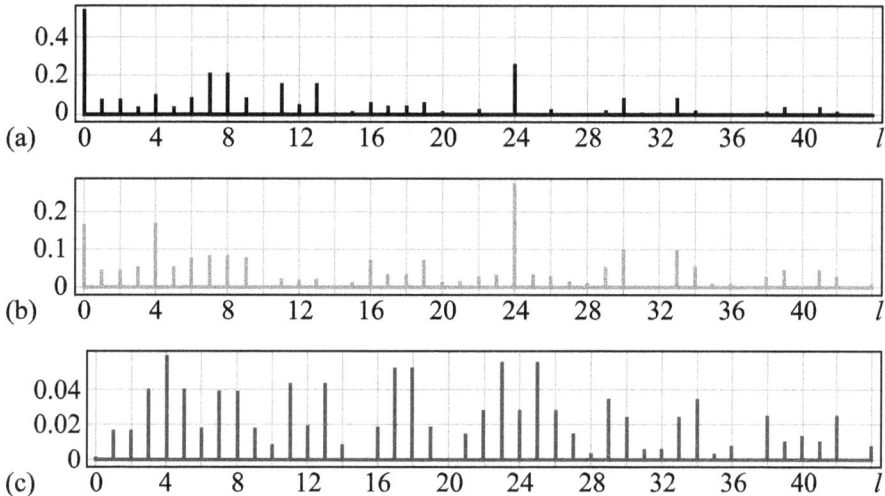

FIGURE 7.11 Wave front expansion coefficients with aberration superposition $Z_3^1(r,\phi) + Z_4^2(r,\phi)$ of various levels: (a) $\alpha = 0.4$, (b) $\alpha = 0.6$, and (c) $\alpha = 1$.

7.2.4 Experimental Results

7.2.4.1 Experimental Investigation of Multi-Order Diffractive Optical Elements Matched with Two Types of Zernike Functions

Figure 7.12 shows the optical setup used in our experiments. A system composed of a lens L_1 ($f_1 = 350$ mm) and a pinhole PH (40 μm aperture) was utilized for expansion and filtration of input laser beam. Then the laser beam incident on the diffractive optical element DOE_1, generating a Hermite–Gaussian mode TEM_{10}. The generated beam illuminated the diffractive optical element DOE_2. In the first series of experiments, we used 20-th order diffractive optical element generating 10 pairs of Zernike functions as the DOE_2; in the second series of experiments we used 25-th order diffractive optical element generating nine pairs of Zernike functions and their combinations. Figure 7.13 shows the phase transmission functions of diffractive optical elements used in experiments. Transmission function of DOE_1 matched with the basis of first type (see Eq. 7.16)). Transmission function of DOE_2 matched with the basis of second type (see Eq. (7.5)). The video camera Cam LOMO TC-1000 (pixel size is 1.67×1.67 μm) was used for shooting the intensity distributions generated in the focal plane of the lens L_2 ($f_2 = 150$ mm).

In experiments, an element DOE_2 was stationary, while an element DOE_1 rotated about its center located on the axis of the beam propagation. Thus, we investigated the influence of the orientation of the illuminating beam relative to the Zernike filter. In the first series of experiments, the 25-th order diffractive optical element with the transmission function shown in Figure 7.13(b) was used as DOE_2. The intensity distribution obtained at different rotation angles DOE_1 relative to this element is shown in Figure 7.14. From these images, you can see that at rotation angles equal to 0 and 180 degrees, peaks of intensity appear in the central parts of some of the diffraction orders. This indicates the presence of the components corresponding Zernike functions encoded in these diffraction orders in the illuminating beam. At angles of 90 degrees and 270, peaks of intensity are not observed in the respective orders. This indicates that the 25-th order Zernike filter is sensitive to rotation. The intensity distribution generated in the zero order allows you to see the rotation of the illuminating beam.

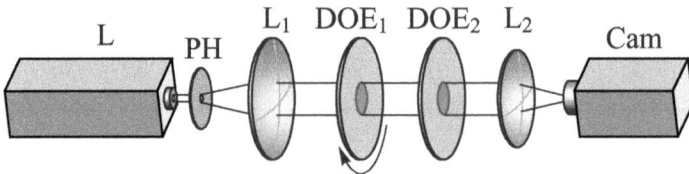

FIGURE 7.12 Experimental optical setup: L is a linearly polarized laser ($\lambda = 633$ nm), L_1 and L_2 are lenses ($f_1 = 150$ mm, $f_2 = 150$ mm), PH is a pinhole (40 μm), DOE_1 is a diffractive optical element to generate the Hermite–Gaussian mode TEM_{10}, DOE_2 is diffractive optical element matched with the basis of Zernike polynomials, Cam is a video camera LOMO TC-1000 (the pixel size is 1.67×1.67 μm).

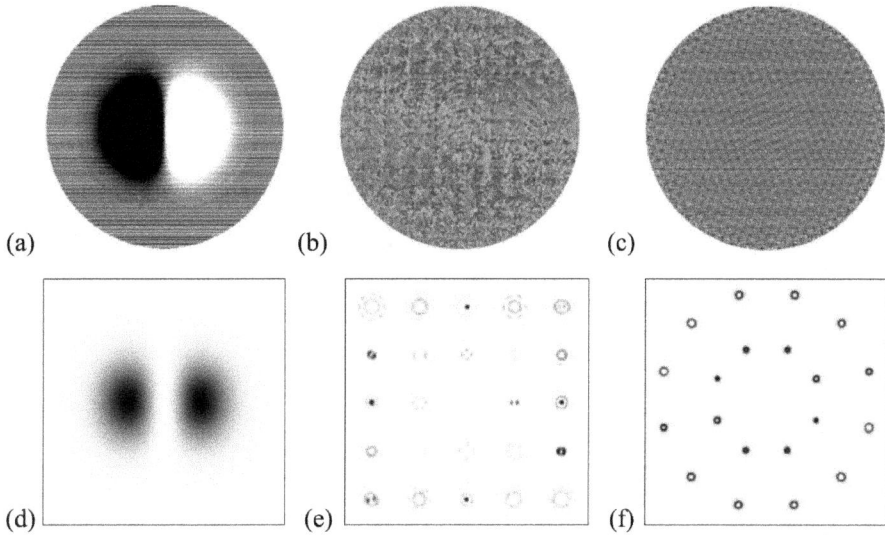

FIGURE 7.13 The phase transmission functions of the diffractive optical elements utilized in experiments (top row), and generated intensity distributions in far-field (negative): (a, d) DOE-generated Hermite–Gaussian mode TEM_{10}; (b, e) 25-th order phase Zernike filter; (c, f) 20-th order phase Zernike filter.

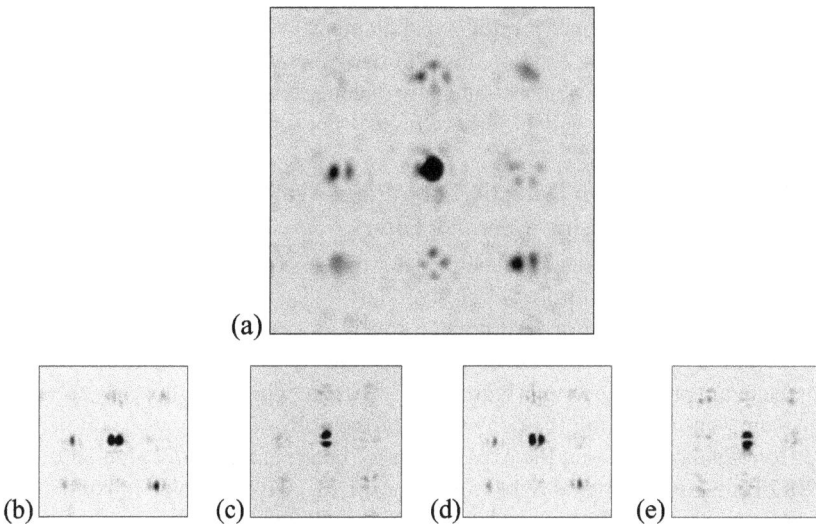

FIGURE 7.14 Intensity distributions (negative) experimentally generated by 25-th order phase Zernike filter illuminated by plane wave (a) and by Hermite–Gaussian mode TEM_{10} at different angles of rotation of the Hermite–Gaussian mode TEM_{10} relative to the propagation axis of the beam: (b) 0 degrees, (c) 90 degrees, (d) 180 degrees, (e) 270 degrees. Only the central part of the generated diffraction pattern is shown.

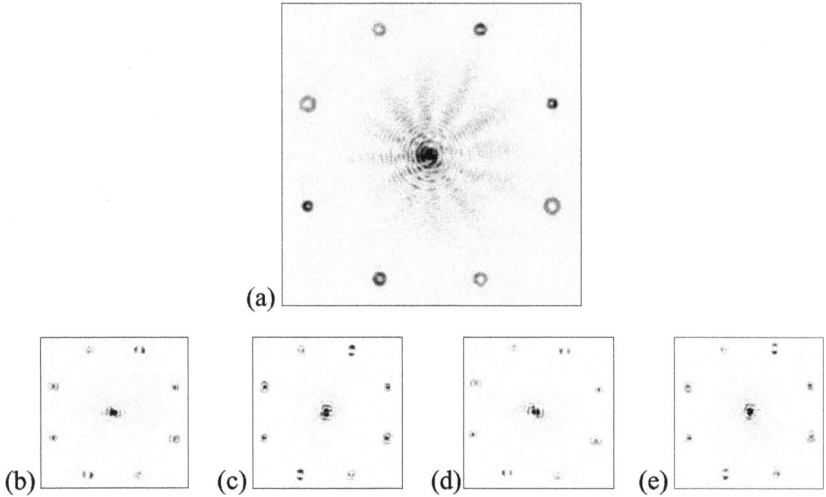

FIGURE 7.15 Intensity distributions (negative) experimentally generated by 20-th order phase Zernike filter illuminated by plane wave (a) and by Hermite–Gaussian mode TEM_{10} at different angles of rotation of the Hermite–Gaussian mode TEM_{10} relative to the propagation axis of the beam: (b) 0 degrees, (c) 90 degrees, (d) 180 degrees, (e) 270 degrees. Only the central part of the generated diffraction pattern is shown.

Figure 7.15 shows the intensity distribution generated in the second series of experiments, when we used 20-th order diffractive optical element with transmission function, as shown in Figure 7.13(c), as an element DOE_2. It is clearly seen in this case, that regardless of the rotation angle of the generated Hermite–Gaussian mode TEM_{10}, distribution of the peaks of intensity in generated diffraction pattern does not change. This indicates that such Zernike filter is not sensitive to rotation.

7.2.4.2 Experiments on Detection of Various Wave Front Aberrations Using a Zernike Filter

Figure 7.16(a) shows the optical scheme used in the experiment. The output from a solid-state laser ($\lambda = 532$ nm) was collimated using a system consisting of a pinhole (PH) with a hole diameter of 40 µm and a spherical lens (L_1) ($f_1 = 250$ mm). In this case, the lens (L_1) was mounted on a linear translation stage and could be moved along the beam propagation axis with a step of 10 µm, which was done for subsequent experiments on measuring the defocusing of the collimated initial beam using our wave front analyzers. Then, the expanded laser beam passed through a HOLOEYE LC 2012 transmissive spatial light modulator (SLM_1) with a 1024×768 pixel resolution and a pixel size of 36 µm, which was used to form a wave front with a required set of aberrations. Lenses (L_2) ($f_2 = 150$ mm), L_3 ($f_3 = 150$ mm), and a diaphragm (D) were used together to ensure spatial filtering of the aberration-distorted beam formed by the first modulator. A HOLOEYE PLUTO VIS reflective spatial light modulator (SLM_2) with a $1,920 \times 1,080$ pixel resolution and a pixel size of 8 µm was used to implement a phase mask of a multi-order analyzing diffractive optical

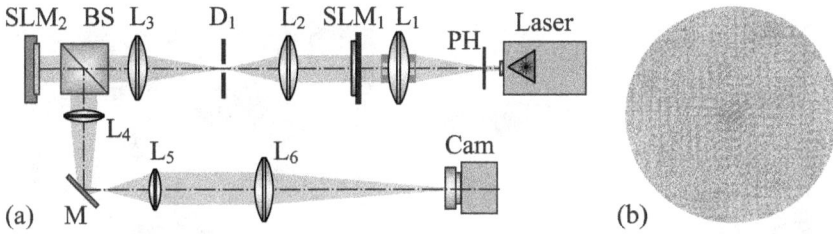

FIGURE 7.16 Detection of wave front aberrations using a multichannel analyzing DOE. (a) Schematic of the experimental setup: laser is a solid-state laser (λ = 532 nm); PH is a pinhole (hole size of 40 μm); L_1, L_2, L_3, L_4, L_5, and L_6 are spherical lenses (f_1 = 250 mm, f_2 = 150 mm, f_3 = 150 mm, f_4 = 150 mm, f_5 = 150 mm, and f_6 = 350 mm); SLM_1 is a transmissive spatial light modulator (HOLOEYE LC 2012); SLM_2 is a reflective spatial light modulator (HOLOEYE PLUTO VIS); D_1 is an aperture; BS is a beam splitter; M is a mirror; Cam is a ToupCam UCMOS08000KPB video camera. (b) Phase mask of a 25-order analyzing DOE, which displays the incident light field in terms of the Zernike polynomials.

element (DOE), which served to decompose the studied light field in terms of the Zernike polynomial basis. The laser beam reflected from the reflective light modulator using a beam splitter (BS), a 4-f optical system of lenses L_4 and L_5 (f_4 = 150 mm, f_5 = 150 mm) and a mirror (M) was directed to lens L_6 (f_6 = 350 mm), which focused it on the matrix of a ToupCam UCMOS08000KPB camera with a 3,264 × 2,448 pixel resolution and a pixel size of 1.67 μm. Part of the scheme, including light modulator SLM_2, lens L_6, and video camera is, in fact, a sensor.

Other elements of the optical system are designed to simulate the studied beam and match the light modulator operating in reflective regime with the rest of the system. The described optical system, in addition to checking the operability of the wave front sensor, is used to calibrate the sensor by forming the studied beams with different aberrations and different α values using a controlled light modulator SLM_1. In the industrial version of the sensor, light modulator SLM_2 will be replaced by a classical transmissive phase DOE made of a transparent material. This version of the sensor is easier to manufacture than the Shack–Hartmann sensor and does not require vibration isolation. In the process of measurements, the intensity patterns were stable both in coordinates and in measured intensity, although the optical system was not vibration-proof.

Phase mask of a 25-order analyzing DOE, which decomposes the incident light field in terms of the basis of Zernike functions (Eq. 7.16) with numbers (n, m) = {(0,0), (1,1), (2,0), (2,2), (3,1), (3,3), (4,0), (4,2), (4,4), (5,1), (5,3), (5,5), (6,0), (6,2), (6,4), (6,6), (7,1), (7,3), (7,5), (7,7), (8,0), (8,2), (8,4), (8,6), (8,8)} is shown in Figure 7.16(b). Although the complete basis (7.10) implies positive and negative values of the index m, taking into account a certain duplication of information in complex conjugate coefficients (with $\pm m$), we used basis (7.10) only with positive m index values to reduce the number of diffraction orders (channels).

Figure 7.17 shows the experimental and numerical (using the Fourier transform) intensity distributions formed in the focal plane of lens L_6 when modulator SLM_2 is illuminated by a laser beam with an aberration less plane wave front.

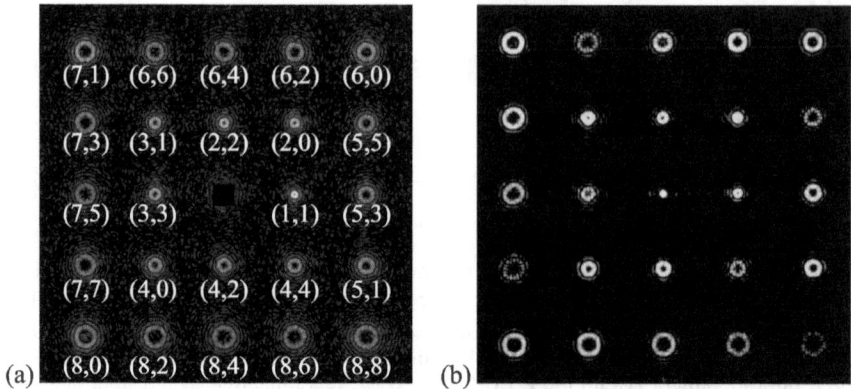

FIGURE 7.17 (a) Numerically and (b) experimentally obtained intensity distributions formed in the focal plane of lens L_6 when SLM_2 is illuminated by a laser beam with a plane wave front. In the images of numerically obtained distributions, the zero diffraction order is cut out.

The absence of any light peaks in the formed diffraction orders is clearly seen in Figure 7.17, which confirms the absence of aberrations in the illuminating beam.

Figures 7.18, 7.19, and 7.20 show the experimentally obtained intensity distributions for various aberrations at $\alpha = 0.4$, 0.6, and 1, respectively. One can see the emergence of correlation peaks in diffraction orders corresponding to additional theoretically predicted aberrations. In all cases, the presence of defocusing (2,0) is clearly visible.

It is convenient to evaluate the value of α in the wave front under study and, accordingly, the applicability of the sensor by the intensity distribution in the zero diffraction order corresponding to the PScF of the beam in question. Using the numerical simulation (see Table 7.2) we showed that for $\alpha \leq 0.4$ corresponding to the range of the sensor applicability, the PScF is close to the Airy pattern of a diffraction-limited system. With $\alpha > 0.4$, the size of the PScF begins to increase due to the appearance of additional petals and rings, and the intensity in the center decreases (see Figure 7.19 and especially Figure 7.20). This serves as a criterion for exceeding the level of aberrations acceptable for the sensor.

Thus, we have experimentally confirmed that at $\alpha \leq 0.4$, the aberration structure can be confidently detected (recognized); however, with a further increase in α, recognition becomes problematic.

7.2.4.3 Experiments on the Collimator Fine-Tuning

One of the most important applications of the developed technique – analysis of wave front aberrations – is the testing and accurate adjustment of various optical components, for example, collimators.

Figure 7.21 shows the intensity distributions obtained in the focal plane of lens L_6 at various displacements of lens L_1 forming the collimator along the beam propagation axis. One can see that when the lens is displaced from the initial plane $z = 0$, which corresponds to an ideal position of the lens (i.e. the position at which a laser

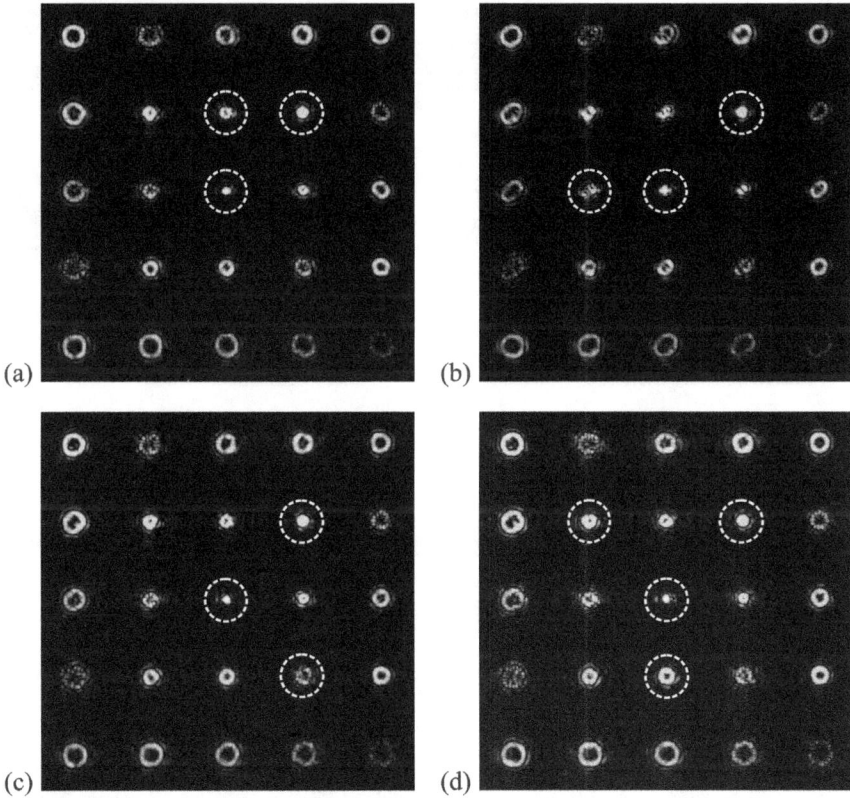

FIGURE 7.18 Experimentally obtained intensity distributions at $\alpha = 0.4$ for (a) $Z_2^2(r,\phi)$ {(0,0), (2,0), (2,2)}; (b) $Z_3^3(r,\phi)$ {(0,0), (2,0), (3,3)}; (c) $Z_4^4(r,\phi)$ {(0,0), (2,0), (4,4)}; and (d) $Z_3^1(r,\phi) + Z_4^2(r,\phi)$ {(0,0), (2,0), (3,1), (4,2)}.

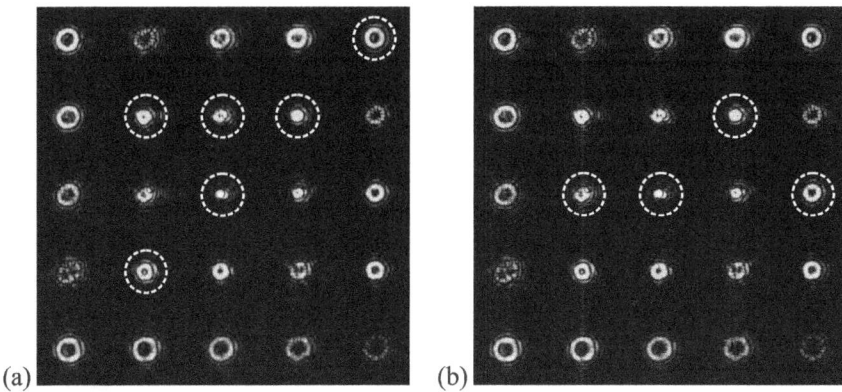

FIGURE 7.19 Experimentally obtained intensity distributions at $\alpha = 0.6$ for (a) $Z_3^1(r,\phi)$ {(0,0), (2,0), (2,2), (3,1), (4,0), (6,0)} and (b) $Z_3^3(r,\phi)$ {(0,0), (2,0), (3,3), (5,3)}.

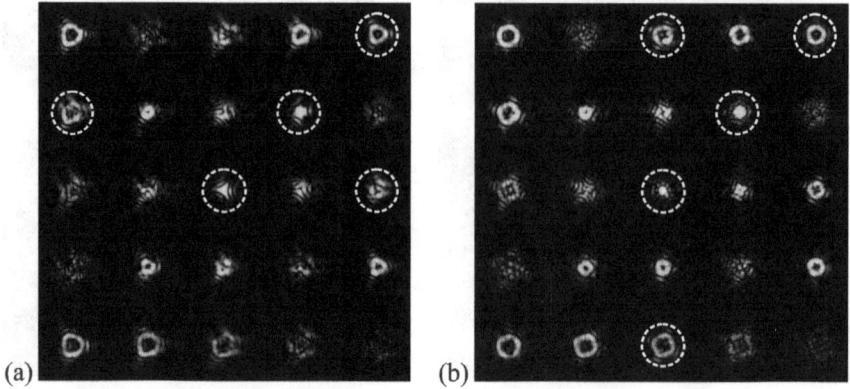

FIGURE 7.20 Experimentally obtained intensity distributions at $\alpha = 1$ for (a) $Z_3^3(r,\phi)$ (0,0), (2,0), (5,3), (6,0), (7,3) and (b) $Z_4^4(r,\phi)$ (0,0), (2,0), (6,0), (6,4), (8,4).

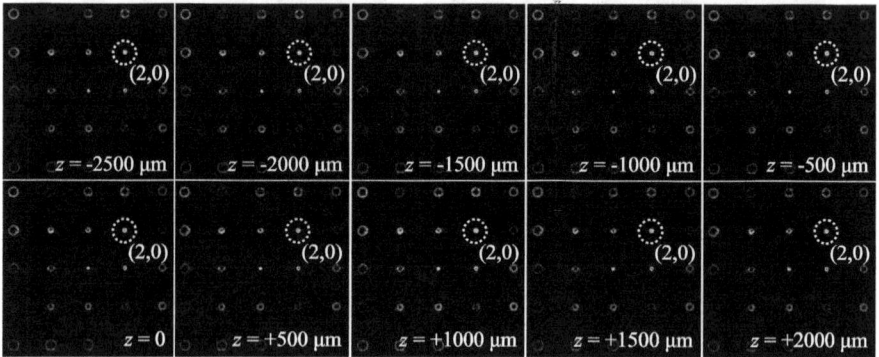

FIGURE 7.21 Intensity distributions obtained in the focal plane of lens L_6 at different displacements of lens L_1 forming the collimator. The dotted circle highlights the diffraction order corresponding to defocusing (Zernike polynomial Z_2^0).

beam with a plane wave front is formed behind the lens) at the center of the diffraction order responsible for the Zernike polynomial with numbers (n, m) equal to (2,0) (i.e. defocusing aberrations), an intensity peak appears.

The intensity of this peak increases equally with distance from the original plane $z = 0$ in both directions. Calculations show that for the used lens with a focal length of 250 mm and the illuminating beam with a diameter of about 5.4 mm, the deviation of the wave front from the plane at the edge of the lens aperture at $z = 0.5$ mm is about $\lambda/20$. Thus, we can conclude that the sensitivity of the proposed sensor to local deviations of the wave front is no worse than $\lambda/20$. The average aberration over entire beam is much less. This sensitivity of the sensor allows one to detect defocusing of less than 0.5 mm for a laser radiation collimator with a relative aperture of about 1:50. Thus, in terms of sensitivity to aberrations, the proposed sensor is quite competitive with the Shack–Hartmann sensor, but the proposed sensor does not require vibration isolation. The main advantage of the proposed sensor is that, unlike the Shack–Hartmann sensor, where the aberration coefficients are calculated by processing mathematically a 2D data array, the values of the coefficients in multichannel DOE sensors are directly proportional to the light intensities measured in individual pixels of the photodetector matrix with fixed coordinates. The entire area of the tested beam is simultaneously involved in the formation of the values of each coefficient, which reduces errors.

7.3 APPLICATION OF WAVE FRONT SENSORS IN OPHTHALMOLOGY

7.3.1 WAVE FRONT SENSORS IN OPHTHALMIC RESEARCH

Over the last decade, aberrometers have been widely employed in many areas of ophthalmology and optometry [57, 58], e.g. for detecting refraction anomalies, diagnosing a dry-eye symptom [59] and keratoconus [60], as well as in refractive eye surgery [61, 62]. Wave front aberrations may be in the form of defocus, astigmatism, and higher-order aberrations. Higher-order aberrations (HOA) occur when the eye surface has minor roughness or imperfections that cannot be compensated for using standard spectacles [63]. As a rule, cornea topography is performed using dedicated algorithms that rely on height data. Intraocular aberrations can be identified by deducting eye aberrations from cornea aberrations. Principles of aberrometer operation may be based on a Shack–Hartmann method, a ray tracing technique, and a Churning principle.

Ophthalmology practices now employ a new technology of ocular pyramid aberrometer. In [64] it was demonstrated that using new pyramid aberrometers *Osiris*, repeatable and consistent measurement results of ocular aberrometry of healthy eyes could be attained.

A wave front sensor with a pyramid-shaped optical element and expanded source has been successfully used to measure human eye aberrations [65]. In this sensor, a significant benefit for the eye is that it can easily be adapted to variations in the aberration range, which may be expected of a human eye, varying from healthy eyes with slight aberration to cornea pathologies causing extreme aberration. The dynamic range of a sensor can be changed by simply changing the length of a light source projected onto the retina. These capabilities may prove useful in practical designs of devices that operate by measuring the human eye wave aberration. A shortcoming of the sensor is that it also collects the light from false reflections from the eye surface.

In addition, pyramid wave front sensors (PWFS) are widely employed in adaptive optics systems, including astronomy, ophthalmology [66], and microscopy. Wave front reconstruction algorithms for pyramid wave front sensors are in high demand. With few exceptions, the majority of currently existing algorithms rely on an assumption of a linear pyramid sensor model. Nonetheless, the use of a simplified nonlinear pyramid operator enables an acceptable quality of wave front reconstruction to be achieved.

Also known is a crossed sinusoidal wave front sensor that is based on a gradient transfer filter placed near the image plane of the system under testing [67]. The theoretical principle behind the sensor operation is comprehensively described in the framework of Fourier optics. Numerical simulation confirms that a root mean square (RMS) measurement accuracy of $\lambda/80$ can be attained, which is much more accurate compared to other types of wave front sensors (WFS). A crossed-sine WFS also offers a benefit of being quasi-chromatic and able to operate with small-length, natural, or artificial light sources.

The crossed sinusoidal WFSs belong to a class of WFSs in which the pupil of the system under testing is imaged on a sensor array, thus providing high spatial resolution. In this aspect, it is closer to a pyramid WFS or an optical differentiation sensor (ODS) [68].

Interesting studies were conducted using Shack-Hartmann aberrometers [69]. Measurements of higher-order aberrations (HOA) conducted using a tracing aberrometer *iTrace* and Shack–Hartmann aberrometer *Topcon KR-1 W* were evaluated for the accuracy and consistency provided. The results arrived at demonstrated excellent repeatability but less reliable reproducibility when measuring HOA. Thus, it may be concluded that due to essentially different results of HOA measurements, the two aberrometers should not be mutually replaced when applied clinically.

A hand-held wave front aberrometer with post-cycloplegic autorefraction (AR) and cycloplegic refraction (CRf) has shown good agreement of measurement results with post-cyclopegic AR and CRf in spherical equivalents but tended to produce slightly myopic results. Considering that the wave front aberrometer has also been shown to overestimate astigmatism, it is recommended that a device for refraction anomaly evaluation be additionally used, which may also be helpful when testing individuals with abnormal posture and bedridden patients [70].

Also popular in ophthalmology are Scheimpflug sensors [71]. Accurate cornea analysis is important for pre-surgery risk evaluation and early detection of ectasia after the refractive surgery [72]. The analyzer DRS (dual rotating Scheimpflug) (Galilei®; Ziemer Ophthalmology, Port, Switzerland) utilizes two rotating Scheimpflug cameras in combination with a Placido topographical system. The analyzer utilizes a Placido disc in order to obtain more accurate topographical data on the anterior cornea curvature as addendum to the data from Scheimpflug cameras. Moreover, the double system generates images on both sides, thus minimizing the decentering effect due to eye movements on the cornea pachymetry and posterior cornea curvature measurements [73]. On the whole, the DRS analyzer enables the anterior cornea curvature for both a healthy eye and an eye after refractive surgery to be measured with good repeatability and reproducibility [74].

To achieve a higher sensitivity and specificity in keratoconus (KC) detection, ophthalmologist practices utilize Scheimpflug-tomograph-type aberrometers, for instance, an Oculyzer (Alcon, USA). The device is intended for doing computer tomography of the cornea and analyzing the anterior segment of the eyeball. With the Oculyzer, the examination results can be presented in a visual form convenient for further use, for instance, making it possible to move in space a 3D cornea model together with all its parameters. Other examination results are visualized as color diagrams, maps, and 3D images, providing an ophthalmologist with a comprehensive picture, making it possible to make a maximally accurate diagnosis, draw up a plan and propose treatment techniques, and prognosticate the treatment outcome.

7.3.2   MEDICAL OPHTHALMOLOGICAL MEASUREMENT DEVICES

Among a multitude of ophthalmological measurement devices, we may single out several devices that are most widely used clinically and described in the specialist literature. The detectors presented here are discussed in high-impact journals and analyzed in research theses, with data from those detectors underlying a decision-making process on the diagnosis and the need for surgical intervention.

7.3.2.1   ALLEGRO Topolyzer VARIO

ALLEGRO Topolyzer VARIO is a medical measurement instrument for the human eye examination, a diagnostic medical device of IIb class according to the Medical Device Directive 93/42/EEC classification. It is used for measuring the cornea topography and was specifically designed for ophthalmologists. The measurement data can be exported and used for refractive surgery with laser systems WaveLight AG (with topographic guidance).

The illumination system, containing a custom reflector, produces concentric intensity rings on the backside of a transparent dome. The dome image is reflected from the eye under examination and a virtual image is captured with a precision lens and an objective of a digital CCD camera. In this way, all distortions in the curvature radii of the eye become accessible for the measurement data processing. An analogous image is then processed with a measuring device, becoming accessible for further processing as a transformed compressed digital image. After getting the relevant data from a diagram, the software uses this information to produce a topographic image of the cornea surface. Finally, the measurement results are displayed on the screen not only in colors but also as a diagram and a 3D image.

This device allows a wave front analysis based on expansion in the Zernike basis to be conducted using data on height measurements by calculating a coefficient that describes the contribution of each Zernike polynomial to the heights measured. The device operates by using index calculus data: data on the curvature and height, Fourier analysis, and an output Zernike analysis. Such a comprehensive use of the key cornea-related data enable an early diagnosis of a wide variety of eye and sight impairments.

7.3.2.2   WaveLight Oculyzer

WaveLight Oculyzer II is a diagnostic instrument intended for the identification and examination of the anterior eye segment, which is specially manufactured for clinical uses in ophthalmologist practices, as well as for optometrist and optician offices.

The device measures the topography of the anterior chamber of the eyeball (including the cornea, aqueous humor, iris, and the front surface of the intraocular lens). The optical power of the cornea is calculated using a Gullstrand model. Other functionalities of the device include the estimation of the cornea shape, the analysis of the intraocular lens condition (lenticular opacity), analysis of the anterior chamber angle, analysis of the anterior chamber depth, analysis of the anterior chamber volume, analysis of the anterior and posterior cortical opacity, analysis of cataract localization (kernel, subcapsular, or cortical), the use of cross images with densitometry, and cornea thickness measurements.

The device also performs cornea measurements and is custom designed for ophthalmologists. The data measured can be exported and utilized jointly with a laser system WAVE ALLEGRETTO (OcuLink) for refractive surgery. The device performs a Zernike analysis of the front and rear cornea surfaces based on height measurements, calculating for each Zernike polynomial a coefficient that describes its contribution to the heights measured.

Using Zernike polynomials, the user can also characterize the front and rear surface of the cornea independently. While mathematically calculating the Zernike coefficients, the polynomials used are being corrected in an optimal way with respect to the heights measured. The larger the number of the polynomials employed, the higher the accuracy of height calculations. If the abnormal Zernike coefficient is absent, the aberration coefficient is 0.0. Values exceeding 1.0 indicate that the cornea surface contains untypical wave components, which may result in an impaired vision. When surgically correcting keratoconus, it is only possible to reform the front cornea surface, with the rear surface remaining astigmatic and thus affecting the retinal image clarity. Moreover, a change in the cornea thickness leads to a change in the entire chamber of the eyeball due to the aqueous humor pressure.

All the factors just mentioned should be accounted for when doing refractive laser surgery. The apparatus WaveLight Oculyzer II makes it possible to represent the front and rear cornea surfaces as Zernike polynomials, also allowing the polynomial weights and deviations from standard "ideal" values to be measured.

7.3.2.3 HD Analyzer

Considering that an objective measurement of image quality is an indispensable step in eye diagnostics, the firm VISIOMETRICS has designed an HD Analyzer – a new instrument that relies on a double passage approach providing an objective clinical evaluation of eyes' optical quality.

The device operates on a principle of refraction of a point light source from the retina. As a result, light passes through the eye medium twice. The HD Analyzer analyzes the size and shape of the reflected light spot. The design of a dual-passage aberrometer is based on a Shack–Hartmann sensor. The device contains all information pertaining to optical properties of the eye, including all higher-order aberrations and scattered light, which are usually overlooked by the majority of aberrometry methods. The said higher-order aberrations may have a significant impact on the refractive surgery outcome due to light scatter in an older eye. Key data provided by the HD Analyzer include OSI (objective scattering index), MTF (modulation transfer function), and PScF (point scattering function).

OSI (objective scattering index) is a parameter that enables the intraocular scattered light to be objectively estimated and calculated by evaluating the amount of light on the periphery of a dual-passage image relative to the amount of light at the center. Thus, the higher the OSI, the higher the intraocular scattering level.

This is the only parameter that allows the intraocular scattered light to be evaluated quantitatively. It is useful in all clinical situations when the amount of scattered light may be of clinical significance, such as the development of a cataract and surgical intervention, refractive surgery, intraocular lens, aging process, dry eye syndrome, and the like.

MTF (modulation transfer function) is a function that allows evaluating the level of detail in the image following the passage through an optical system, evaluating the ratio of the contrast between a system-generated image and the source image. In the human eye, the MTF represents the loss of contrast following the eye passage.

Like in any optical system, a decrease in contrast in the human eye is higher for high spatial frequencies (fine details in the image). Thus, MTF is a function of spatial frequency. It is possible to simulate an ordinary scene imaged in the patient's retina. In particular, the program presents an image of an infant put 1 m away from the viewer. The modeling is conducted through the convolution of the original scene with the eye PScF measured by the device. Thus, it shows how aberrations and intra-ocular scattering in the patient's optical system affect the image acquired. This does not mean, however, that an individual sees the same image as that shown on the screen, because the modeling takes into account only an optical image quality rather than neural network processing of the retinal image.

Thus, using the HD Analyzer it is possible to obtain quantitative and objective estimates of (1) the intraocular scattered light, (2) optical quality of the eye, (3) optical quality loss due to tear film destruction, (4) accommodation and pseudo-accommodation, and (5) maps of a dual-passage retinal image.

7.3.2.4 Tomey TMS-4

Tomey TMS (Japan) is a diagnostic instrument based on Placido discs and intended for kerato topography measurements, allowing one to obtain a high-quality photograph of the cornea surface that is undistorted with shadows cast by superciliary arches and nose. With 25 discs used in the optical scheme and 256 data points being analyzed on average on each disc, the device makes it possible to evaluate the cornea refractivity at 6,400 points (TMS-4, User manual).

With the kerato topograph TMS-4, abnormal astigmatism can be diagnosed using different types of maps – sagittal and tangential ones, which use different curvature radii when measuring the refractive power of the cornea. The sagittal map is calculated using a cornea radius related to the vision (sagittal) axis, whereas the tangential map is calculated using an instantaneous curvature radius unrelated to the vision axis. The sagittal map offers more information relating to the cornea refractivity, while the tangential map provides more data on the cornea shape and local irregularities.

7.3.2.5 Pentacam HR

The diagnostic instrument *Pentacam HR* manufactured by the company Oculus (Germany) belongs to a class of projection mappers or kerato mappers and is intended for analyzing the anterior eye segment with a Scheimpflug camera.

Unlike Placido disc-based topography operating in reflection and only measuring parameters of the front cornea surface, this instrument forms an optical cross section of the front eye segment by projecting a light slit in a similar way to a narrow light beam in a slit lamp. Images of the structures under analysis (cornea, anterior chamber of the eyeball, iris, intraocular lens) are acquired based on a Scheimpflug principle.

Using the resulting Scheimpflug images, the instrument software then outlines contours of the front and back cornea surfaces. With this approach, the cornea shape can be directly determined by measuring the elevation of its points, based on which all other parameters are then calculated (tilt, curvature, and optical power). This approach also allows the reconstructed 3D image of the front eye segment to be mapped, and data on the cornea thickness to be obtained at each point. The differential maps can also be used for a comparison of the cornea refractive power (sagittal map) and corneal pachymetry.

7.3.2.6 OPD-Scan ARK-10000 (Nidek)

The OPD-Scan aberrometer comprises a wave front analyzer (seven measurement zones, 2,520 points at the eye pupil projection), a Placido disc-based topographer with a keratometry interval ranging from 33.75 to 67.5 dioptres (D), 11,880 measurement points, and a refraction analyzer. Unlike a standard refractometry method employed in autorefractometers, which measure refraction in a 3-mm region, the OPD-Scan aberrometer measures refraction in 3-mm and 5-mm regions. Besides, the OPD-Scan features a wider range of measurements of the cylindrical refraction component from 0 to ±16D, as compared with a standard autorefractometer (from 0 to ±10D), which is a significant benefit for patients with irregular corneal astigmatism.

Among the multitude of parameters the aberrometer measures, the most important and informative parameter of image quality in an eye optical system is the point spread function (PSF). The size and spatial distribution of the light energy from a point light source focused on the retina after light has traveled through the eye refractive medium determine the PSF.

In addition, the instrument calculates an important parameter enabling the evaluation of eye optical quality, namely, the Strehl coefficient [75–76]. It describes the ratio of the PSF maximum in a diffraction-free (ideal) optical system to the PSF maximum in a system with aberrations. The Strehl coefficient takes values in the interval $0 < St < 1$. If an optical system is aberration-free, the Strehl coefficient equals one. For a healthy human eye, the Strehl coefficient with regard to higher-order aberration is about 0.05.

7.3.3 Analysis of Wave Front Sensors in Ophthalmological Studies

Potentialities of wave front sensors for solving tasks in applied ophthalmology were experimentally studied in eye clinics around the world. Among most recent studies (2018–2020), some publications concerned with an analysis of different types of aberrometers are worth special mentioning.

Moorfields Eye Hospital (UK, London) reported results of a study that examined human eyes following treatment under a wave front sensor control that involved the use of a pyramid aberrometer (SCHWIND eye-tech-solutions GmbH) [77]. Similar studies were conducted at the eye clinic Maja Clinic (Serbia, Nish), where the instruments used included a WaveLight Allegro Oculyzer, a WaveLight Allergo Biograph, and an ultrasound pachymeter DGH Pachette 3. Average values of the central corneal thickness measured using different instruments were put under comparison [78].

Studies conducted at the eye clinic Al Vatani (Egypt, Cairo) looked into detecting keratoconus (KC) with higher sensitivity and specificity by means of a Scheimpflug sensor called Oculyzer [79]. At Beijing Tongren Eye Center (China, Beijing), researchers conducted a comparative study of aberrometers Oculyzer and Topolyzer Vario for eye measurements prior and after a corneal refractive surgery [80]. In addition, Branchevsky Eye clinic (Samara, Russia) conducted a comparative analysis of optical instruments based on Placido discs, Scheimpflug camera, and optical coherence tomography (OCT) for kerotometry measurements in patients after laser eyesight correction [81].

7.3.3.1 Study of a Pyramid Aberrometer Peramis

Moorfields Eye Hospital (UK, London) published results [77] of the examination of eyes following myopia laser-assisted in situ keratomileusis (LASIK) treatment, under wave front control, executed with an excimer laser Amaris 1050RS and a pyramid aberrometer Peramis (SCHWIND eye-tech-solutions GmbH). Repeatability limits of the results were calculated for spherical refraction and higher-order aberration through consecutive eye scans made prior and after the surgical treatment for the first 100 measurements. The repeatability limits were found to be 95% for pyramid aberration measurements. This data suggests good repeatability of measurements, alongside safety and effectiveness of pyramid aberrometry upon a routine myopic LASIK procedure.

7.3.3.2 Analysis of the Use of Aberrometers Oculyzer and Biograp

Measurements of the central corneal thickness (CCT) is an important step of the eye examination preceding any corneal refractive surgery procedures and post-surgery patient follow-up, observation of patients with corneal diseases including keratoconus and Fuchs corneal dystrophy, as well as in CCT-dependent intraocular pressure control. Such studies were conducted at a specialist eye clinic "Maja Clinic" (Nish) using instruments such as WaveLight Allegro Oculyzer, WaveLight Allergo Biograph, and ultrasound pachymeter DGH Pachette 3. A comparison was conducted of average CCT values measured with the different instruments [78].

With no statistically meaningful differences revealed between CCT measurements conducted using an Oculyzer, BioGraph, and ultrasound pachymetry, the CCT measurement results from any of the three instruments can be considered reliable.

The authors of the research came to a conclusion that any of the measurement results obtained by the instruments may independently be considered as reliable and form a basis for selecting therapy tactics. Each of the instruments may find its place in ophthalmology depending on the patient's eye condition and required treatment procedures.

7.3.3.3 Analysis of the Use of the Aberrometer Oculyzer

Research by ophthalmologists at the eye clinic Al-Vatani (Egypt, Cairo) [79] was aimed at identifying indices for detecting keratoconus (KC) with higher sensitivity and specificity compared to those currently available with the use of the software for rotational Scheimpflug visualization (Oculyzer I, Pentacam). Note that while Placido disc-based systems provide accurate data in many cases of KC, they have essential limitations associated with the curvature map. The cornea coverage area is limited to about 60% of the entire cornea surface, thus excluding critical data relating to many peripheral or paracentral pathologies and providing no information relating to the rear corneal surface. Moreover, it appears impossible to construct pachymetry maps that present the cornea thickness distribution. In addition, there are limited possibilities to reconstruct the cornea topography from curvature measurements data [82].

In the study at the eye clinic Al-Vatani (Cairo, Egypt), patients' eye visualization was conducted by the Pentacam company under a trademark Allegro Oculyzer (WaveLight, GmbH, Erlangen, Germany) with the aid of the software 1.16r12. The limitations of the said study are that it was conducted using just a base Pentacam model (Allegro Oculyzer I), which is currently utilized by many eye centers. Further studies of Pentacam HR are recommended because when conducting refractive surgery screening, conventional high-resolution Scheimpflug visualization instruments produce different objective values and the two instruments are not mutually replaceable. It is also recommended that the accuracy of the indices should be compared with a new biochemical index that is based on deformation parameters and cornea thickness profile because it is shown to be highly accurate in distinguishing between a normal cornea and a KC-affected cornea.

7.3.3.4   Comparative Analysis of Sensors Oculyzer and Topolyzer

A study conducted at Beijing Tongren Eye Center (China) was concerned with a consecutive comparison of data from aberrometers Oculyzer and Topolyzer Vario [80] prior and during the corneal refractive surgery conducted with an excimer laser WaveLight EX500.

The Oculyzer and Topolyzer Vario map the correlation between the Cartesian coordinates XY of the apex and the eye pupil center. When the patient's eye focuses on a target, the (reproducible) cornea apex is the highest point most close to the vision ray, irrespective of the pupil size. Should a human-eye optical system be coaxial, the optical axis and the cornea would intersect exactly at the cornea apex. For the correct ablation adjustment, the surgeon needs to analyze the pre-surgery corneal topography in combination with intra-surgery condition. It was experimentally revealed that the effect of a manual adjustment of Kapp's tilt angle based on the Oculyzer topography manual is analogous to the automated tilt angle adjustment based on the Topolyzer Vario topography when using an excimer laser WaveLight EX500. The manual adjustment based on the Oculyzer topography is recommended in the case of discrepancy between the Topolyzer Vario topography and a live image during surgery.

Ophthalmologists managed to demonstrate a high level of similarity between patterns of Oculyzer topography and Topolyzer Vario topography. The data of Oculyzer topography becomes significant when Topolyzer Vario topography images do not coincide with a live image in the course of surgery with an excimer laser WaveLight EX500.

7.3.3.5   Analysis of the Diagnostic Instruments Oculyzer,
Tomey AO-2000, and IOL Master

Ophthalmologists at Branchevsky Eye Clinic (Samara, Russia) conducted a comparison [81] of topography maps of a group of patients who underwent excimer-laser-aided myopia correction. The parameters were measured using three different instruments: IOL Master 700, Tomey AO-2000, and Oculyzer.

Schleimpflug tomography was implemented using an Oculyzer (Alcon, USA) by making 25 meridian scans, with the analysis conducted using SimK indices. Keratometry data obtained using Placido disc-based topography were recorded on an optical biometer with built-in Placido topographer AO-2000 (Tomey, Japan). With this instrument, keratometry of corneal areas of different diameters can be done. For a comparative analysis, keratometry data from a 2.5-mm zone were utilized. An optical biometer IOL Master 700 (Carl Zeiss Meditec AG, Germany) operates by using an OCT principle with a tunable wavelength.

No statistically meaningful differences between the average values of keratometry of a steep meridian, keratometry of a flat meridian, and an average value of keratometry obtained from the three instruments analyzed were revealed. However, statistically meaningful differences were revealed between the instruments Tomey AO-2000 (Placido) and a Schleimpflug tomograph Oculyzer when measuring astigmatism. Although the study did not intend to identify the instrument providing the most reliable keratometry data, finding essentially different keratometry patterns could be useful for the interpretation of the examination data.

7.3.4 ATTEMPTS OF AN ANALYSIS OF EYE ABERRATION

Our studies aimed to solve a problem of differential diagnostics of eye cornea optical structures through an analysis of focusing system aberrations. In [20, 83], an analysis of eye cornea aberrations was conducted using data presented by Branchevsky Eye Clinic. As a result of the analysis, basis Zernike functions most characteristic of some particular eye pathologies were revealed.

An analysis of weight coefficients most informative in terms of the classification of patient diagnoses was conducted. As a result, basis Zernike functions most informative in terms of diagnosing certain eye pathologies were identified.

7.3.4.1 Analysis of a Human Eye Corneal Aberrations

People are known to obtain a predominant proportion of information about the outside world with their eyes; hence, the quality of life deteriorates with deteriorating vision. Science has accumulated enormous expertise fighting eye diseases, from which it became clear that a major problem is age-related changes in the eye optical system, which can be evaluated on the basis of wave aberrations.

The human eye can be described as a lens array composed of three major components: a cornea, a pupil, and a crystalline lens [84]. All optical characteristics of a healthy eye are defined by a combination of aberrations in the cornea and intraocular optics.

Studies [85–87] have shown that the total amount of aberration in the whole eye is always lower than that in the front corneal segment or internal optics. This is explained by the fact that aberrations of the cornea and the crystalline lens compensate for each other [88]. There is convincing proof that the aberration compensation between the cornea and intraocular optics occurs in the case of astigmatism (along horizontal/vertical axis), horizontal coma, and spherical aberration [89]. The general result of aberration compensation consists in a decrease of aberration amount in the retina plane, potentially leading to a higher-quality optical image in the foveola. In young people, aberrations of the cornea and crystalline lens are normally higher than those in the whole eye. This fact lends support to the hypothesis that the crystalline lens of the human eye plays an important role in the compensation of the cornea aberration, resulting in an improved retina image.

Wave front analysis using the Zernike basis expansion facilitates making a diagnosis in clinical ophthalmology, also making it possible to design almost perfect lenses enabling the best eyesight correction possible to date.

Data on aberration of the human eye optical system was obtained using Branchevsky Eye Clinic aberrometers WaveLight Oculyzer II and HD Analyzer (a fragment of the data is shown in Figure 7.22).

For a better visual representation, Figure 7.23 depicts a "Zernike pyramid" composed of several first Zernike functions. From top to bottom, radial numbers are varying from $n = 0$ to 4 and from left to right – azimuthal numbers vary from $m = -n$ to $m = n$.

Patterns of the point spread function (PSF) in the presence of typical aberrations were numerically constructed using a simplest Fourier correlator scheme in the Zemax programming environment [90].

The optical scheme in Figure 7.24 consists of two identical lenses made from BK8 glass and an aperture of radius 3 mm at the center, with the numerical simulation

ID	Age	Eye	Doctor's Diagnosis	Z00	Z11	Z1-1	Z22	Z20
1	27	OS	Low myopia	1.34E-01	−4.18E-04	−1.33E-03	6.35E-04	7.83E-02
2	27	OD	Low myopia	1.33E-01	−1.52E-03	8.90E-04	6.94E-04	7.78E-02
3	20	OS	Mediate myopia	1.23E-01	3.75E-04	−2.10E-04	9.59E-04	7.25E-02
4	20	OD	Mediate myopia	1.24E-01	−7.31E-04	6.93E-04	3.21E-04	7.28E-02
5	24	OS	Mediate myopia	1.26E-01	−9.28E-04	6.60E-04	4.99E-04	7.40E-02
6	24	OD	Mediate myopia	1.26E-01	−8.69E-04	2.50E-05	5.42E-04	7.39E-02
7	27	OS	Low myopia with compound myopic astigmatism	1.31E-01	−1.33E-03	−1.36E-03	5.34E-03	7.73E-02
8	27	OD	Low myopia with compound myopic astigmatism	1.30E-01	−8.83E-04	1.77E-03	1.15E-03	7.65E-02
9	23	OS	Low myopia	1.41E-01	2.25E-04	−3.55E-04	7.54E-04	8.33E-02
10	24	OD	Low myopia	1.42E-01	1.82E-03	−4.10E-05	8.81E-04	8.41E-02
11	24	OS	Mediate myopia with compound myopic astigmatism	1.37E-01	1.07E-03	−1.07E-03	8.68E-04	8.06E-02
12	24	OD	Mediate myopia with compound myopic astigmatism	1.38E-01	4.81E-04	5.31E-04	1.13E-03	8.09E-02
13	22	OS	Mediate myopia with compound myopic astigmatism	1.25E-01	−1.38E-03	1.05E-03	1.01E-03	7.33E-02

FIGURE 7.22 A fragment of the data obtained in the studies (Zernike weight coefficients).

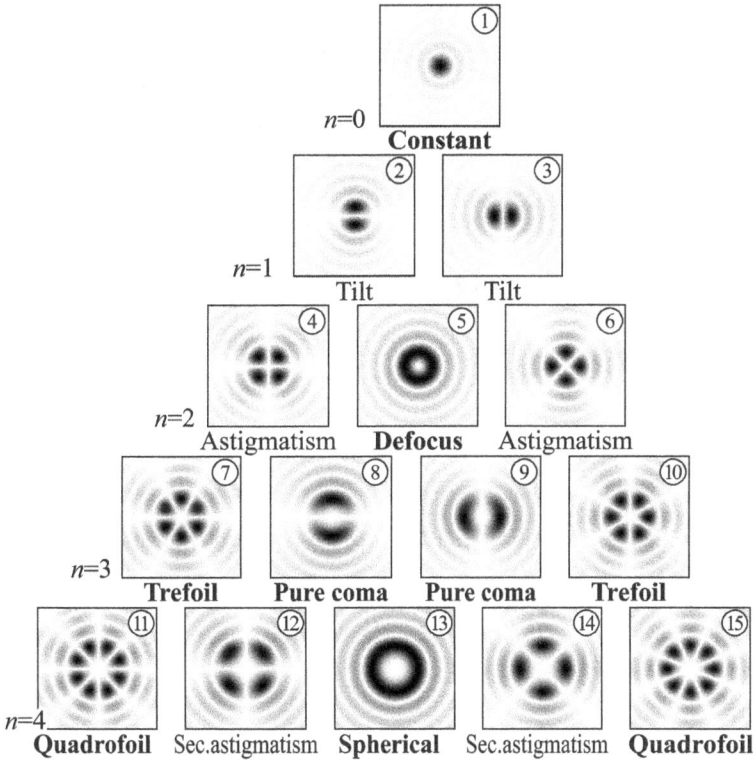

FIGURE 7.23 Several examples of Zernike functions.

conducted for a 780-nm wavelength. Rather than using the lens radius, the Zernike polynomials were normalized relative to the output pupil radius calculated using Zemax, which was found to be 8.37 mm. The first surface of the second lens was consecutively subjected to third-order coma aberrations (Figure 7.25) corresponding to the seventh Zernike function (the coefficient assumed to equal 1).

The system just described can be numerically simulated using the following algorithm. The wave front can be constructed as superposition of Zernike polynomials using relevant formulae before calculating the PSF as a Fourier transform of the derived distribution. For a more visual representation, Figure 7.27 depicts a PSF pyramid corresponding to the Zernike pyramid. The results obtained in this way

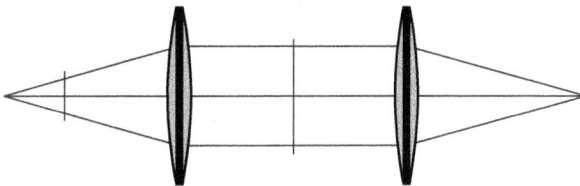

FIGURE 7.24 An optical scheme of a Fourier correlator.

(a)   (b)   (c)   (d)

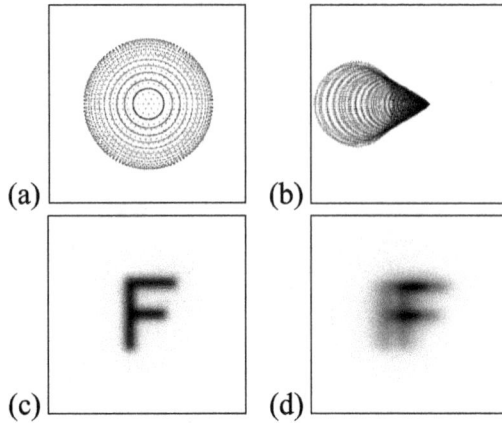

FIGURE 7.25   PSF (a) without aberration, (b) with coma, and (c, d) corresponding test images.

agree well with the numerical simulation performed in the Zemax programming environment. The possibility of rotating the pattern by angle φ_0 (Figure 7.26) is made possible through the use of the transform:

$$\psi(r,\varphi) = R_{nm}(r)\left[a\sin(m\varphi) + b\cos(m\varphi)\right], \quad (7.31)$$

where

$$a = \sin(|m|\varphi_0), b = \cos(|m|\varphi_0), \varphi_0 \in [0; 2\pi].$$

Let us perform the numerical simulation of image distortion following the introduction of aberration into a wave front incident on an ideal lens of an imaging system.

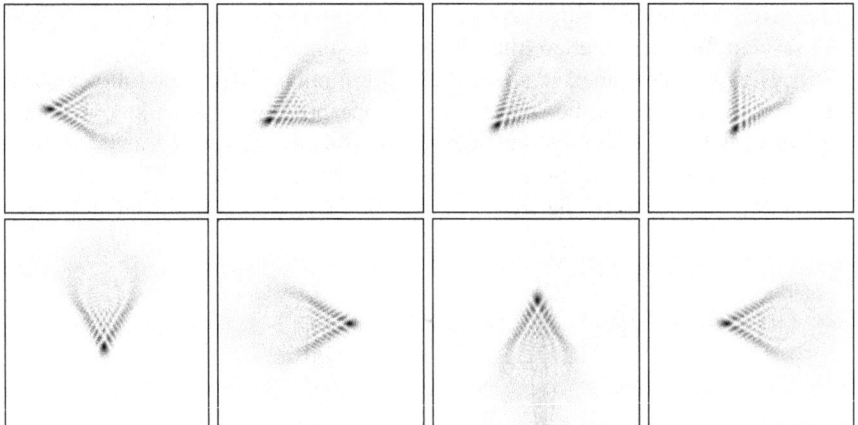

FIGURE 7.26   Rotation of the PSF of the Zernike function at $n = 3$, $m = 1$, $\varphi_0 = \{0, \pi/6, \pi/4,$ $\pi/3, \pi/2, \pi, 3\pi/2, 2\pi\}$.

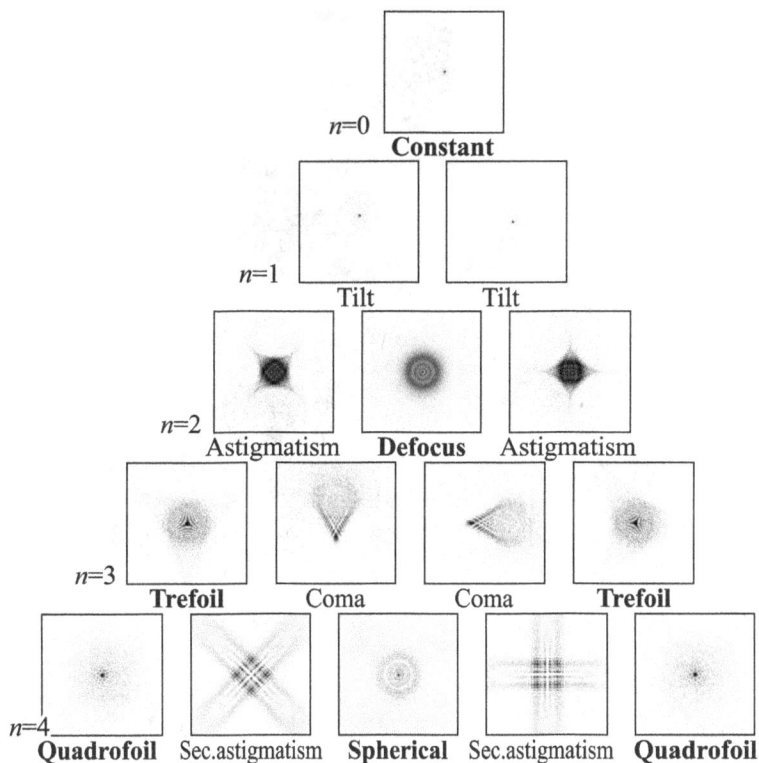

FIGURE 7.27   Patterns illustrate PSFs corresponding to typical aberrations.

For the acquisition of an ideal image $f(x, y)$, an imaging optical system containing an ideal lens was utilized. A distorted image $f_a(x, y)$ was numerically simulated using the following algorithm: a Fourier image $F(u, v)$ of an ideal image was multiplied by an aberration-distorted wave front $w(x, y)$, before performing an inverse Fourier transform $f_a(x,y) = \Im^{-1} \left[\Im \left[f(x,y) \right] w(x,y) \right]$. Thus, convolution of an ideal image with a distorted PSF was numerically simulated.

The aforementioned algorithm was utilized to study the impact of all Zernike polynomials up to a fourth order inclusively on image distortion (as shown in Figure 7.27). Some types of the same-order aberrations with the same weight coefficients but with different signs of the meridional index m were found to introduce a different degree of distortion into the same test image (a cross, see Figures 7.28 and 7.29). The reason is that some types of aberration are orthogonal to the test image utilized. This

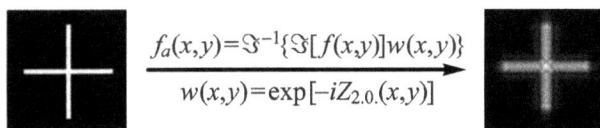

FIGURE 7.28   A test image is distorted by defocus-type aberration.

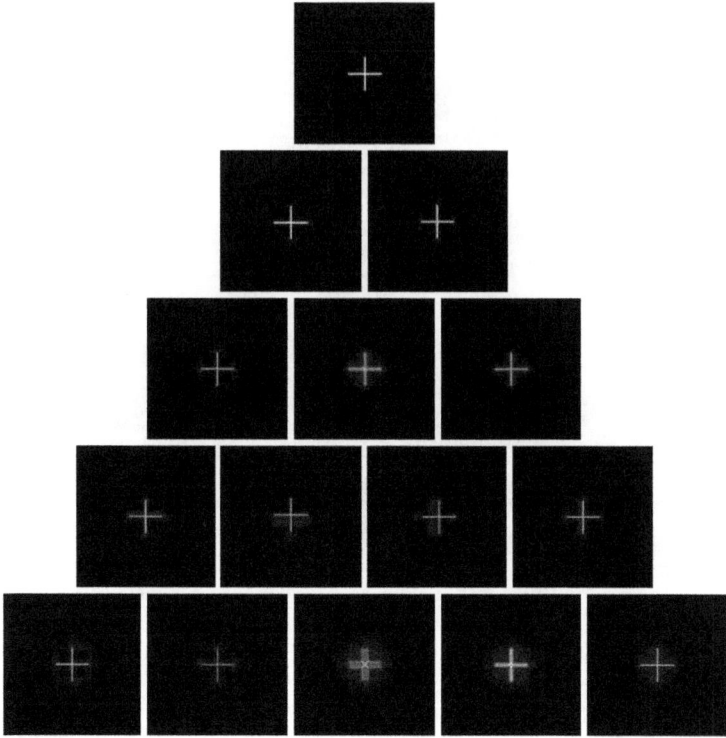

FIGURE 7.29   Distortion of a test image "cross."

property may be useful when specific phase patterns of DOEs intended to compensate for some aberrations need to be generated.

By way of illustration, aberration in the form of astigmatism Z_{22} Z_{44} deteriorates the image quality in a greater degree when compared with analogous aberration Z_{2-2} Z_{4-4}. In the meantime, aberration in the form of defocus (Z_{20}) and spherical aberration (Z_{40}) introduce similar image distortion but differ in the impact degree.

It is worth noting that apart from numbering and normalization, Zernike functions may have different angular dependence. In particular, multichannel diffractive optical elements [31, 53, 91, 92] utilized for wave front analysis operate by using an exponential and trigonometric angular dependence. Note that in the former case the Zernike basis is rotation invariant [31]. In the latter case, the rotation of the pattern leads to a change in coefficients by the basis functions.

The human eye can be described as a lens array composed of three major components: a cornea, a pupil, and a crystalline lens. For the image to be formed on the retina, the performance of the optical components is being mutually coordinated. The improper functioning of the eye system leads to vision defects, ranging from a minor blur of the image to complete blindness. Being the predominant component of an eye's optical power (~70%), the cornea (tear film included) is the main source of optical aberrations. The anterior corneal surface has an elongated profile with the steeper central region compared to the periphery. This shape serves to reduce the number of spherical aberrations in the whole eye. However, corneas of different individuals significantly differ in shape, possibly causing astigmatism and higher-order asymmetric aberrations (e.g. coma). An ideal eye focuses an image of the outside world specifically in the foveola (a region containing only photoreceptors), irrespective of the field angle. However, a perfect image cannot be formed in a real eye as some amount of aberration deteriorates optical characteristics of the eye.

Lower-order aberrations are predominant in the human eye, accounting for up to 90% of all wave aberrations [91]. At the top of the aberration list is defocus, with positive defocus causing long-sightedness and negative – short-sightedness. The second from the top is astigmatism. Correction of low-order aberration with the aid of spectacles, contact lenses, or laser surgery essentially improves visual acuity in the majority of cases. On the other hand, lower-order aberrations known to occur in the human eye currently are unable to be corrected by means of popular optical or surgical techniques. The eye also has chromatic aberration, which is the result of dispersion in optical elements with particular values of the refractive index. Different eye parts are affected by different wavelengths. Thus, the incident white light is decomposed into a color spectrum in the eye. Putting it simply, chromatic aberration is wavelength-dependent spherical refraction. Types of chromatic aberration (dispersion) are traditionally divided into longitudinal and transverse ones. The former is characterized by varying on-axis optical power and wavelength and is relatively the same in different individuals. Transverse aberration causes the image to be shifted across the entire image plane depending on the wavelength and is significantly varying in different individuals. With the real world being polychromatic, chromatic aberration undoubtedly poses limitations on the image quality.

Diffraction is a fundamental property of light waves, which occurs when light passes through a pinhole comparable in size with its wavelength. In the eye, diffraction takes place when light passing through the eye interacts with the iris edge. Even with no aberrations, an infinitesimally small point cannot be imaged in the retina due to diffraction. In general, considering that all optical systems (including the eye) have a rigid diaphragm, it is impossible to design a diffraction-free optical device. Theoretically, it is possible to improve the image quality in an optical system by minimizing aberrations. Nonetheless, diffraction-related limits on the image quality cannot be overcome, and the highest image quality attainable is called diffraction-limited as it is limited solely by diffraction. Thus, even "perfect" optical systems cause unavoidable blur due to diffraction. With no aberration, a perfect eye would transform incoming wave fronts into converging spherical waves. Thus, due to diffraction, an image of a point has a finite size. The image of a point in a diffraction-limited optical system is called an Airy spot.

Scattered light also hampers the performance of an optical system (of an eye) in terms of image processing. This results in a lower image contrast, causing halos and flares. Scattering in the eye medium mostly occurs as a result of diffusion of light and deterioration of the transparency of the cornea (including the tear film) and the crystalline lens [53]. Retinal scattering may depend on the fraction of light traveling along photoreceptors and the retina layer from which the light is reflected (depending on the wavelength used). Higher-order aberrations are comparatively low, with predominant aberrations represented by third-order comatic aberrations (vertical coma, horizontal coma, slanted trefoil, horizontal trefoil) and spherical aberration.

The study discussed herein was based on data sets presented by Branchevsky Eye Clinic, which contained data on conditionally healthy eyes (with no pronounced pathologies in either the pupil or the crystalline lens) as well as those with pathologies, like low and medium myopia. A reference model of the cornea can be described by a sphere or an ellipse of eccentricity 0.75. The reference model is always axisymmetric, with its shape affecting only the coefficients of axisymmetric Zernike functions (1) with the numbers $(n, 0)$. The central radius of the reference model is always taken to be equal to an average central radius of measurements. In the absence of a reference model, the largest contribution to the cornea shape comes from the component corresponding to the Zernike function $(2,0)$ (a paraboloid), because this component is closest to the cornea in shape. Shown in Figure 7.30 is an optical setup enabling the simplest human eye model to be implemented in Zemax [90, 93, 94]. The numerical simulation was conducted for a wavelength of 780 nm (the wavelength used by the ophthalmological instrument HD Analyzer).

A number of polynomials were consecutively applied to the cornea surface. Figure 7.31 depicts the performance of an optical system after coma described by the Zernike function $(3,1)$ with the coefficient 1 introduced to the anterior cornea surface.

When specifying a normalization radius for the Zernike polynomials, instead of using the lens radius, an output pupil radius calculated using Zemax was utilized, amounting to 100 mm.

Within this study, a simplest statistical analysis was conducted, with the (anterior and posterior) cornea surfaces of individuals with healthy eyes being independently tested. The individuals were grouped by age (20–29 years and 30–39 years) and diagnosis (weak myopia and medium myopia).

Using an aberrometer WaveLight Oculyzer II, Zernike coefficients for the anterior and posterior cornea surface can be measured separately. The averaged values are given in Figures 7.32–7.35. For a better visualization of the Zernike coefficients, coefficients with the numbers $(n, 0)$, $n = 0, 2, 4$ were excluded from consideration (by putting them equal to zero). Although they are by an order of magnitude higher than

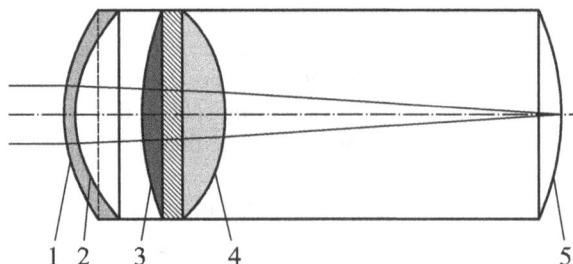

FIGURE 7.30 Schematic view of key eye refractive surfaces: 1 is an anterior cornea surface, 2 is a posterior cornea surface, 3 is a front crystalline lens surface, 4 is a back crystalline lens surface, and 5 is retina.

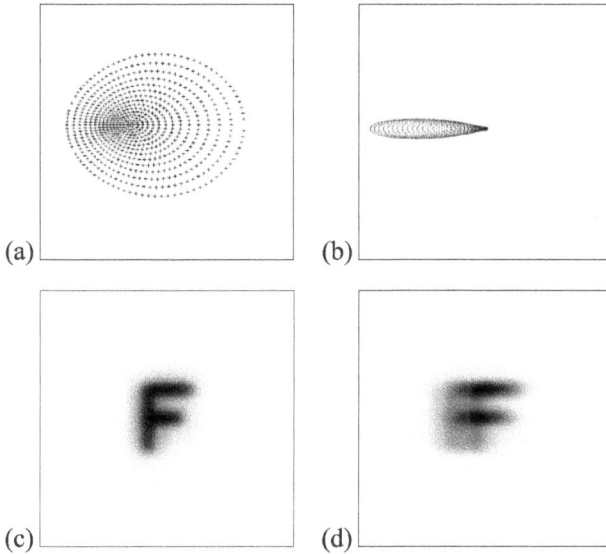

FIGURE 7.31 PSF (a) without aberration and (b) with coma and (c, d) corresponding test images.

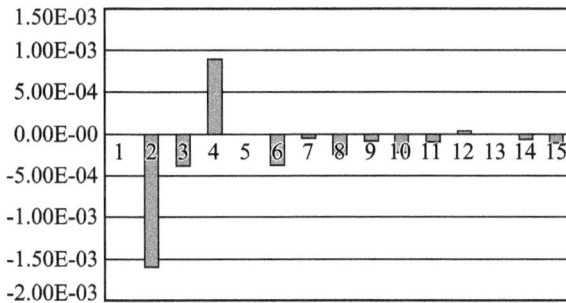

FIGURE 7.32 Values of weight coefficients of the Zernike polynomials for the anterior cornea surface of healthy eyes.

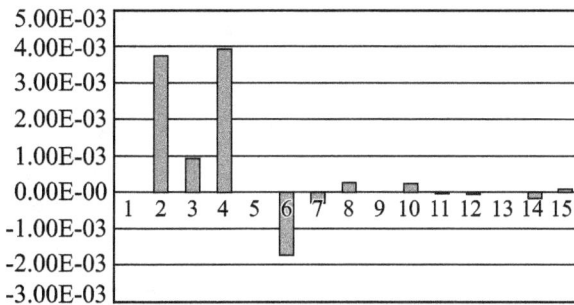

FIGURE 7.33 Values of weight coefficients for the Zernike polynomials for the posterior cornea surface of healthy eyes.

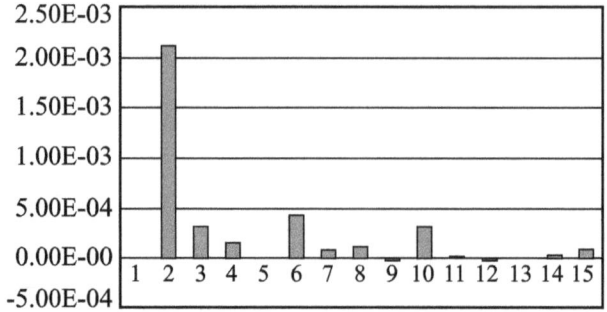

FIGURE 7.34   Values of weight coefficients for the Zernike polynomials for the anterior cornea surface in patients with weak myopia.

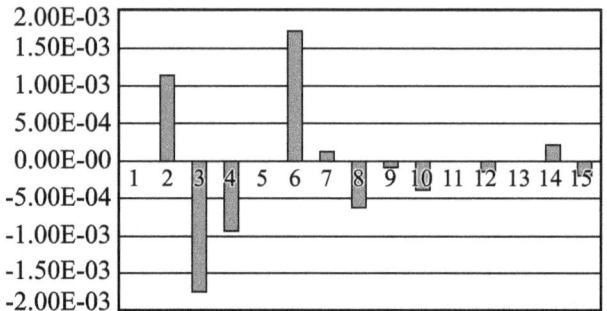

FIGURE 7.35   Values of weight coefficients for the Zernike polynomials for the posterior cornea surface in patients with weak myopia.

all the other coefficients, weight coefficients for these functions carry no information whatsoever on any deviations.

Figure 7.36 depicts patterns corresponding to averaged aberrations in the anterior cornea surface in patients with weak myopia diagnosis and corresponding PSFs.

Similar patterns for the posterior cornea surface are shown in Figure 7.37.

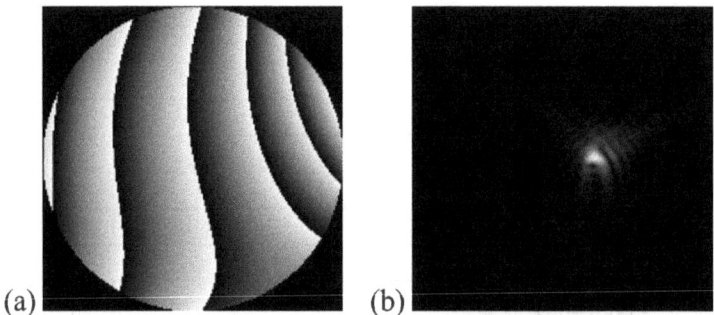

FIGURE 7.36   (a) Averaged aberrations over the anterior cornea surface in patients with weak myopia and (b) PSF.

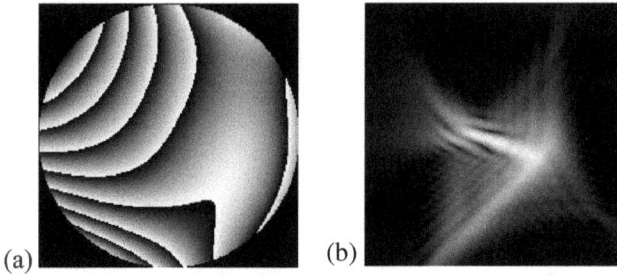

FIGURE 7.37   (a) Averaged aberrations of the posterior cornea surface in patients with weak myopia and (b) PSF.

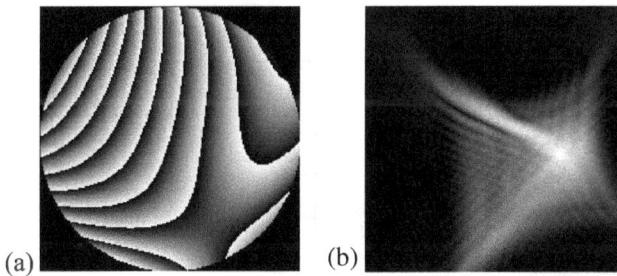

FIGURE 7.38   Combined averaged aberrations (of the anterior and posterior cornea surface) in patients with weak myopia: (a) wave front and (b) PSF.

To evaluate the impact of whole-cornea aberrations, wave front aberrations were summed up, with the result shown in Figure 7.38.

In a similar way, results for medium myopia have been derived (Figure 7.39).

Thus, the most characteristic Zernike functions for particular eye pathologies have been revealed (see for details Figure 7.40). In particular, in a sampling for patients with weak myopia, a tilt both on the anterior and posterior cornea surfaces was observed, but astigmatism was revealed on a larger scale (especially for the

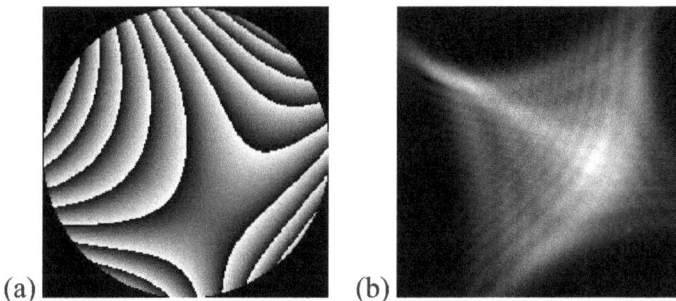

FIGURE 7.39   Combined averaged aberrations (of the anterior and posterior cornea surface) in patients with medium myopia: (a) wave front and (b) PSF.

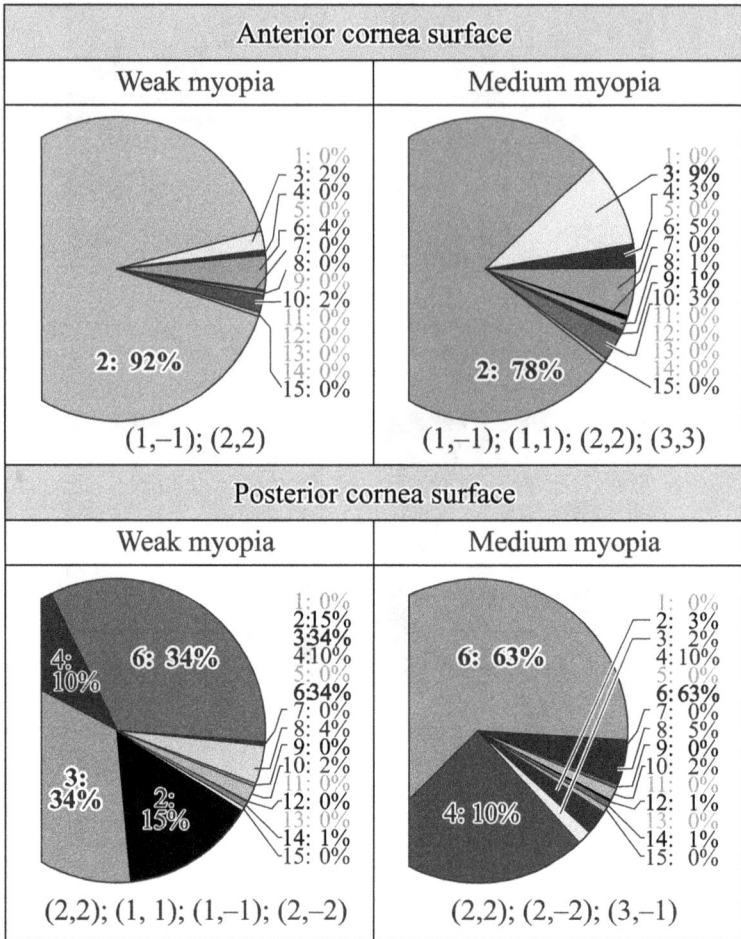

FIGURE 7.40 Summary table for the anterior and posterior cornea surface, correlation between the Zernike polynomials, most pronounced weight coefficients, and the diagnosis.

posterior surface). An analysis of samplings for medium-myopic eyes has shown that the influence of the third-order coma comes to the forefront. Though the aberrations typical of low myopia are still retained.

7.3.4.2 Extracting Informative Features Based on Zernike Polynomial Coefficients for Different Pathologies of the Human Eye

Research conducted in [83] was concerned with an analysis of wave front aberrations linked with changes in the curvature of the human eye cornea. The analysis was conducted by representing the aberration of the anterior and posterior cornea surface as a superposition of Zernike functions. The study aimed to examine weight coefficients of the Zernike polynomials, with the data collected in the course of clinical tests at Branchevsky Eye Clinic. Most informative weight coefficients were analyzed in terms of their use for grouping patients according to particular diagnoses. A comparison was conducted of the classification results obtained on the basis of 30 features derived both on the anterior and posterior cornea surface and using most informative features. The features were ranked in terms of their information content for a concrete classification task. The information content was evaluated based on the values of a separability criterion and additionally evaluated by calculating a classification error with the aid of a K-means method. As a result of the statistical analysis, basis Zernike functions most informative for particular eye pathologies were revealed.

Then, the most informative weight coefficients needed to be extracted from the presented set in order to reduce the false classification probability. The task was implemented using a programming language R intended for intelligent data analysis, with built-in mathematical tools of the language used in the study. The input data for the analysis included n objects, each being characterized by 30 features, and classes defined by the following binary relations: weak myopia/medium myopia (I), weak myopia/weak myopia + astigmatism (II), medium myopia/medium myopia + astigmatism (III), weak myopia/medium myopia + astigmatism (IV), and medium myopia/weak myopia + astigmatism (V).

When analyzing the classification quality, a learning sample was constructed using a new classifier that was calibrated based on a reference vector method [95]. Within an approach called a U-method [96], the classifier was synthesized using just those objects of the learning sampling that were not found in the test sampling. True distribution objects can be replaced by objects not used for the classifier synthesis and independent of the objects involved in the classifier synthesis. While the U-method can be implemented in a number of ways, the probability of false classification was evaluated using a method of excluding a single object. Objects from the test sampling were classified using object-class related data from the training sampling using a nearest-neighbor algorithm. In the course of experimental studies, an average rate of classification error was found to be 0.319 when using the total set of features. To conduct an analysis of weight coefficients, binary relations for the average vectors for each of the aforementioned diagnoses were chosen. For convenience sake, a continuous numbering of Zernike basis functions was proposed, where $N = 1{:}15$ are basis functions describing the curvature of the anterior cornea surface and $N = 16{:}30$ are basis functions describing the curvature of the posterior cornea surface (the correlation between the ordinal number N and the basis function Z_{nm} is shown on the abscissa of the plots in Figures 7.41–7.45).

Let a vector $\mathbf{a}_{avg}(x_1, x_2, \ldots, x_{30})$ be the center of class a of a 30-dimensional feature space and a vector $\mathbf{b}_{avg}(x_1, x_2, \ldots, x_{30})$ be the center of class b. Then, the most informative features in terms of the classification of the two classes can be revealed from the difference of coordinates of the two vectors, $\mathbf{c} = |\mathbf{a}_{avg} - \mathbf{b}_{avg}|$, enabling the proximity of each weight coefficient to be revealed independently for different pathologies. Results for the binary relation of interest are presented in Figure 7.41.

Bar charts in the plots enable the most informative features to be revealed. The larger is the absolute difference for each feature c_i, the more informative it is. Afterwards, the most informative features among the vectors $\mathbf{a}_{avg}$ and $\mathbf{b}_{avg}$ were

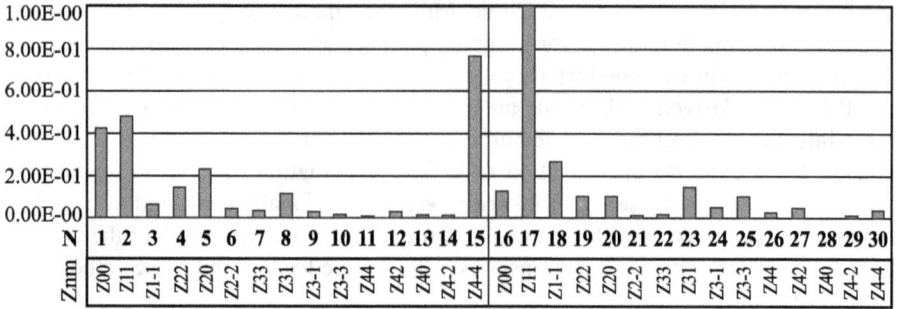

FIGURE 7.41 The most informative weight coefficients for the binary relation – weak myopia/medium myopia.

clustered, with corresponding error rates shown in Table 7.5. Table 7.5 suggests that the most informative features that occur in three and more binary relations (I–V) are represented by polynomials with the following indices (N): 1, 5, 15, 16, 17, 18, 19, and 20. They describe aberrations including constant (Z_{00}), defocus (Z_{20}), quatrefoil (Z_{4-4}) on the anterior cornea surface and constant (Z_{00}), distortion (Z_{11}, Z_{1-1}),

TABLE 7.5
Informative Weight Coefficients for Binary Relations (I–V)

Binary Relations	I	II	III	IV	V
Numbers of Zernike Weight	1	–	1	1	1
Coefficients (N)	2	2	–	–	2
	–	–	4	4	–
	5	–	5	5	–
	–	–	–	–	6
	–	–	–	–	8
	15	15	–	15	–
	–	–	16	16	16
	17	17	–	17	17
	18	18	18	18	–
	–	–	19	19	19
	–	–	20	20	20
	–	–	–	–	21
	–	–	–	–	22
	–	–	23	–	23
	–	–	–	–	24
	–	–	–	–	25
	–	–	–	–	26
	–	–	–	–	27
	–	–	–	–	28
	–	–	–	–	30
Error	0.233	0.160	0.241	0.142	0.239

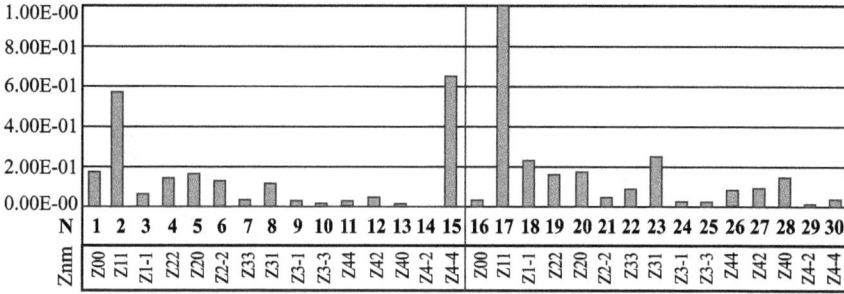

FIGURE 7.42 The most informative Zernike weight coefficients for the binary relation weak myopia/weak myopia + astigmatism.

astigmatism (Z_{22}), and defocus (Z_{20}) on the posterior cornea surface. Table 7.5 contains only features that were found to be informative for distinguishing between at least one pair of classes.

Similarly, an information content of the coefficients for the binary relations II–V can be evaluated (Figures 7.42–7.45).

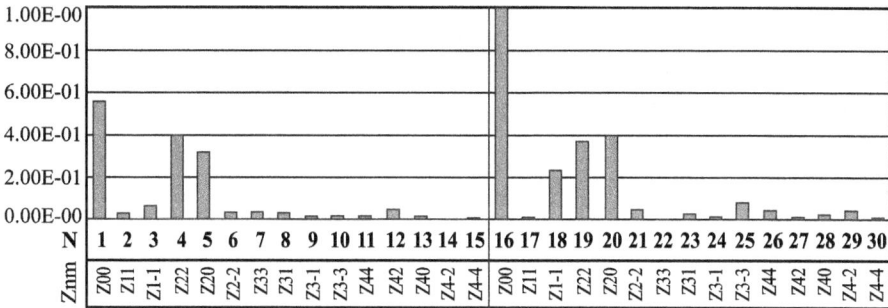

FIGURE 7.43 The most informative Zernike weight coefficients for the binary relation medium myopia/medium myopia + astigmatism.

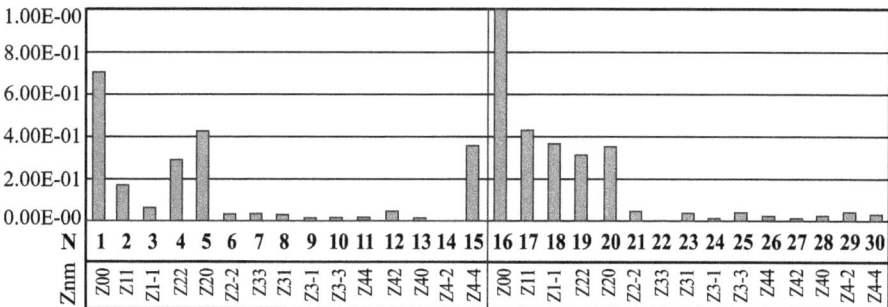

FIGURE 7.44 The most informative Zernike weight coefficients for the binary relation weak myopia/medium myopia + astigmatism.

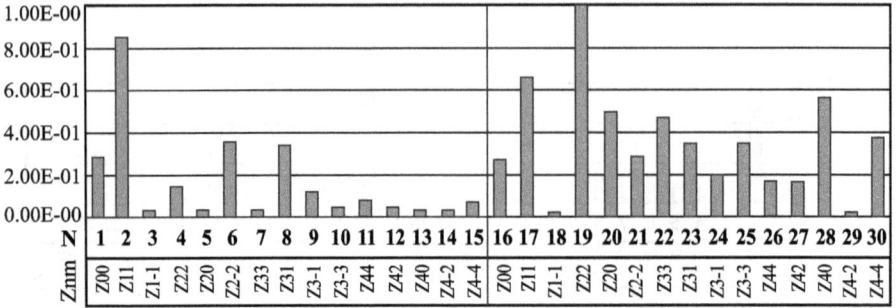

FIGURE 7.45 The most informative Zernike weight coefficients for the binary relation medium myopia/weak myopia + astigmatism.

From Table 7.5, the clustering error rates are seen to exceed values acceptable for making a diagnosis [97], therefore a decision was made to use a more complex procedure for selecting most effective features for illness diagnosis [98–100].

The approach relies on statistical analysis techniques when constructing an informative feature space. In [99, 100], similar approaches to selecting informative features for the analysis of medical-biological objects were described.

Thus, the selection procedure aims to find the most informative features for a particular classification task. In the course of study, the information content was evaluated based on a separability criterion [96, 97] $J_1 = tr((T)^{-1}B)$, where $T = B + W$, B is an intergroup scattering matrix whose elements are calculated using the formula:

$$b_{ij} = \sum_{k=1}^{g} n_k \left(\overline{x_{ik}} - \overline{x_i} \right) \left(\overline{x_{jk}} - \overline{x_j} \right), \quad i,j = \overline{1,p}, \tag{7.32}$$

where n is the number of elements of a sampling containing p features divided into g classes. The intergroup scattering matrix W is described by its elements:

$$w_{ij} = \sum_{k=1}^{g} \left(x_{ikm} - \overline{x_i} \right) \left(x_{jkm} - \overline{x_j} \right), \quad i,j = \overline{1,p}, \tag{7.33}$$

where x_{ikm} is the value of the i-th feature of the m-th element in the class k, $\overline{x}_{ik} = 1/n_k \sum_{m=1}^{n_k} \overline{x}_{ikm}$ is the average value of the i-th feature in the class k, $\overline{x}_i = (1/n)\sum_{k=1}^{g} n_k \overline{x}_{ik}$ is the average value of the i-th feature for all classes, and n_k is the number of elements in the class k. The larger the value of the criterion, the better the separability of the classes. The information content of features was additionally evaluated by calculating a clustering error based on a K-means technique. When performing clustering, K equals the number of the classes under analysis, and the error rate is defined as a proportion of wrongly clustered sampling vectors.

As far as the task under analysis is concerned, evaluating the information content based on the values of an individual separability criterion would be of low efficiency because the value of the criterion is low. For instance, if all four classes of interest are evaluated simultaneously, the criterion value is below 0.1, meanwhile when analyzing each pair of classes, the criterion value is not higher than 0.2.

TABLE 7.6

Values of Group Separability Criteria for Each Pair of Classes

Binary Relation	Values of Separability Criteria	
	All Features	Informative Features
I	0.909	0.891
II	0.880	0.840
III	0.765	0.728
IV	0.810	0.762
V	0.820	0.778

Because of this, when seeking to discern each pair of classes, the information content was evaluated by evaluating the contribution of each feature into the separability of the entire feature space. Thus, features with the lowest contribution to the total information content were excluded from the set of the features used to discern two definite classes (binary relations I, II, III, IV). The features continued to be excluded until the value of the all-group separability criterion was reduced by more than 0.05. Finally, we extracted five sets of features sufficiently informative to discern each pair of classes. Values of the group separability criterion and the clustering error obtained for all 30 initial features and resulting sets of informative features are shown in Table 7.6.

The analysis of Table 7.6 suggests that all features chosen are informative enough to discern different pathology classes. Importantly, the feature space dimensionality has been essentially reduced.

In an experimental study, we established a number of comparisons for which the preceding features were informative, evaluating their efficiency for discerning the aforementioned classes. The features were broken down as follows: recommended ones – those utilized in at least five pairs of comparisons; non-recommended ones – those utilized in no more than two pairs of comparison; with the remaining features termed as conventional.

Using the derived features as a basis, clustering was performed, with the clustering errors shown in Table 7.7. From an analysis of Table 7.7, it can be inferred that the most informative features that are found in all the binary relations of interest (I–V) are polynomials with indices N: 1, 3, 7, 8, 13, 20, 21, 22, 23, 26, 28. These polynomials describe aberrations, which include constant (Z_{00}), distortion (Z_{1-1}), trefoil (Z_{33}), coma (Z_{31}), and spherical aberration (Z_{40}) for the anterior cornea surface, and defocus (Z_{20}), astigmatism (Z_{2-2}), trefoil (Z_{33}), coma (Z_{31}), quatrefoil (Z_{44}), and spherical aberration (Z_{40}) for the posterior surface.

Hence, the conclusion is that for the classes under analysis to be discerned, the use of the following features may be recommended (Table 7.8): distortion (Z_{1-1}) and coma (Z_{31}) – the anterior cornea surface; defocus (Z_{20}), astigmatism (Z_{2-2}), trefoil (Z_{33}), quatrefoil (Z_{44}), and spherical aberration (Z_{40}) – the posterior surface.

It is interesting to note that highly informative features contain higher-order polynomials Z_{31}, Z_{33}, Z_{40}, Z_{44} (higher-than-second order). In classical ophthalmology,

TABLE 7.7

Summary Table Containing the Most Informative Features on the Anterior and Posterior Cornea Surfaces

Binary Relations	I	II	III	IV	V
Numbers of Zernike Weight Coefficients (N)	1	1	1	–	1
	–	2	2	2	–
	3	3	3	3	–
	–	–	4	4	4
	5	5	–	–	–
	–	6	6	–	6
	7	7	7	7	–
	8	8	8	8	–
	9	9	–	–	9
	10	–	–	–	–
	11	11	–	11	–
	–	12	–	–	12
	13	–	13	13	13
	14	14	14	–	–
	15	–	–	–	–
	–	16	16	–	–
	17	–	–	17	–
	18	18	–	–	18
	–	19	–	–	–
	20	20	20	20	20
	21	21	21	21	21
	22	22	–	22	22
	23	23	23	–	23
	24	24	–	–	24
	25	–	–	25	–
	26	26	26	26	–
	27	27	–	27	–
	–	28	28	28	28
	29	–	–	–	29
	30	–	30	–	30
Error	0.030	0.024	0.049	0.047	0.026

TABLE 7.8

Recommended Features for the Anterior and Posterior Cornea Surface

Recommended Features	
Anterior Cornea Surface	Posterior Cornea Surface
$Z_{1-1}\ Z_{31}$	$Z_{20}\ Z_{2-2}\ Z_{33}\ Z_{44}\ Z_{40}$

specialists take into consideration just the first- and second-order polynomials, which account for the so-called cylinder and sphere. However, the study presented here has shown that the account of higher-order aberrations (third-order coma, trefoil, quatrefoil, and fourth-order spherical aberration) has enabled a threefold reduction in the classification error, leading to a more accurate computer-aided diagnosis.

A listing of software codes written in the C++ programming language for the Visual Studio development environment for calculating a distorted image is provided in Appendix B.

7.4 CONCLUSION

A review of recent studies conducted by professional ophthalmologists has shown that there are no wave front sensor designs, currently utilized or proposed, that would be capable of simultaneously providing high spatial resolution in the pupil of the optical components under testing and a measurement precision comparable with that of laser interferometers. Over the last decade, aberrometers have found applications in many areas of ophthalmology, including detection of refraction anomalies, keratoconus diagnosis, and refractive surgery. Practicing ophthalmologists use a wide variety of instruments for the eye examination, such as ocular pyramid and tracing aberrometers, Shack–Hartmann sensors, and Scheimpflug cameras in combination with Placido disc-based topographers.

For the ophthalmologist to be able to make a maximally accurate diagnosis, draw up an adequate plan, select treatment procedures, and prognosticate the treatment outcome, the patient's eyesight needs to be thoroughly examined. Independent studies [77–82] conducted in a number of eye clinics show that popular wave front sensors, such as ALLEGRO Topolyzer VARIO, WaveLight Oculyzer, HD Analyzer, Tomey TMS-4, Pentacam HR, OPD-Scan ARK-10000, and others, generally provide consistent data. However, in specific cases, measurements of the human eye optical system may produce different results, which may be due to both technological reasons (different indirect technologies and data processing algorithms) and varying approaches to data interpretation.

The analysis of the human eye aberrations was conducted using measurements data provided by Branchevsky Eye Clinic [20]. Aberrations were represented as a superposition of Zernike functions. As a result of the analysis [which excluded from consideration coefficients corresponding to the functions with the numbers $(n,0)$, $n = 0,2,4$], Zernike basis functions most typical of some eye pathologies were identified. In particular, a sampling for weakly myopic eyes revealed a tilt for both anterior and posterior cornea surfaces, but astigmatism was found to be prevailing (especially for the posterior surface). In patients with medium myopia, the sampling analysis found that the third-order coma had an essential effect, meanwhile aberrations typical of weak myopia were also retained.

A study reported in [83] looked into wave front aberrations caused by variations in the human eye corneal topography. The analysis relied on the representation of anterior and posterior cornea surface aberrations as a superposition of Zernike functions. The study focused on weight coefficients of the Zernike functions, with the data obtained in a series of clinical studies conducted at Branchevsky Eye Clinic.

The study analyzed most informative weight coefficients in terms of the classification of patients by definite diagnoses. We proposed two approaches to extracting informative coefficients, with the feature extraction technique based on separability criteria shown to be most efficient for solving a particular classification task. Interestingly, among highly informative features there are higher-than-second order polynomials Z_{31}, Z_{33}, Z_{40}, Z_{44}. In classical ophthalmology, specialists take into consideration just the first- and second-order polynomials, which account for the so-called "cylinder" and "sphere." However, the study presented here has shown that the account of higher-order aberrations (third-order coma, trefoil, quatrefoil, and fourth-order spherical aberration) has enabled a threefold reduction in the classification error, leading to a more accurate computer-aided diagnosis. Therefore, the said features need to be used when correcting the eye cornea surface curvature to improve vision acuity.

Thus, based on the experience gained when studying wave front aberrations in the human eye, we have demonstrated the relevance of the development of new types of aberrometers and wave front sensors for professional ophthalmology, enabling the human eye optical system to be better evaluated and the accuracy of computer-aided diagnosis to be enhanced, making possible the transition to personified hi-tech medicine.

ACKNOWLEDGEMENT

"The work was supported by the Russian Science Foundation grant No. 22-12-00041"

REFERENCES

1. David, F. 2015. *Buscher practical optical interferometry*. Cambridge: Cambridge University Press. ISBN: 978-1-107-04217-9.
2. Malacara, D. 2007. *Optical shop testing*. Hoboken, NJ: John Wiley & Sons Inc. ISBN: 978-0-471-48404-2.
3. Vasil'ev, L. A. 1971. *Schlieren methods*. New York, Jerusalem, London: Israel Program for Scientific Translations. ISBN: 978-0-7065-1100-0.
4. Hartmann, J. 1900. Bemerkungen über den Bau und die Justierung von Spektrographen. *Zeitschrift für Instrumentenkunde*. 20:17–27, 47–58.
5. Artzner, G. 1992. Microlens arrays for Shack-Hartmann wavefront sensors. *Opt. Eng.* 31(6):1311–22. https://doi.org/10.1117/12.56178.
6. Minin, I., and Minin, O. 2015. *Diffractive optics and nanophotonics. Resolution below the diffraction limit*. Cham: Springer International Publishing. ISBN: 978-3-319-24251-4.
7. Zernike, F. 1955. How I discovered phase contrast. *Science*. 121(3141):345–9. https://doi.org/10.1126/science.121.3141.345.
8. Situ, G., Warber, M., Pedrini, G., and Osten, W. 2010. Phase contrast enhancement in microscopy using spiral phase filtering. *Opt. Commun.* 283(7):1273–7. https://doi.org/10.1016/j.optcom.2009.11.084.
9. Zernike, F. 1934. Beugungstheorie des schneidenver-fahrens und seiner verbesserten form, der phasenkontrastmethode. *Physica*. 1(7–12):689–704. https://doi.org/10.1016/S0031-8914(34)80259-5.

10. Nijboer, B. R. A. 1943. The diffraction theory of optical aberrations: Part i: General discussion of the geometrical aberrations. *Physica.* 10(8):679–92. https://doi.org/10.1016/S0031-8914(43)80016-1.

11. Nijboer, B. R. A. 1947. The diffraction theory of optical aberrations: Part II: Diffraction pattern in the presence of small aberrations. *Physica.* 13(10):605–20. https://doi.org/10.1016/0031-8914(47)90052-9.

12. Nienhuis, K., and Nijboer, B. R. A. 1949. The diffraction theory of optical aberrations: Part III: General formulae for small aberrations; Experimental verification of the theoretical results. *Physica.* 14(9):590–604. https://doi.org/10.1016/0031-8914(49)90002-6.

13. Born, M., and Wolf, E. 1999. *Principles of optics: Electromagnetic theory of propagation, interference and diffraction of light.* 7th ed. Cambridge: Cambridge University Press. ISBN: 978-0-521-64222-4.

14. Roddier, N. 1990. Atmospheric wavefront simulation using Zernike polynomials. *Opt. Eng.* 29(10):1174–80. https://doi.org/10.1117/12.55712.

15. Neil, M. A. A., Booth, M. J., and Wilson, T. 2000. New modal wave-front sensor: A theoretical analysis. *J. Opt. Soc. Am. A.* 17(6):1098–107. https://doi.org/10.1364/JOSAA.17.001098.

16. Thibos, L. N., Applegate, R. A., Schwiegerling, J. T., and Webb, R. 2002. Standards for reporting the optical aberrations of eyes. *J. Refract. Surg.* 18:S652–S660. https://doi.org/10.3928/1081-597X-20020901-30.

17. Evans, C. J., Parks, R. E., Sullivan, P. J., and Taylor, J. S. 1995. Visualization of surface figure by the use of Zernike polynomials. *Appl. Opt.* 34(34):7815–9. https://doi.org/10.1364/AO.34.007815.

18. ANSI Z80.28. 2004. Methods for reporting optical aberrations of eyes. American National Standards Institute Inc.

19. Lombardo, M., and Lombardo, G. 2010. Wave aberration of human eyes and new descriptors of image optical quality and visual performance. *J. Cataract Refr. Surg.* 36(2):313–31. https://doi.org/10.1016/j.jcrs.2009.09.026.

20. Khorin, P. A., Khonina, S. N., Karsakov, A. V., and Branchevskiy, S. L. 2016. Analysis of corneal aberration of the human eye. *Comput. Opt.* 40(6):810–7. https://doi.org/10.18287/0134-2452-2016-40-6-810-817.

21. Martins, A. C., and Vohnsen, B. 2019. Measuring ocular aberrations sequentially using a digital micromirror device. *Micromachines.* 10(2):117. https://doi.org/10.3390/mi10020117.

22. Soifer, V. A., ed. 2002. *Methods for computer design of diffractive optical elements.* New York: John Wiley & Sons Inc. ISBN: 978-0-471-09533-0.

23. Soifer, V. A., ed. 2013. *Computer design of diffractive optics.* Cambridge: Cambridge International Science Publishing Limited & Woodhead Publishing Ltd. ISBN: 978-1-84569-635-1.

24. Picart, P. 2015. *New techniques in digital holography.* John Wiley & Sons Inc. ISBN: 978-1-84821-773-7.

25. Golub, M. A., Karpeev, S. V., Krivoshlykov, S. G., Prokhorov, A. M., Sisakyan, I. N., and Soifer, V. A. 1984. Spatial filter investigation of the distribution of power between transverse modes in a fiber waveguide. *Sov. J. Quantum Electron.* 14(9):1255–6. https://doi.org/10.1070/QE1984v014n09ABEH006201.

26. Golub, M. A., Karpeev, S. V., Kazanskiĭ, N. L., Mirzov, A. V., Sisakyan, I. N., Soifer, V. A., and Uvarov, G. V. 1988. Spatial phase filters matched to transverse modes. *Sov. J. Quantum Electron.* 18(3):392–3. https://doi.org/10.1070/QE1988v018n03ABEH011528.

27. Almazov, A. A., Khonina, S. N., and Kotlyar, V. V. 2005. Using phase diffraction optical elements to shape and select laser beams consisting of a superposition of an arbitrary number of angular harmonics. *J. Opt. Technol.* 72(5): 391–9. https://doi.org/10.1364/JOT.72.000391.

28. Khonina, S. N., and Ustinov, A. V. 2019. Binary multi-order diffraction optical elements with variable fill factor for the formation and detection of optical vortices of arbitrary order. *Appl. Opt.* 58(30):8227–36. https://doi.org/10.1364/AO.58.008227.

29. Ha, Y., Zhao, D., Wang, Y., Kotlyar, V. V., Khonina, S. N., and Soifer, V. A. 1998. Diffractive optical element for Zernike decomposition. *Proc. SPIE.* 3557:191–7. https://doi.org/10.1117/12.318300.

30. Booth, M. J. 2003. Direct measurement of Zernike aberration modes with a modal wavefront sensor. *Proc. SPIE.* 5162:79–90. https://doi.org/10.1117/12.503695.

31. Porfirev, A. P., and Khonina, S. N. 2016. Experimental investigation of multi-order diffractive optical elements matched with two types of Zernike functions. *Proc. SPIE.* 9807:98070E. https://doi.org/10.1117/12.2231378.

32. Garitchev, V. P., Golub, M. A., Karpeev, S. V., Krivoshlykov, S. G., Petrov, N. I., Sissakian, I. N., Soifer, V. A., Haubenreisser, W., Jahn, J.-U., and Willsch, R. 1985. Experimental investigation of mode coupling in a multimode graded-index fiber, caused by periodic microbends using computer-generated spatial filters. *Opt. Commun.* 55(6):403–5. https://doi.org/10.1016/0030-4018(85)90140-3.

33. Karpeev, S. V., Pavelyev, V. S., Khonina, S. N., and Kazanskiy, N. L. 2005. High-effective fiber sensors based on transversal mode selection. *Proc. SPIE.* 5854:163–9. https://doi.org/10.1117/12.634603.

34. Karpeev, S. V., Pavelyev, V. S., Khonina, S. N., Kazanskiy, N. L., Gavrilov, A. V., and Eropolov, V. A. 2007. Fibre sensors based on transverse mode selection. *J. Mod. Opt.* 54(6):833–44. https://doi.org/10.1080/09500340601066125.

35. Khonina, S. N., Kotlyar, V. V., Soifer, V. A., Paakkonen, P., Simonen, J., and Turunen, J. 2001. An analysis of the angular momentum of a light field in terms of angular harmonics. *J. Mod. Opt.* 48(10):1543–57. https://doi.org/10.1080/09500340108231783.

36. Moreno, I., Davis, J. A., Pascoguin, B. M. L., Mitry, M. J., and Cottrell, D. M. 2009. Vortex sensing diffraction gratings. *Opt. Lett.* 34(19):2927–9. https://doi.org/10.1364/OL.34.002927.

37. Lei, T., Zhang, M., Li, Y., Jia, P., Liu, G. N., Xu, X., Li, Z., Min, C., Lin, J., Yu, C., and Niu, H. 2015. Massive individual orbital angular momentum channels for multiplexing enabled by Dammann gratings. *Light Sci. Appl.* 4(3):e257. https://doi.org/10.1038/lsa.2015.30.

38. Karpeev, S. V., Pavelyev, V. S., Soifer, V. A., Khonina, S. N., Duparre, M., Luedge, B., and Turunen, J. 2005. Transverse mode multiplexing by diffractive optical elements. *Proc. SPIE.* 5854: 163–9. https://doi.org/10.1117/12.634547.

39. Moreno, I., Davis, J. A., Ruiz, I., and Cottrell, D. M. 2010. Decomposition of radially and azimuthally polarized beams using a circular-polarization and vortex-sensing diffraction grating. *Opt. Express* 18(7):7173–83. https://doi.org/10.1364/OE.18.007173.

40. García-Martínez, P., Sánchez-López, M. M., Davis, J. A., Cottrell, D. M., Sand, D., and Moreno, I. 2012. Generation of Bessel beam arrays through Dammann gratings. *Appl. Opt.* 51(9):1375–81. https://doi.org/10.1364/AO.51.001375.

41. Lyubopytov, V. S., Tlyavlin, A. Z., Sultanov, A. K., Bagmanov, V. K., Khonina, S. N., Karpeev, S. V., and Kazanskiy, N. L. 2013. Mathematical model of completely optical system for detection of mode propagation parameters in an optical fiber with few-mode operation for adaptive compensation of mode coupling. *Comput. Opt.* 37(3):352–9.

42. Ni, B., Guo, L., Yue, C., and Tang, Z. 2017. A Novel measuring method for arbitrary optical vortex by three spiral spectra. *Phys. Lett. A.* 381(8):817–20. https://doi.org/10.1016/j.physleta.2016.12.050.

43. Khonina, S. N., Karpeev, S. V., and Paranin, V. D. 2018. A technique for simultaneous detection of individual vortex states of Laguerre–Gaussian beams transmitted through an aqueous suspension of microparticles. *Opt. Laser Eng.* 105:68–74. https://doi.org/10.1016/j.optlaseng.2018.01.006.

44. Abderrahmen, T., Park, K.-H., Zghal, M., Ooi, B. S., and Alouini, M.-S. 2019. Communicating using spatial mode multiplexing: Potentials, challenges, and perspectives. *IEEE Commun. Surv. Tut.* 21(4):3175–203. https://doi.org/10.1109/COMST.2019.2915981.
45. Wilby, M. J., Keller, C. U., Haert, S., Korkiakoski, V., Snik, F., and Pietrow, A. G. M. 2016. Designing and testing the coronagraphic Modal Wavefront Sensor: a fast non-common path error sensor for high-contrast imaging. *Proc. SPIE.* 9909:990921. https://doi.org/10.1117/12.2231303.
46. Lyu, H., Huang, Y., Sheng, B., and Ni, Z. 2018. Absolute optical flatness testing by surface shape reconstruction using Zernike polynomials. *Opt. Eng.* 57(9):094103. https://doi.org/10.1117/1.OE.57.9.094103.
47. Bereznyi, A. E., Prokhorov, A. M., Sisakyan, I. N., and Soifer, V. A. 1984. Bessel-optics. *Soviet Phys Dokl.* 29(2):115–7.
48. Khonina, S. N., Kotlyar, V. V., Shinkaryev, M. V., Soifer, V. A., and Uspleniev, G. V. 1992. The phase rotor filter. *J. Mod. Opt.* 39(5):1147–54. https://doi.org/10.1080/09500349214551151.
49. Degtyarev, S. A., Porfirev, A. P., and Khonina, S. N. 2017. Zernike basis-matched multi-order diffractive optical elements for wavefront weak aberrations analysis. *Proc. SPIE.* 10337:103370Q. https://doi.org/10.1117/12.2269218.
50. Koltyar, V. V., and Khonina, S. N. 2003. Multi-order diffractive optical elements to process data. In Perspectives in engineering optics. Delhi: Anita Publications. 47–56.
51. Khonina, S. N., Kazanskiy, N. L., and Soifer, V. A. 2012. Optical vortices in a fiber: Mode division multiplexing and multimode self-imaging. In Recent progress in optical fiber research. London: IntechOpen Limited. 327–52. ISBN: 978-953-307-823-6.
52. Kotlyar, V. V., Khonina, S. N., Soifer, V. A., Wang, Y., and Zhao, D. 1997. Coherent field phase retrieval using a phase Zernike filter. *Comput. Opt.* 17:43–8.
53. Khonina, S. N., Kotlyar, V. V., and Kirsh, D. V. 2015. Zernike phase spatial filter for measuring the aberrations of the optical structures of the eye. *J. Biomed. Photonics Eng.* 1(2):146–53. https://doi.org/10.18287/JBPE-2015-1-2-146.
54. Guo, H., Korablinova, N., Ren, Q., and Bille, J. 2006. Wavefront reconstruction with artificial neural networks. *Opt. Express.* 14(14):6456–62. https://doi.org/10.1364/OE.14.006456.
55. Paine, S. W., and Fienup, J. R. 2018. Machine learning for improved image-based wavefront sensing. *Opt. Lett.* 43(6):1235–8. https://doi.org/10.1364/OL.43.001235.
56. Nishizaki, Y., Valdivia, M., Horisaki, R., Kitaguchi, K., Saito, M., Tanida, J., and Vera, E. 2019. Deep learning wavefront sensing. *Opt. Express.* 27(1):240–51. https://doi.org/10.1364/OE.27.000240.
57. Maeda, N. 2009. Clinical applications of wavefront aberrometry – A review. *Clin. Exp. Ophthalmol.* 37(1):118–29. https://doi.org/10.1111/j.1442-9071.2009.02005.x.
58. Mello, G. R., Rocha, K. M., Santhiago, M. R., Smadja, D., and Krueger, R. R. 2012. Applications of wavefront technology. *J. Cataract. Refract. Surg.* 38(9):1671–83. https://doi.org/10.1016/j.jcrs.2012.07.004.
59. Montés-Micó, R., Cáliz, A., and Alió, J. L. 2004. Wavefront analysis of higher order aberrations in dry eye patients. *J. Refract. Surg.* 20(3):243–7. https://doi.org/10.3928/1081-597X-20040501-08.
60. Maeda, N., Fujikado, T., Kuroda, T., Mihashi, T., Hirohara, Y., Nishida, K., Watanabe, H., and Tano, Y. 2002. Wavefront aberrations measured with Hartmann-shack sensor in patients with keratoconus. *Ophthalmology.* 109(11):1996–2003. https://doi.org/10.1016/s0161-6420(02)01279-4.
61. Gobbe, M., Reinstein, D. Z., and Archer, T. J. 2015. LASIK-induced aberrations: Comparing corneal and whole-eye measurements. *Optom. Vis. Sci.* 92(4):447–55. https://doi.org/10.1097/OPX.0000000000000557.

62. Denoyer, A., Ricaud, X., Van Went, C., Labbé, A., and Baudouin, C. 2013. Influence of corneal biomechanical properties on surgically induced astigmatism in cataract surgery. *J. Cataract. Refract. Surg.* 39(8):1204–10. https://doi.org/10.1016/j.jcrs.2013.02.052.

63. Cerviño, A., Hosking, S. L., Montes-Mico, R., and Bates, K. 2007. Clinical ocular wavefront analyzers. *J. Refract. Surg.* 23(6):603–16. https://doi.org/10.3928/1081-597X-20070601-12.

64. Plaza-Puche, A. B., Salerno, L. C., Versaci, F., Romero, D., and Alio, J. L. 2019. Clinical evaluation of the repeatability of ocular aberrometry obtained with a new pyramid wavefront sensor. *Eur. J. Ophthalmol.* 29(6):585–92. https://doi.org/10.1177/1120672118816060.

65. Iglesias, I., Ragazzoni, R., Julien, Y., and Artal, P. 2002. Extended source pyramid wave-front sensor for the human eye. *Opt. Express.* 10(9):419–28. https://doi.org/10.1364/OE.10.000419.

66. Hutterer, V. et al. 2019. Real-time adaptive optics with pyramid wavefront sensors: Part i. A theoretical analysis of the pyramid sensor model. *Inverse Probl.* 35(4):045007. https://doi.org/10.1088/1361-6420/ab0656.

67. Hénault, F., Spang, A., Feng, Y., and Schreiber, L. 2020. Crossed-sine wavefront sensor for adaptive optics, metrology and ophthalmology applications. *Eng. Res. Express.* 2(1):015042. https://doi.org/10.1088/2631-8695/ab78c5.

68. Oti, J. E., Canales, V. F., and Cagigal, M. P. 2003. Analysis of the signal-to-noise ratio in the optical differentiation wavefront sensor. *Opt. Express.* 11(21):2783–90. https://doi.org/10.1364/OE.11.002783.

69. Xu, Z., Hua, Y., and Qiu, W. et al. 2018. Precision and agreement of higher order aberrations measured with ray tracing and Hartmann-Shack aberrometers. *BMC Ophthalmol.* 18:18. https://doi.org/10.1186/s12886-018-0683-8.

70. Han, J. Y., Yoon, S., Brown, N. S., Han, S.-H., and Han, J. 2020. Accuracy of the hand-held wavefront aberrometer in measurement of refractive error. *Korean J. Ophthalmol.* 34(3):227–34. https://doi.org/10.3341/kjo.2019.0132.

71. Kim, B. K., Mun, S. J., and Yang, Y. H. et al. 2019. Comparison of anterior segment changes after femtosecond laser LASIK and SMILE using a dual rotating Scheimpflug analyzer. *BMC Ophthalmol.* 19:251. https://doi.org/10.1186/s12886-019-1257-0.

72. Randleman, J. B., Woodward, M., Lynn, M. J., and Stulting, R. D. 2008. Risk assessment for ectasia after corneal refractive surgery. *Ophthalmology.* 115(1):37–50. https://doi.org/10.1016/j.ophtha.2007.03.073.

73. Salouti, R., Nowroozzadeh, M. H., Zamani, M., Fard, A. H., and Niknam, S. 2009. Comparison of anterior and posterior elevation map measurements between 2 Scheimpflug imaging systems. *J. Cataract. Refract. Surg.* 35:856–62. https://doi.org/10.1016/j.jcrs.2009.01.008.

74. Kim, E. J., de Oca, I. M., Wang, L., Weikert, M. P., Koch, D. D., and Khandelwal, S. S. 2015. Repeatability of posterior and total corneal curvature measurements with a dual Scheimpflug–Placido tomographer. *J. Cataract. Refract. Surg.* 41(12):2731–8. https://doi.org/10.1016/j.jcrs.2015.07.035.

75. Güell, J. L., Pujol, J., and Arjona, M. et al. 2004. Optical quality analysis system: Instrument for objective clinical evaluation of ocular optical quality. *J. Cataract. Refract. Surg.* 30(7): 1598–9. https://doi.org/10.1016/j.jcrs.2004.04.031.

76. Santamaría, J., Artal, P., and Bescós, J. 1987. Determination of the point-spread function of human eyes using a hybrid optical-digital method. *J. Opt. Soc. Am. A.* 4(6):1109–14. https://doi.org/10.1364/JOSAA.4.001109.

77. Frings, A., Hassan, H., and Allan, B. D. 2020. Pyramidal aberrometry in wavefront-guided myopic LASIK. *J. Refract. Surg.* 36(7):442–8. https://doi.org/10.3928/108159 7X-20200519-03.

78. Zlatanović, M., Živković, M., Hristov, A., Stojković, V., Novak, S., Zlatanović, N., and Brzaković, M. 2019. Central corneal thickness measured by the Oculyzer, BioGraph, and ultrasound pachymetry. *Acta Medica Medianae.* 58(2):33–7. https://doi.org/10.5633/AMM.2019.0206.

79. Roshdy, M. M., Wahba, S. S., and Fikry, R. R. 2018. New corneal assessment index from the rational thickness and other OCULUS values (CAIRO index). *Clin. Ophthalmol.* 12:1527–32. https://doi.org/10.2147/OPTH.S171827.

80. Sun, M.-S., Zhang, L., Guo, N., Song, Y.-Z., and Zhang, F.-J. 2018. Consistent comparison of angle Kappa adjustment between Oculyzer and Topolyzer Vario topography guided LASIK for myopia by EX500 excimer laser. *Int. J. Ophthalmol.* 11(4):662–7. https://doi.org/10.18240/ijo.2018.04.21.

81. Branchevskiy, S. L., and Branchevskaya, E. S. 2018. Comparative analysis of devices based on Plasido, Scheimpflug and OCT for measuring keratometry in patients after laser vision correction. *Modern Technologies Ophthalmol.* 5:185–7. https://doi.org/10.25276/2312-4911-2018-5-185-187.

82. Belin, M. W., and Zloty, P. 1993. Accuracy of the PAR corneal topography system with spatial misalignment. *CLAO J.* 19(1):64–8. https://doi.org/10.1097/00140068-199301000-00012.

83. Khorin, P. A., Ilyasova, N. Y., and Paringer, R. A. 2018. Informative feature selection based on the Zernike polynomial coefficients for various pathologies of the human eye cornea. *Comput. Opt.* 42(1):159–66. https://doi.org/10.18287/2412-6179-2018-42-1-159-166.

84. Lombardo, M., and Lombardo, G.. 2010. Wave aberration of human eyes and new descriptors of image optical quality and visual performance. *J. Cataract. Refract. Surg.* 36:313–31. https://doi.org/10.1016/j.jcrs.2009.09.026.

85. Artal, P., Guirao, A., Berrio, E., and Williams, D. R. 2001. Compensation of corneal aberrations by the internal optics in the human eye. *J. Vis.* 1(1):1–8. https://doi.org/10.1167/1.1.1.

86. Artal, P., Berrio, E., Guirao, A., and Piers, P. 2002. Contribution of the cornea and internal surfaces to the change of ocular aberrations with age. *J. Opt. Soc. Am. A.* 19(1):137–43. https://doi.org/10.1364/josaa.19.000137.

87. He, J. C., Gwiazda, J., Thorn, F., and Held, R. 2003. Wave-front aberrations in the anterior corneal surface and the whole eye. *J. Opt. Soc. Am. A.* 20(7):1155–63. https://doi.org/10.1364/JOSAA.20.001155.

88. Mrochen, M., Jankov, M., Bueeler, M., and Seiler, T. 2003. Correlation between corneal and total wavefront aberrations in myopic eyes. *J. Refract. Surg.* 19(2):104–12. https://doi.org/10.1117/12.470593.

89. Kelly, J. E., Mihashi, T., and Howland, H. C. 2004. Compensation of corneal horizontal/vertical astigmatism, lateral coma, and spherical aberration by internal optics of the eye. *J. Vis.* 4(4):262–71. https://doi.org/10.1167/4.4.2.

90. ZEMAX®. Optical design program. User's manual. July 8, 2011. https://neurophysics.ucsd.edu/Manuals/Zemax/ZemaxManual.pdf.

91. Khonina, S. N., Kotlyar, V. V., and Wang, Y. 2001. Diffractive optical element matched with Zernike basis. *Pattern Recognit. Image Anal.* 11(2):442–5.

92. Kirilenko, M. S., Khorin, P. A., and Porfirev, A. P. 2016. Wavefront analysis based on Zernike polynomials. *CEUR Workshop Proc.* 1638:66–75. https://doi.org/10.18287/1613-0073-2016-1638-66-75.

93. Tocci, M. How to model the human eye in Zemax. Zemax knowledgebase. 2007. https://support.zemax.com/hc/en-us/articles/1500005575002-How-to-model-the-human-eye-in-OpticStudio.

94. Guirao, A., Redondo, M., and Artal, P. 2000. Optical aberrations of the human cornea as a function of age. *J. Opt. Soc. Am. A.* 17(10):1697–702. https://doi.org/10.1364/josaa.17.001697.

95. Vapnik, V. N. 2000. *The nature of statistical learning theory.* 2nd ed. New York: Springer Science+Business Media. ISBN: 978-1-4419-3160-3.

96. Fukunaga, K. 1990. *Introduction to statistical pattern recognition.* San Diego: Academic Press. ISBN: 978-0-12-269851-4.

97. Ilyasova, N. Y. 2013. Methods for digital analysis of human vascular system. Literature review. *Comput. Opt.* 37(4):517–41.

98. Ilyasova, N., Paringer, R., and Kupriyanov, A. 2016. Regions of interest in a fundus image selection technique using the discriminative analysis methods, 408–417. In *Computer vision and graphics*, ed. L. J. Chmielewski, A. Datta, R. Kozera, K. Wojciechowski. Springer International Publishing AG. https://doi.org/10.1007/978-3-319-46418-3_36.

99. Kutikova, V. V., and Gaidel, A. V. 2015. Study of informative feature selection approaches for the texture image recognition problem using Laws' masks. *Comput. Opt.* 39(5):744–50. https://doi.org/10.18287/0134-2452-2015-39-5-744-750.

100. Gaidel, A. V., and Krasheninnikov, V. R. 2016. Feature selection for diagnosing the osteoporosis by femoral neck X-ray images. *Comput. Opt.* 40(6):939–46. https://doi.org/10.18287/2412-6179-2016-40-6-939-946.

8 Optical Computing
Key Problems, Achievements, and Perspectives

Nikolay Lvovich Kazanskiy[1,2], Muhammad Ali Butt[1,3], and Svetlana Nikolaevna Khonina[1,2]
[1]Image Processing Systems Institute – Branch of the
Federal Scientific Research Centre "Crystallography and
Photonics" of Russian Academy of Sciences, Samara, Russia
[2]Samara National Research University, Samara, Russia
[3]Warsaw University of Technology, Institute of
Microelectronics and Optoelectronics, Warszawa, Poland

8.1 INTRODUCTION

The topic of optical computing (OC) dates to the early 1960s, or perhaps before, when the military became interested in utilizing the Fourier transform (FT) relationships essential in coherent optical imaging systems to perform processes like convolution and correlation. On data supplied in imaging optical format to a bulk optical system, it was easily demonstrated that these processes could be performed with considerable speed. Such processors were fundamentally analog in operation, and, as a result, they constantly struggled to maintain adequate dynamic range and signal-to-noise ratios, severely restricting their use. Despite several spectacular demonstrations, silicon digital electronic processing appears to have nearly always been chosen for final manufacturing equipment [1]. Nonetheless, there is still a lot of interest in such specialized machines, and it is certainly true that they can attain very high comparable digital processing speeds. The use of nonlinear optical tools to establish the basic digital processing functions of AND, OR, NAND, NOR, etc., sparked even more interest in the early 1980s, this time in imaging optical format, where a single lens could, in theory, image a very large number of parallel channels from a two-dimensional (2D) array of devices. As a result, assertions have been made that those future high-speed computers would use vastly parallel digital-optical processing to attain speeds much above those conceivable with electronics [2].

Following that, a slew of large R&D initiatives were aimed at capitalizing on the potential. Such statements are established on basic expectations about prospective digital throughput, but easily disregard the enormous practical issues that come with putting them into practice. Because the analog light wave level represents the digital state, practically all optical logic systems addressed in the literature use threshold logic, meaning highly tight control of the optical power level across a complicated multichannel system. Given how readily unanticipated 3-dB insertion losses may be obtained in complicated systems of mirrors, holograms, lenses, and other components, one would wonder if this had any chance of being realized. Nevertheless, employing a dual-rail optical signaling system, the symmetric-SEED technique established at AT&T Bell Labs and explored elsewhere does give a sophisticated resolution to this problem [3]. Other important concerns include the fact that the finest optical logic systems are optically activated electrical equipment, and huge ones at that because light must enter them, and optical wavelengths are extremely long by electronic norms. The practical concerns of compiling high-resolution imaging optical systems, the shallow depth of field, the accuracy inferred in the lenses (in terms of focal length), and the mechanical tolerances and massive dimensions of constructed structures are all simple issues that are conveniently overlooked by many. Nevertheless, none of these disadvantages indicates that it cannot be done; rather, they raise the challenge, and as some of the results will demonstrate, genuinely spectacular experimental procedures established on free-space optics may be created in the research laboratory.

The switches, optical logic gates (OLGs), and memory components that regulate the flow of electrons in an electronic computer would be replaced by optical systems in a digital optical computer. Optical modulators, which come in a variety of shapes and sizes, may perform the roles of the OLGs and switches. A 1×1 switch is a simple on–off switch with the ability to link two lines. A 1×2 switch connects one line to one of two lines. Two lines are connected by a 2×2 switch. It can be in one of two states: input (I/P) 1 linked to output (O/P) 1 and I/P 2 connected to O/P 2, or I/P 1 connected to O/P 2 and I/P 2 connected to O/P 1, known as the bar state, or I/P 1 connected to O/P 2 and I/P 2 connected to O/P 1, known as the cross-state. As a result, it's known as a crossbar switch. A crossbar switch may be stretched to $n \times n$ arrangement, which allows any of n I/P lines to be linked to any of n O/P lines at any time without causing interference.

Optical switches can be made from electro-optic modulators, acousto-optic modulators, magneto-optic modulators, and liquid crystals, among other forms of optical modulators. A crossbar switch can also be made from an array of modulators. An array of $n \times n$ light valves, for instance, might be used to create $n \times n$ crossbar switch. The light source would be a vertical linear array of n laser diodes. The light from the diodes is dispersed horizontally by a cylindrical lens, such that each diode illumines one row of the $n \times n$ array. Then, one lens per row of the modulator array, a set of n cylindrical lenses, positioned perpendicular to the first lens, is employed to direct the signal emitted by the array onto a horizontally oriented linear array of n photodetectors. This setup allows light from any laser diode to be connected to any detector without interruption. Switches that are operated both electrically and optically have been designed. Optically regulated apparatuses would be used in an all-optical computer. The switches should also be tiny, high-speed apparatuses that can be produced in vast arrays and consume very little switching energy. Several alternative technologies have been used to illustrate the needed switching functionalities. However, massive, high-density arrays of OLGs are currently being developed.

A multinational research team lead by Russia's Skolkovo Institute of Science and Technology has developed an exceptionally energy-efficient optical switch in conjunction with IBM [4]. The switch is extremely rapid and does not involve any cooling. It might serve as the foundation for a next generation of computers that manage photons instead of electrons. A 35-nanometer semi-conducting polymer consisting of organic material is placed between highly reflecting surfaces to form the switch. This results in the formation of a tiny chamber that traps light beam. The gadget is powered by two lasers: a pump laser and a seed laser. When the pump laser shines on the switch, thousands of indistinguishable quasiparticles develop in the same area, generating a Bose–Einstein condensate – a collection of particles that individually act like a single atom. The seed beam is utilized to alter this condensate between two measured states that act as binary codes "0" and "1." The new technology can do 10^{12} calculations per second, which is 1,000 times quicker than today's finest commercial transistors. Furthermore, it requires significantly less energy to transition states than transistors. This is due to the optical switch's ability to be actuated by a single photon of light. Equivalent electrical transistors that employ single electrons typically need enormous quantities of cooling device, which uses a lot of electricity. The new switch, on the other hand, will function at room temperature. Before it can be deployed, the technology has a long way to go. The first electronic transistor literally takes years to make its way into a personal computer. The researchers face a hurdle in that, despite requiring relatively little energy to switch, the device still requires continual input from the pump laser. The team is researching ways to get around this by employing superfluorescent perovskite supercrystals to aid in reducing power usage. Despite the difficulties, the researchers expect that the novel switch will be employed in various types of optical computing systems in the near future, maybe as a method to ramp up supercomputer processing [4].

The urgent demand for optical technology relies on the fact that the time response of electrical circuits limits today's computers [5]. A solid transmission medium restricts signal speed and volume while also generating heat that destroys components. A 1-foot length of wire, for instance, creates roughly 1 nanosecond (billionth of a second) of time delay. The extreme downsizing of microscopic electronic apparatuses also causes "cross talk," or signal mistakes that compromise the system's dependability. These and other challenges have prompted researchers to look for solutions in light itself. Light doesn't even have the time domain restrictions of electronics, does not require insulators, and may even deliver dozens or hundreds of photon communication streams at distinct color frequencies at the same time, those that are not susceptible to electromagnetic (EM) interference and do not experience electrical short circuits. They feature low-loss transmission and a broad bandwidth; that is, they can communicate with numerous channels simultaneously without interference. They may transmit signals inside the same or neighboring optical fibers (OFs) with little to no interference or cross talk. They are smaller, lighter, and less costly to produce, and they work better with stored data than magnetic materials. Scientists want to construct a new class of computers that run 100 million times quicker than today's apparatuses by restoring electrons and cables with photons, OFs, crystals, thin films, and mirrors [6].

Optical processing hits at a critical juncture in computing history. The need for AI is growing at the same time as Moore's Law is breaking down in silicon-based computing. Optalysys is a game-changing technology startup that uses groundbreaking optical processing techniques to address this challenge, allowing new levels of AI capability for high-resolution image and video-based applications [7].

The technology is emerging to speed up several of the most demanding processor-intensive operations while using a fraction of the energy that silicon processors consume. It may be configured to execute optical correlation and convolution operations using the API or Tensorflow interface to unleash new levels of AI and pattern recognition potential. Optalysys employs photons instead of electrons, but it also does high-resolution computations, allowing huge image/pattern-based data to be handled at speeds significantly quicker than silicon. The Optalysys technology can deliver something distinctive in the AI space: a scalable processor that can accomplish end-to-end, full-resolution processing of multi-megapixel image and video data, or contextually pre-process data for enhancing the effectiveness of existing Convolutional Neural Network (CNN)-type models for high-resolution data applications, thanks to the characteristics of diffractive optics.

8.2 TYPES OF COMPUTING AND NECESSITY OF OPTICS FOR COMPUTING

OC may be divided into two types: digital optical computing (DOC) and analog optical computing (AOC). For more than 30 years, DOC based on Boole logics has been established, employing a technique comparable to general-purpose computing realized on transistors. Nonetheless, given reduced optical device integration density, it is impossible to beat traditional digital computing. AOC, on the other hand, makes use of the physical characteristics of light, including amplitude and phase, as well as the interactions between light and optical apparatuses, to perform specific computations. Due to the obvious unique mathematical portrayal of computational processes in one AOC system, it is considered dedicated computing. AOC can achieve higher data processing speed in specialized applications, including pattern recognition and numerical calculation, as compared to standard digital computing. As a result, AOC has received a lot of academic attention as one of the most potential pervasive computing systems in the post-Moore age.

Optical technology offers tremendous improvements in computational efficiency and speed, along with considerable size and cost reductions [8, 9]. Since many processes may be conducted at the same time, an optical desktop computer might process data 100,000 times quicker than existing versions. Low production costs, resilience to EM interference, low-loss transmission tolerance, independence from short electrical connections, and the capacity to deliver wide bandwidth and transport signals without interference within the same or neighboring OFs are all features of optics. The distinction between optical and electrical synchrony may be shown with a simplistic example. Imagine an imaging system with $1,000 \times 1,000$ distinct units per mm^2 in the object plane that are optically coupled to a comparable number of points per mm^2 in the image plane; the lens essentially conducts a real-time FFT (fast Fourier transform) of the image plane. A million processes are necessary to do this electronically. If parallelism is combined with rapid switching speeds, startling computing speeds are possible. Consider there are only 100 million OLGs on a chip, which is significantly fewer than the figure indicated earlier. Furthermore, imagine that each OLG has a switching period of just 1 nanosecond (organic optical switches may switch at sub-picosecond rates, but electronic switching has a maximum picosecond switching time). Well over 1,017-bit operations per second may be performed by such a device. When compared to the gigabits (109) or terabits (1,012) per second speeds that electronics are now restricted to or aiming for, this is a significant difference. In other words, an operation that would take a normal computer a hundred thousand hours (>11 years) might just be completed in less than an hour by an optical computer.

As photons are uncharged and do not react with each other as easily as electrons, light does have another benefit. As a result, in full-duplex functioning, beams of light can travel across each other without altering the information transmitted. Loops in electronics produce noisy voltage spikes anytime the EM fields pass through the loop change. Furthermore, switching pulses with a high frequency or a short duration will generate cross talk in surrounding lines. Signals in nearby OFs or optical-integrated channels are unaffected by each other, and they do not pick up noise from loops. Ultimately, optical materials outperform magnetic materials in terms of storage density and availability. The subject of OC is growing quickly and offers many exciting possibilities for transcending the limits of today's electrical computers. Optical apparatuses are already being integrated into a variety of computing systems. Resulting in mass manufacture, the cost of laser diodes as coherent light sources has reduced dramatically. Optical CD-ROM discs have also been widely used in both home and office PCs [10].

8.3 OPTICAL PROCESSING ARCHITECTURE

Optical information processing is introduced and established by utilizing all of light's attributes of speed and parallelism to handle data at a high rate. The data is represented as a light wave or graphic. The inherent parallel processing of OC was frequently noted as a fundamental benefit of optical processing over electronic processing employing predominantly serial processors. As a result, optics holds significant promise for interpreting enormous amounts of data in real time. OC is established on the FT feature of a lens. When employing coherent light, a lens executes the FT of a 2D transparency in its front focal plane in its rear focal plane. The lens computes the precise FT with amplitude and phase in an analog method [11].

Figure 8.1 depicts the topology of a general-purpose optical processor for data processing. The I/P plane, the processing plane, and the O/P plane make up the system.

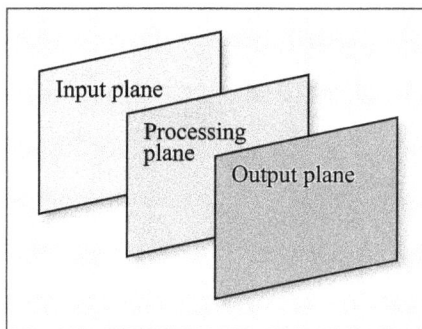

FIGURE 8.1 An optical processor's topology.

The I/P plane displays the data to be processed; this plane will usually implement an algorithm which helps in the conversion of signals from electricity to light. The conversion is done via a Spatial Light Modulator (SLM). A 1D or 2D signal can be used as an I/P. In the context of a 1D I/P signal, an acousto-optic cell is commonly utilized, whereas 2D SLMs are used for 2D signals. Due to the lack of SLMs in the early years, the I/P plane consisted of a stationary slide. As a result, the concepts and possibilities of optical processors could be proved, but no real-time applications could be presented, rendering the processor worthless for most real-world applications. Lenses, holograms (optically recorded or computer produced), and nonlinear components can all be found in the processing plane. This is perhaps the most important component of the processing, and it can be done at the speed of light in most optical processors. The O/P plane, which detects the processing results, is made up of a photodetector, a photodetector array, or a camera. Since the bulk of them are running at video rates, Figure 8.1 indicates that the speed of the entire process is restricted by the speed of its slowest component, which is usually the I/P plane SLM. The SLM is an essential component in the creation of realistic optical processors, but it's also one of their weakest. Likewise, the poor performance and expensive price of SLMs have slowed the development of a real-time optical processor.

Real-time pattern recognition was first thought to be one of the most potential technologies of optical computers, hence the following two optical correlator designs were developed. Because the separation between the I/P plane and the O/P plane is four times the focal length of the lenses, the basic correlator is termed 4-f in Figure 8.2(a).

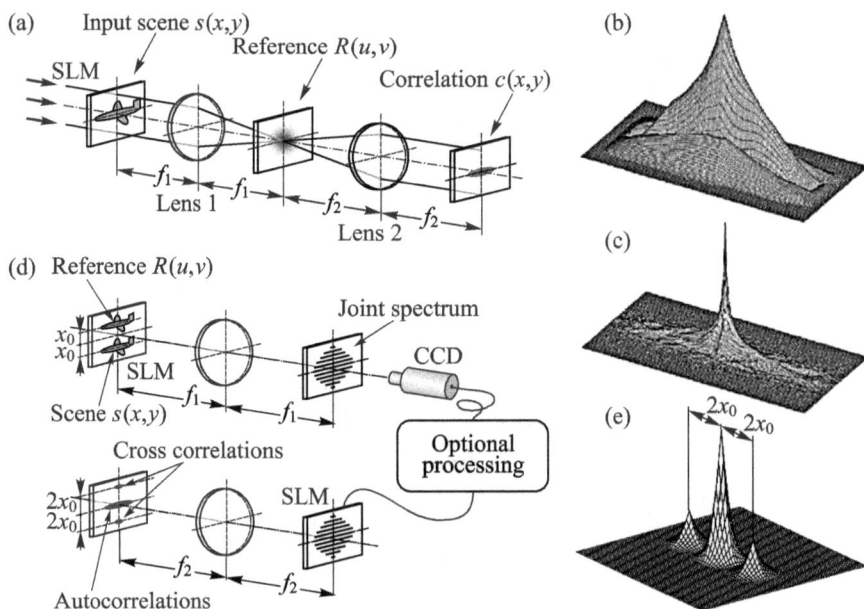

FIGURE 8.2 (a) Optical setup of basic 4-f correlator [22], (b) autocorrelation peak for a matched filter [22], (c) autocorrelation peak for a phase-only filter [22], (d) optical setup of JTC [22], (e) O/P plane of the JTC [22].

This extremely simple design is established on Maréchal and Croce's spatial filtering work in 1953 [12] and was further refined by numerous researchers over the years [13, 14]. The I/P scene is projected in the I/P plane, and Lens 1 performs the FT. The reference's actual conjugated FT is placed in the Fourier plane and hence multiplied by the I/P scene's FT. Lens 2 uses a second FT to determine the relationship between the I/P scene and the reference in the O/P plane. The fundamental issue of this arrangement was developing a sophisticated filter using the reference's FT, and A. V. Lugt suggested in 1964 to utilize a Fourier hologram of the benchmark as a filter [15]. Figure 8.2(b and c) depicts the O/P correlation peak for autocorrelation when the correlation filter is a matched filter and when the correlation filter is phase-only, respectively.

Weaver and Goodman [16] proposed a new optical correlator layout in 1966, the joint transform correlator (JTC), which is shown in Figure 8.2(d). The two pictures, the reference $r(x, y)$ and the scene $s(x, y)$, are arranged beside each other on the I/P plane, which is the first lens FTs. After detecting the intensity of the combined spectrum, the FT is used.

The cross correlations between the scene and the reference are among the terms that make up the second FT. This FT may be done optically using an SLM, as seen in Figure 8.2(d). When the reference and scene are similar, the O/P plane of the JTC is shown in Figure 8.2(e). The two cross-correlation peaks are the only ones worth looking at. The CCD camera can be exchanged with an optical component such as an optically addressed SLM or a photorefractive crystal to create a completely optical processor. Because the JTC does not need the computation of a correlation filter, it is the appropriate design for real-time applications, including target tracking where the benchmark must be revised at a rapid rate. Coherent optical processors are illustrated in Figure 8.2. Wave intensities, rather than complex wave amplitudes, are used to carry information in incoherent optical processors. Incoherent processors aren't affected by phase changes in the I/P plane and don't produce coherent noise. The non-negative real value of the information, on the other hand, necessitates the employment of a variety of methods to execute some image-processing applications [17, 18]. Linear optical processing may be broken down into space-invariant activities like correlation and convolution, as well as space-variant operations like coordinate transformations [19] and the Hough transform [20]. Logarithm transformation, thresholding, and analog-to-digital conversion are examples of nonlinear processing that may be done optically [21].

8.4 TYPES OF OPTICAL COMPONENTS USED TO IMPLEMENT OPTICAL FUNCTIONS

Deep learning has an unquenchable thirst for processing resources. Classical deep learning has silently developed a bottleneck due to interference of electrical impulses, energy usage, and physical constraints [23], even though electronic parts established on silicon can still sustain it presently. Academic and corporate circles are attempting to find new approaches for resolving electrical flaws that are less computationally intensive. It has huge benefits in information transmission and OC due to its high speed of 300,000 km per second, which is 300 times faster than that of an electron, and its data-carrying capacity and variety, which is 2×10^4 times greater than that of electric channels, as well as high parallelism and strong anti-interference [24]. Switching electricity with light has emerged as a viable and sustainable work style, following the current trend.

8.4.1 SPATIAL LIGHT MODULATOR (SLM)

Light wave processing has provided useful techniques for converting data into spatially modulated coherent light waves using SLM apparatuses, allowing the creation of digital holographic images [11]. The capacity to alter the phase and amplitude of light in the far-field is one of the hologram's most valuable qualities. The FT illustrates how a hologram (near field) interacts with its replay field (far-field). In free space, the far-field might develop at the focal point of a positive lens or an indefinite distance from the near field plane [25]. Waveforms from an existent item can be reproduced using holograms. With developments in digital technology and light wave processing, it is now feasible to compute interference patterns computationally and construct synthetic wave fronts of any shape. Computer-generated holograms (CGH), diffractive optical elements (DOE), phase/amplitude masks, diffractive grating, and other terms can be used to describe these interference patterns. Because they all work on the concept of diffraction, the nomenclature is rather arbitrary.

CGHs and DOEs have historically been used to implement various optical operations in correlators as spatial filters [26–29]. Despite the advent of dynamically controlled SLMs, the use of DOEs in Fourier correlators remains relevant in high-energy applications [30] since they have a significantly higher damage threshold than SLM. Research by V. A. Soifer's team have designed multichannel spatial optical elements that allow coherent light fields to be optically decomposed into a series of orthogonal functions. Using angular harmonics to calculate the light field's angular momentum [31–35], detection and analysis of wave front aberrations using Zernike polynomials [36–40], and using the optical Karhunen–Loeve decomposition, we may derive decorrelated image characteristics [41, 42]. Relying on segmented spatial filters in the sequence of diffraction gratings, an optical approach for producing a directions field for fringed/contour pictures such as interferograms and fingerprints has been established [43–46]. To deepen focus and adjust for defocusing and chromatic aberrations, imaging systems undergo phase apodization and optical wave front coding [47–50]. The spiral phase plate as a phase rotor filter is suggested to be used to visually realize the m-th order Hankel transform as well to optically discriminate the light field with rotational symmetry [51]. In contrast to the commonly used 2D Hilbert transforms in image processing, the vortex spatial phase filter is utilized to execute the radial Hilbert transform, which is isotropic [52–54]. Figure 8.3 shows the operation of the optical correlator with a spatial filter in the form of the spiral phase plate for the implementation of the radial Hilbert transform to contrast the image of the fundus vessels [52]. In this way, a correlation filter can be used not only for the classical task of optical detection (recognition) (Figure 8.2) but also for the optical implementation of a mathematical operation such as the Hilbert transform (Figure 8.3).

The SLM is an electrically programmable device that may modify light in compliance with a fixed spatial (pixel) pattern. It may typically be used to adjust the phase

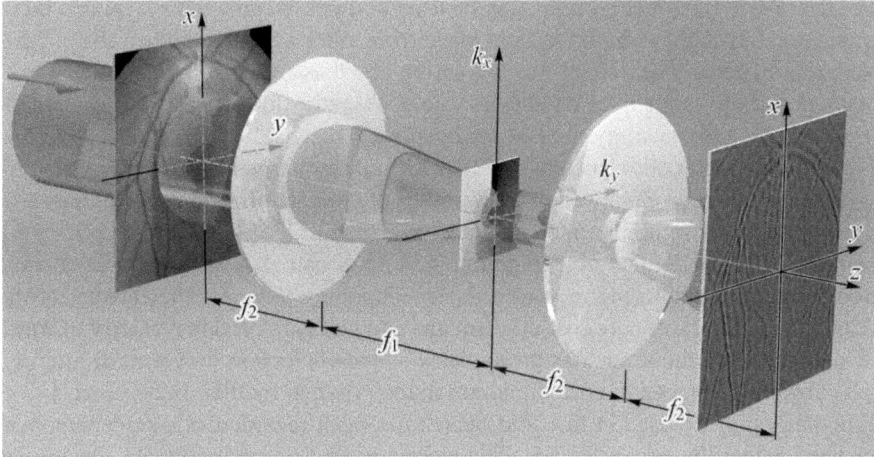

FIGURE 8.3 Illustration of the optical implementation of the radial Hilbert transform using the spiral phase plate as a spatial filter to contrast the image of the fundus vessels [52].

and/or amplitude of incoming light. As a result, SLM may easily realize phase-only, amplitude-only, or a combination of phase-amplitude [55]. Commercially available SLMs are quite "slow," although faster ones are also known [56]. A digital micromirror device (DMD) is considered a fast analog of SLM. DMDs show switching speed ranging from several kHz to tens of kHz (settling time for full-scale angle change is around 10 μs) [57], however, high speed comes at the expense of limited modulation depth and diffraction efficiency [58]. Thus, in applications where high energy efficiency is not required, it is possible to use DMDs instead of SLMs. There are a variety of modulation techniques to choose from. EO-SLM (electro-optic spatial light modulator) is one of the most appealing and commonly utilized. Liquid crystal is used as the modulation material in EO-SLM. A micro-display is used for incident light manipulation and collecting in a liquid crystal SLM, as well. This may be done in two ways: as a transmissive display using LCD SLM technology or as a reflective display using liquid crystal on silicon (LCoS) SLM technology. The liquid crystal molecular alignment is one of the modulator's most prominent characteristics. This might take the form of a vertical, parallel, or twisted structure. As a result, with the right polarizing optics, the incident beam of light parameters, such as amplitude, phase, or their combination, may be efficiently adjusted.

   The nematic LCoS technology is a form of SLM that allows for phase-only modulation [59]. Furthermore, it belongs to the electrically addressed reflection modulator class, in which the liquid crystal is controlled by a direct and precise voltage, and the beam of the light wave front may be adjusted as well. The LCoS SLM may be used to rebuild pictures from CGH as a diffractive modulator [60]. CGHs may be used for a variety of communication applications, and it is increasingly being used in indoor visible light communication networks [61]. Additionally, a range of optimization approaches, including the iterative Fourier transform algorithm (IFTA), linear Fourier transform (i.e. linear phase mask), simulated annealing, and the Gerchberg–Saxton algorithm, can be used to quickly construct acceptable holograms [62, 63]. The beam of the light wave front can be modified when the SLM is used as a diffractive device for reconstructing graphics from CGH.

As previously stated, LCoS displays have gained a lot of traction as potential micro-displays for a variety of SLM applications. Likewise, they have appealing and important characteristics, such as excellent spatial resolution and light efficiency. As a result, they've been employed in a wide range of optical utilizations, including communication, reconfigurable interconnects [64], storage [65], diffractive optics [66], metrology [67], and quantum computing [68]. They can also be used for light wave processing and monitoring in wave shaper technology [69]. Another feature of the LCoS is that it is extremely cost-efficient and can be configured in a variety of ways. Several functionalities, such as group delay ripple correction, wavelength filtering, and chromatic dispersion correction, are made possible by this. In addition, LCoS technology may be used in flex grid, which has been regarded as a key component for next-generation networks [65]. The classic fixed grid with 50 GHz spacing, as regulated by the International Telecommunication Union (ITU) Telecommunication Standardization Sector (ITU-T), has a variety of drawbacks. The fixed grid has been reported to result in poor usage of the optical spectrum. Furthermore, it severely limits the system's transmission capacity. The flex grid implementation, on the other hand, allows for the use of several modulation formats and their coexistence on a common infrastructure. They can also be multiplexed densely and effectively, which helps optical networks not only extend their reach but also increase the per-channel data rate. It is also expected that the deployment of WSS and SDM would considerably aid in the expansion of the network's coverage and capacity [65].

Electrically addressed SLMs (EASLMs) and optically addressed SLMs (OASLMs) are two types of SLMs that differ in how data is loaded into the apparatuses [70]. The EA has been the most prominent approach in modern commercial SLMs because of the growth of electronic information technology over the last few decades. However, because data must be converted back and forth between the optical and electrical domains, EASLMs are not the greatest solution for future all-optical information processing systems. The OASLMs, on the other hand, enable light to be modulated directly by light without having to go through an electronic-optical transformation [71, 72]. Additionally, OASLMs are required for a variety of all-optical purposes that EASLMs cannot handle, such as coherent to incoherent image conversion, real-time optical correlation, and parallel all-optical processing [8, 73–76]. OASLMs may be built-in theory using material nonlinearities, with modulation over the read light achieved by spatially precisely modifying the characteristics of the materials through nonlinear optical stimuli [77]. Natural materials, on the other hand, have insufficient nonlinearities to permit efficient "light-control-by-light" inside nanoscale volumes. This renders the apparatuses exceedingly bulky or necessitates a lot of pumping power to collect big enough nonlinear modulations, making them unsuitable for the nano-era.

The recent remarkable advancement in dynamic optical metasurface (MS) technology provides a chance to overcome challenges and proposes a unique framework for nanoscaled SLM [78, 79]. The MS can also enhance optical engagements by focusing light on nanoscale volumes, allowing for greater control of light fields in response to external mechanical, chemical, and magnetic stimuli, for example. The ultracompact EASLMs established on MS have recently been observed relying on the active manipulation of beams of light by additional electrical fields [80]. An optically addressed spatial light modulator (OASLM) established on metasurface-based OASLM (MS-OASLM) is proposed in [81], with a nonlinear polarization manipulation of read light by another write light at the nanoscale as the operating mechanism. It delivers 500-line pairs/millimeter (equivalent to a pixel size of just 1 µm), which is more than ten times greater than a standard commercial SLM. The MS-OASLM has exceptional compactness and a thickness of just 400 nm. MS-OASLMs like this might pave the way for next-generation all-optical data processing and high-resolution display technologies.

8.4.2 PLASMONIC SWITCHES

Moore's law is expected to surpass its physical constraints, putting modern computer systems centered on von Neumann architecture at a physical limit [82, 83]. Potential alternate prospect next-generation computing approaches are necessary due to the rapidly rising development of bandwidth needs with lower power utilization. As a result of the usage of optical interconnects in high computing chips rather than electrical interlocks to minimize power consumption and increase speeds, photonic computing is expected to have a competent replacement for existing electronic computing systems. Furthermore, Si photonics is considered a leading field owing to its capacity to provide effective light modulation, light confinement, and compliance with today's CMOS production techniques [84, 85]. A decade ago, the shift from electronic to photonic systems began. Related to current material discoveries, fabrication technique advances, and more ongoing research efforts in this field, photonic interconnects and circuits have witnessed significant advancements in recent years.

Photonic logic circuits rely heavily on optical switching and modulation. Optical computations are established on photonic logic circuits. Utilizing the thermo-optic effect, free carrier dispersion effect, and Pockels electro-optic effect, intensive research in the direction of efficient optical switches, optical modulators, and optical logic circuits has been presented [86–89]. All these tools are volatile, requiring constant voltage to manage, and, as a result, they consume a lot of power. Furthermore, due to their low electro-refractivity, they have long interaction durations. As a result, these apparatuses have substantial footprints, making ultra-compact designs difficult to achieve [90, 91]. A novel type of material known as phase change materials (PCMs) has found considerable application in both the electronics and photonics domains in recent years [92, 93]. With high-speed switching between two steady states, PCMs reveal changes in electrical and/or optical characteristics. Several photonic and plasmonic apparatuses have been proposed and investigated, including on-chip optical modulators and optical switches established on PCMs [94, 95]. One of the widely used PCMs in photonic apparatuses is $Ge_2Sb_2Te_5$ (GST) [96].

Even though effective energy nonvolatile (NV) memory and OLGs based on PCMs have been thoroughly researched and established over the last 20 years, PCM-based photonic NV memories and photonic logic circuits have received little attention. Novel NV combinational and sequential logic circuit designs are investigated, as well as NV hybrid electro-optic plasmonic circuits. The electro-optic devices are made up of a plasmonic waveguide (WG) with a mono PCM layer. Changing the phase of the PCM between amorphous to crystalline likewise changes the optical losses in the WG. Electrical threshold flipping or thermal conduction heating via externally applied radiators or the plasmonic WG metal itself as an integrated heater can be used to generate phase shift in the PCM. All OLGs, a half adder circuit, and sequential circuits may be built employing plasmonic switches as active components, as illustrated. Furthermore, the plasmonic switches and logic functions have minimal extinction ratios larger than 20 dB, are small, have little operational power, and operate at high speeds. To develop an effective architecture for logic processes, photonics, plasmonics, and electronics are merged on the same platform.

A plasmonic slot WG with PCM as an active material coating the slot WG's surface is used in the NV hybrid electro-optic plasmonic switch [97]. Figure 8.4(a) depicts the suggested plasmonic slot WG's construction, which includes Si-Au tapered WGs,

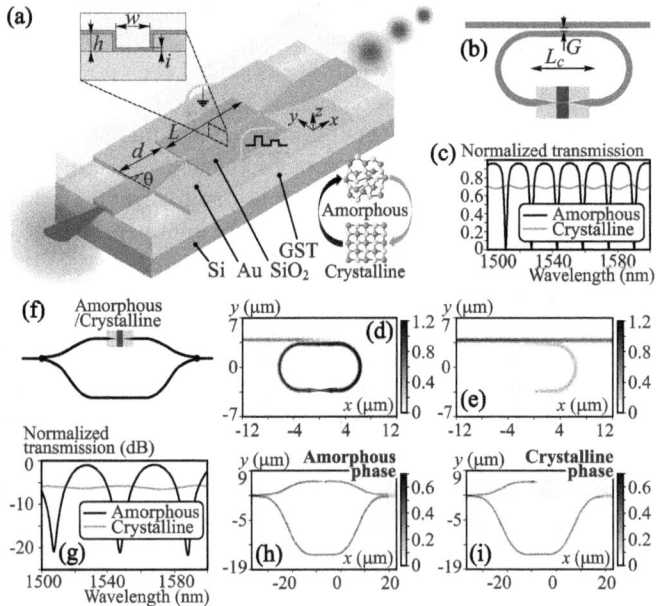

FIGURE 8.4 (a) The broadband NV hybrid EO plasmonic switch is shown schematically. The side view picture of the plasmonic slot WG with a thin coating of GST is shown in the inset [97]. (b) The hybrid NV EO plasmonic switch serves as the central unit in the racetrack μ-RR-based EO switch [97]. (c) Normalized transmission spectra for the amorphous and crystalline phases of the PCM layer of the racetrack μ-RR-based EO switch. (d, e) The E-field mapping of the EO switch for the amorphous phase and the crystalline phase of the PCM layer at the resonance wavelength of 1,558 nm, respectively [97]. (f) Asymmetric MZI-based EO switch with hybrid NV EO plasmonic switch as an active component in one arm. (g) Normalized transmission band for the crystalline and amorphous phases of the PCM layer in the plasmonic switch for the asymmetric MZI based EO switch. The E-field distribution of the EO switch for the (h) amorphous phase and (i) crystalline phase of the PCM layer at the resonance wavelength of 1,548 nm [97].

a slot in an Au film to form a plasmonic slot WG, and a PCM coating for instance GST on the slot WG's surface. The arrangement is made up of three primary components, the first of which is a dielectric to plasmonic mode converter in the I/P region of a hybrid Si-Au tapered WG. The major optical modulation happens in the second portion, which is a MIM plasmonic slot WG covered with a thin coating of GST, and the third part is a plasmonic to dielectric mode converter in the O/P section, which uses a hybrid Au-Si tapered WG. The cross-sectional image (in the y–z plane) of the GST-coated plasmonic slot WG is shown in the left top inset of Figure 8.4(a) [97].

As illustrated in Figure 8.4(b), the EO switch is established on a racetrack microring resonator (μ-RR) with an NV hybrid EO plasmonic switch in the ring WG as an active component. The optical mode is linked to the ring through a Si WG with the same cross section as the hybrid plasmonic switch's Si WG. The transmission spectra for both phases are shown in Figure 8.4(c). The optical loss in the ring is minimal when GST is in the amorphous phase. Consequently, light can pass through the RR and propagate via the optical mode. For specific wavelengths, resonance is obtained in the RR, and, after a roundtrip, light coupled back to the I/P WG's through-port undergoes a 180° out-phasing in comparison to light arriving from the I/P WG's I/P point.

As a result, destructive interference develops between light flowing from the I/P point and light coupled back to the I/P WGs through the port. As a result, light departing the O/P port at resonance wavelengths is eliminated as shown in Figure 8.4(d). Optical mode, on the other hand, cannot travel via the ring WG due to momentous optical loss in the crystalline phase. Consequently, there is no intrusion, and optical power is sent via the O/P port as shown in Figure 8.4(e) [97].

Established on the hybrid plasmonic switch anticipated in [97], an asymmetric MZI is depicted in Figure 8.4(f). The asymmetric MZIs were designed using compact and low-loss Y-junction-based splitters and combiners. The I/P segment's splitter divides the I/P power evenly between the MZI's two arms, while the O/P section's combiner merges the optical powers of the two arms. The asymmetric MZI's hybrid plasmonic switch is incorporated in the sorter arm. Figure 8.4(g–i) shows the transmission spectra and field mappings of the MZI for GST amorphous and crystalline phases, respectively.

8.4.3 NEURAL NETWORKS

Artificial intelligence (AI), as among the most effective areas in computer science, focuses on simulating the framework of the nervous system by constructing artificial neural networks (ANNs), which maintain connections between neurons in multiple layers of the neural networks (NN) and give it higher accuracy and robustness. ANN's research has advanced significantly since the 1980s. It has also effectively solved many functional challenges for modern computers in the fields of pattern recognition, intelligent robots, automatic control, prediction and estimation, biomedicine, economy, and other fields, all while retaining better intelligence attributes. Advanced machine learning techniques, which include ANNs [98, 99], have gotten a lot of interest because of their practical uses in important tasks like image identification and speech processing [100]. NNs employ a lot of multiply-accumulate (MAC) operations, which puts a huge amount of pressure on current electronic computing technology (e.g. CPU, GPU, FPGA, ASIC). For MAC activities, application-specific apparatuses are recommended. Most NNs depend solely on real-valued arithmetic, even though complicated arithmetic might provide a large benefit. For example, a single complex-valued neuron with orthogonal decision boundaries can help resolve the symmetry challenge and the XOR issues, while a single real-valued neuron cannot [101]. Research indicates that complex-valued arithmetic [102] might help NNs function better by providing extensive representational capacity, quick convergence, powerful applicability, and noise-resistant memory mechanisms. Since complex numbers must be formed by two real numbers, which fuels the growth of MAC operations – the most commonly frequently utilized computationally expensive components of NN algorithms-conventional digital electronic computing platforms experience considerable slowdown when implementing algorithms utilizing complex-valued operations [103, 104]. To circumvent these difficulties, it has been suggested that the computationally demanding process of building NNs be delegated to OC, which is proficient in genuinely complex-valued arithmetic [75].

Low power consumption, fast processing time, huge data storage, and intrinsic parallelism are all benefits of OC that cannot be matched by its electrical cousin. Numerous optical NN algorithms have been suggested. Photonic chip-based optical NNs, for example, have grown extremely popular due to their remarkable adaptability, scalability, and durability. This system has previously shown neuromorphic photonic weight banks [105], all-optical NNs [106], and optical reservoir computing with tremendous results [107]. On an embedded silicon photonic device, a typical fully connected NN was realized experimentally [108]. Although this optical device is established on light interference, the NN algorithms used are real-valued, negating the merits of sophisticated NNs. The optical impulses were already transformed to photocurrents before entering the accumulator, resulting in a highly parallelized optical NN accelerator established on photoelectric multiplication, which was also built for real arithmetic. On-chip training, optical nonlinear activations [109], and different NN designs are further subjects linked to optical NNs [110].

Apart from OC platforms, analog electronic apparatuses have effectively exhibited multilayer perceptrons [111, 112] and convolutional NNs, as compared to more prevalent digital electronic apparatuses [113]. Some earlier research has looked at complex-valued NNs on analog electrical apparatuses [114–116]. Complex-valued reservoirs also result in enhanced system dynamics and enhanced efficiency in reservoir computation. Although optical NNs can handle information in multiple degrees of freedom (e.g. magnitude and phase) using complex-valued numerical methods and acquire more effective information processing and analysis, there have been few investigations in OC platforms for integrating general-purpose and complex-valued NNs [117]. Strongly dependent on traditional deep learning algorithms developed for real-valued arithmetic on traditional electronic computers, existing optical solutions have not walked into this prospective flatland. These real-valued optical NNs are built exclusively on the intensity information of the light waves, ignoring the phase information, which eliminates one of OC's primary advantages.

These problems were answered by designing and demonstrating an optical neural chip (ONC) that performs complex-valued arithmetic, demonstrating the benefits of chip-based complex-valued networks via OC. It has been proved that an optical neural chip (ONC) can build fully complex-valued NNs [118]. The complex-valued ONC's system is analyzed in four contexts: basic Boolean tasks, species classification of an Iris data set, nonlinear data set classification (Circle and Spiral), and handwriting recognition. When referred to its real-valued equivalent, the complex-valued ONC achieves strong skillsets (i.e. precision, quick resolution, and the capacity to generate nonlinear decision boundaries). Figure 8.5(a) depicts

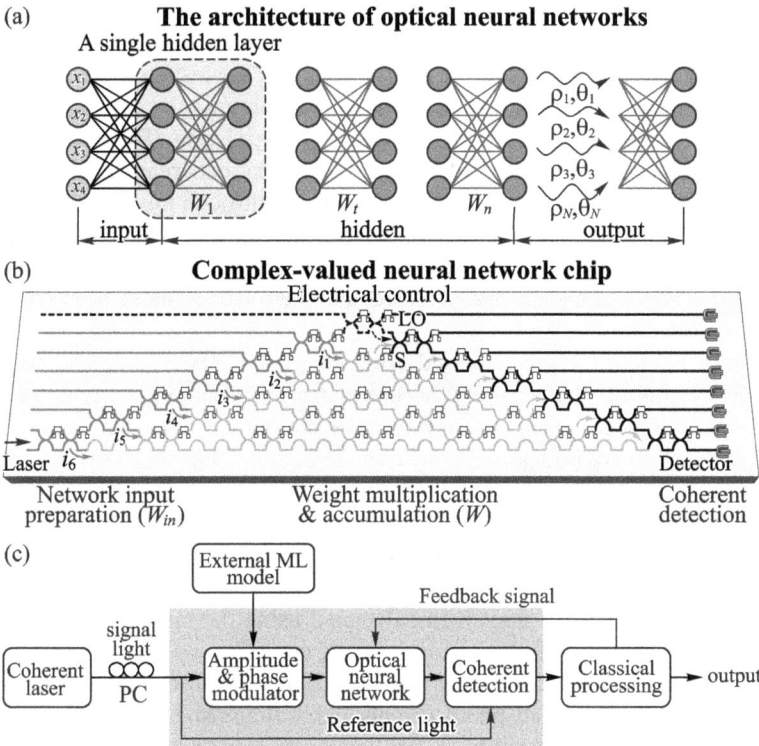

FIGURE 8.5 A complex-valued coherent optical NN's composition. (a) An I/P layer, many hidden layers, and an O/P layer make up an optical NN. During the initial I/P preparation and network evolution, the light signals are encoded and modulated by both amplitude and phase in our complex-valued architecture [118]. (b) The ONC's diagram for implementing complex-valued networks. On a single chip, I/P preprocessing, weight multiplication, and coherence recognition are all combined. The MZIs in gray are responsible for the division and modulation of the light signals ($i_1 - i_6$). The black dot MZI distinguishes the reference light that will be utilized for coherent detection later. Light gray indicates the MZIs that were utilized to build the 6×6 complex-valued weight matrix. On-chip coherent tracking is established on the remaining gray designated MZIs [118]. (c) The ONC system's process. Signal and reference light is generated using a 1,550-nm coherent laser. The amplitude and phase of the signal light on each path are modified by the machine learning (ML) job. The light inference is used to perform the weighted sum process passively. The measurement findings are transferred to the electrical interface for processing, which includes activation function application and cost function computation. The modified weight matrices are then used to reprogram the ONC chip [118].

the optical NN's architecture, consisting of an I/P layer, many hidden layers, and an O/P layer [118]. During the initial I/P signal preparations and network growth in the complex-valued architecture, light signals are encoded and modulated both by optical magnitude and phase. Figure 8.5(b) depicts the ONC design for implementing complex-valued NNs [118].

A single-chip handles I/P preprocessing, weight multiplication, and coherence detection. The I/P signals are generated using a coherent laser ($\lambda = 1,550$ nm). The ONC is simply a multiport interferometer with a special arrangement of Mach–Zehnder interferometers (MZIs). Each MZI is made up of two beam splitter–phase shifter (BS–PS) pairs. The BS has a stable transmittance of 50:50, while the PS is thermally regulated to modify the phase. Different MZIs presented in the figure have distinct capabilities. The bottom point couples the light into the chip. The one chain of MZIs is responsible for I/P light splitting and modulation. The other MZI label distinguishes the baseline light that will be utilized for coherent detection. The on-chip light division ensures that light signals flowing through multiple optical routes are polarized similarly and have a constant relative phase. The machine learning job determines the modulation of the I/P. For jobs involving real-valued I/Ps, the amplitude of the light signals is modulated, and the relative phases between distinct pathways are reduced to 0.

On-chip coherent tracking is based using the gray MZIs. The optical chip's O/P light signals contain both magnitude and phase information, whereas traditional intensity monitoring systems merely retrieve magnitude data. The intensity and coherent detection are both possible with the integrated chip. The purpose of coherent tracking is to find the phase angles between both the reference and signal light. The O/P current achieved by linking photodiodes at both O/Ps in a balanced manner is $I_1 \alpha 2 A_s A_1 \cos\phi_s$, where A_s and A_1 are the signal and reference light amplitudes, respectively. Likewise, if the baseline light is phase shifted by $\pi/2$, the O/P current is $I_Q \alpha 2 A_s A_1 \sin\phi_s$. The ϕ_s is then calculated using the ratio of I_1 and I_Q, which also aids in the removal of physical noise from optical components. The activation analysis determines which detecting approach is used. A transimpedance amplifier (TIA) converts the recorded photocurrents into voltage signals, which are subsequently gathered and analyzed by a traditional processor with an analog-to-digital converter (ADC). As demonstrated in Figure 8.5(c), feedback signals may be created and routed back to the ONC to alter chip layouts [118].

8.4.4 DIFFRACTIVE NN

Prior works to OLGs focused primarily on constructive/destructive interference effects between the I/P light signals, encompassing linear [119–122] and nonlinear interference [123]. The reported works are heavily reliant on accurate positioning of the basic characteristics of two I/P light signals, the control light, and/or the pump light, such as phase difference, polarization, and intensity; if the two nanowires are close to each other, as in the plasmonic logic gate, there is also a mandatory rule on the size of I/P beams of light to prevent a big false I/P. Consequently, greater tight control of I/P light may more fully actualize constructive or destructive interference, resulting in a bigger intensity contrast ratio between the two O/P optical logic states "1" and "0," which is a critical quality to evaluate an OLG's performance. The heavy dependency on precise I/P light management has two negative effects on the development of miniaturized OLGs. First, the substantial optical components required to perform these controls are considered, and downsizing becomes challenging. Second, because of the complexities of achieving perfect I/P light control, their performance may be unstable, and the intensity contrast ratio between two O/P logic states may become rather low in practical circumstances. It is therefore very desired for compact OLGs to eliminate these important I/P light needs. Due to the necessity of developing innovative designs for all-optical apparatuses and systems, such a goal remains an open problem that has long been sought after.

All seven fundamental optical logic operations (OLOs) are realized in a small system utilizing just plane waves as the I/P signal, thanks to a simple yet universal design method called a diffractive NN [106]. A compound Huygens' metasurface (MS) implements the diffractive NN, which may somewhat imitate the functions of an ANN [124]. After training, the compound MS can disperse or focus the I/P encoded light in one of two tiny areas/points, one representing logic state "1" and the other representing logic state "0." Three basic OLGs – NOT, OR, and AND – are experimentally confirmed at microwave frequency utilizing a two-layer high-efficiency dielectric MS as a conceptual example. There are two significant advantages to the design technique. First, the implementation of OLOs here eliminates the need for sophisticated and exact control of I/P light characteristics, which sets this technique apart from earlier work. Furthermore, the I/P layer's architecture is quite broad and strong, and it can be easily changed into numerous user-friendly and programmable formats. Second, if the transmittance state of the I/P layer is dynamically tunable, for example, electrically tunable if the optical mask is generated by a spatial light modulator (SLM) [125], the suggested technique can enable comprehensive logic functionality in a single optical network.

The I/P layer is a common optical mask that is designed to generate numerous zones, as shown in Figure 8.6(a) [126]. Each optical mask zone is set to have two alternative states for optical transmission without sacrificing generality, and its high (low) transmittance state signals whether it is (is not) selected for OC. Then, merely by allocating each of the seven fundamental optical logic operators and the I/P logic states to a specific region, it is simple and efficient to directly specify all seven basic optical logic operators and the I/P logic states in the optical mask. The hidden layers are responsible for decoding the encoded I/P light and rendering the computed result at the O/P layer. The pattern of the I/P layer is shown in Figure 8.6(b). For the sake of simplicity, each region's high (low) transmittance condition is considered to have a transmittance of 100% (0%) [126]. A cascaded two-layer transmission MS with an axial spacing of $170\lambda_0$ is used to create the concealed layers (one of the tunable parameters in the training process of diffractive NN). Each MS is made up of 30×42 meta-atoms [inset in Figure 8.6(c)] [126], each of which has a $0.570\lambda_0$-width square cross section. Taking use of its unique qualities such as high transmittance and polarization insensitivity, a simple yet practical high-efficiency dielectric MS is developed. The calculated field intensity after training is depicted in Figure 8.6(d–m) [126]. Most of the fields are appropriately concentrated within one of two tiny, specified zones, as intended.

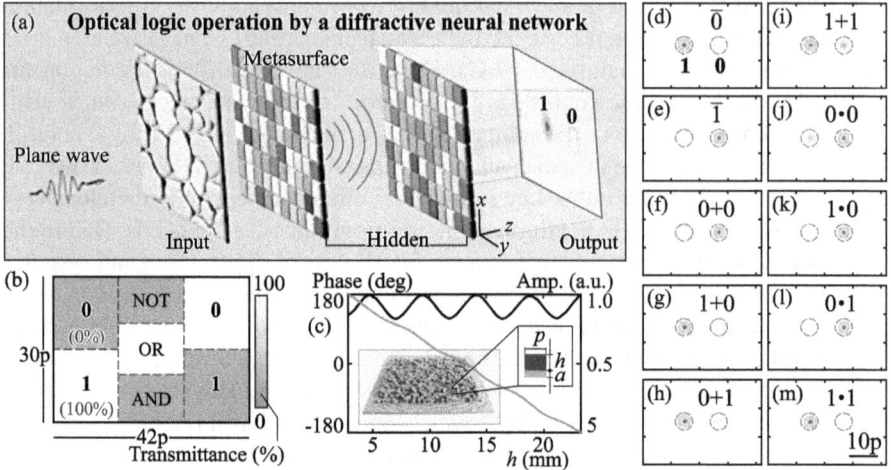

FIGURE 8.6   (a) A diffractive NN is laid out for photon-based logic functions. A diffractive NN is used to numerically demonstrate three basic logic operations: NOT, OR, and AND. Two levels of MSs make up the hidden layers here [126]. (b) The I/P layer's diagram. The white (gray) region's light transmittance is set to 100 percent (0 percent). (c) The MS's transmittance response, which is made up of a 2D array of subwavelength meta-atoms. Each meta-atom may adjust the incoming light's phase (gray line) and amplitude (black line) locally [126]. (d–m) Intensity distribution for three logic operations with random I/P logic states at the O/P layer. The O/P optical logic state is specified as "1" ("0") if the field is concentrated on the tiny, prescribed areas on the left (right). Two dashed circles in each panel highlight the selected regions [126].

8.4.5   PHOTONIC CRYSTAL ALL-OPTICAL LOGIC GATES

With the progress of science, there seems to be a significant change in computing in many ways over the years. Mechanical methods were used to create the earliest computers (1623 to 1945) [127]. Researchers attempted to create apparatuses that could readily answer mathematical problems in the early 17th century. Several scientists, including Gottfried Leibniz, Wilhelm Schickard, and Blaise Pascal, attempted to develop a calculator that could handle addition, subtraction, multiplication, and division. George Schertz and Edward Schertz created a system that could handle 15-digit numbers using a 4-bit difference engine. One of the institutions that employed the mechanical computer for punch card technology for the enumeration was the U.S. Census Bureau, which was designed by Herman Hollerith of the International Business Machines [127].

A wide range of components, including optical gates, optical switches, optical interconnects, and optical memory, are required to form an optical computer. Because of its applicability in ultrafast information processing [128] and the ways to carry out various logical operations in OC systems, all-optical logic gates (LGs) have become popular recently [129]. As a result, building all-optical LGs is the first step in achieving advanced digital functionality in optical computers. Electronic LGs were previously employed, but the highest switching speed obtained was 50 ps with an average power of 0.5 mW per switching [130]. The reduced capacitance of p-n junctions in semiconductor-based LGs is the explanation for this.

Despite today's electronic LGs being tiny, switching is still restricted by interlinking capacitance; on the other side, optical LGs have switching speeds in the femtosecond range and are only restricted by the speed of light traveling through them [131]. The prototype of all-optical LGs can be done in a variety of methods. The first technique employs a semiconductor optical amplifier (SOA), which has a high gain owing to refractive index variations. The original way to make all-optical LGs was to use one of three approaches to introduce nonlinearity: cross-gain modulation [132, 133], cross-phase modulation [132], or four-wave mixing [133, 134]. SOA was also utilized to construct SOA-assisted interferometer-based gates, which are all-optical LGs [135]. However, SOA-based gates have several drawbacks, including SOA-based apparatuses being constrained by SOA's slow carrier recovery time, unsteady gates owing to polarization sensitivity, and the SOA Mach–Zehnder interferometer (MZI) technique, which necessitates more than two SOAs and complicates the scheme by needing the proper tuning of the filter for SOA with the help of fiber LGs, as detuning of the filter. Nonlinear WGs, in which localized nonlinear media have been used by adjusting the control power, is another way for developing all-optical LGs [136–138]. Nonlinear WG-based gates have several drawbacks, including a substantial I/P signal power need and polarization dependency, which pose production issues.

PCs are periodic structured dielectric or EM media with photonic bandgaps (PBGs) that prevent light from propagating through them [139–142]. John was the one who introduced PC [143]. Unlike semiconductor crystals, which alter the characteristics of electrons, these crystals impact the characteristics of photons. Light has various benefits over electrons, including the ability to move faster in dielectric material than electrons in the conductive metal and a larger data capacity in the dielectric than electrons. One of the most significant functions in the realm of high data transmission that 2D PCs may achieve is all-optical LGs. Furthermore, there are two main types of PC-based gates: PBG-based gates and non-PBG-based LGs.

8.4.5.1 Non-Photonic Bandgap-Based All-Optical Logic Gates

Instead of establishing the PBG of the apparatuses, the I/P beam of any wavelength is injected at the I/P, and a logic function is conducted utilizing a self-collimated beam (SCB) in non-PBG-based PC gates. In this case, incoming light transmits to a device in a given direction without diffraction. The phenomena of total internal reflection are utilized to construct all-optical LGs utilizing an SCB. Total internal reflection refers to the angle of incidence being larger than the critical angle, according to the relation $\theta > \arcsin(n_L/n_H)$, where n_L represents the low refractive index and n_H represents the high refractive index. In a 2D PC, a device for the photonic-integrated circuit (PIC) was formulated and established on an SCB [144]. The structure is useful for making optical switches and LGs, both of which are important parts of a PIC. The device with Si rods in the air was designed using square lattice geometry. To achieve the out-phasing at the O/P, a low refractive index medium was created such that one-half of the beam was allowed to propagate while the other was reflected, suggesting the existence of transmitted and reflected beams for the I/P signals. By altering the phase difference between the reflected and transmitted beams, the OR and XOR gate structures were developed. When the phase difference between the I/P beams was $2k\pi + \pi/2$, O/P O_1 functioned as an OR gate and O/P O_2 as an XOR

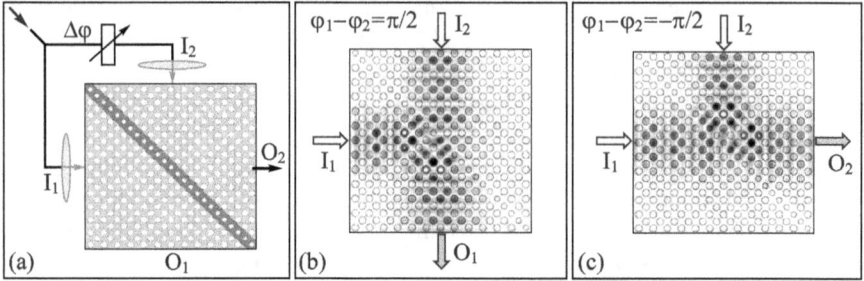

FIGURE 8.7 (a) The switch's schematic representation. On the I/P faces I_1 and I_2, two beams with different phases collide. The computed steady-state field pattern of the E-polarized mode at $0.194(a/\lambda)$ when incoming beams propagate in the Γ-M direction in (b) and (c). The phase difference between the two incident beams is adjusted to $\pi/2$ and $-\pi/2$, respectively, by the phase difference $\varphi_1 - \varphi_2$ [144].

gate, as illustrated in Figure 8.7(a–c) [144]. O_1 worked as an XOR gate and O_2 acted as an OR gate when the phase difference was $2k\pi - \pi/2$ owing to a phase difference of $-\pi/2$ between the I/P beams. The device's frequency range was 0.188 to 0.199, with a 17-dB extinction ratio.

For the PIC implementation, an all-OLG architecture of NOT, OR, AND, and XOR gates centered on the SCB was presented [145]. The structure was constructed using 2D square lattice geometry with air holes in Si as the foundation material. The framework has been developed and established on the phase difference between the I/P signals, following the phenomena of the SCB. When light is launched at both the I/P and reference points of the same intensities, there was partial reflectance of the I/P light wave while the reference light wave was completely mirrored, and when these light waves interfered with each other, there was an O/P that relied on the phases at the I/Ps and interfered in a constructive manner or destructively. It was proposed to use SCB-based logic gates for AND, NAND, XNOR, and NOR OLGs. The framework has been developed with triangular lattice topology and rods made of Si material in an air background [146]. Table 8.1

TABLE 8.1
Evaluation of Different Kinds of Non-PBG-Based All-OLGs

Logic Operation	Operational Wavelength (nm)	Lattice Type	Polarization	CR (dB)	Ref.
OR, XOR	–	Square (rods in air)	TE	17	[144]
WG for OLGs	1,550	Square (air holes in Si)	TE/TM	–	[147]
NOT, OR, AND, XOR	1,550	Square (rods in Si)	TE	30	[145]
AND, NAND, XNOR, NOR	1,555.1	Square (rods in air)	TE	6	[146]
XOR, OR	1,550	Triangular (rods in air)	–	–	[148]
AND	1,550	Triangular (rods in air)	–	–	[149]

shows that different studies offered alternative architectures for creating various gates. Most OLGs were created in [145], which implemented AND, NAND, XNOR, and XOR OLGs. Si rods in an air environment are the structure utilized by all researchers. However, in the case of contrast ratio (CR), the structure developed in [146] for NOT gate produced the greatest results, i.e. 30 dB. Furthermore, OLGs constructed with a self-collimated light wave have a few drawbacks, including a low CR, a wide area, a high cost owing to the big size, and signal guiding in only the vertical and horizontal directions.

8.4.5.2 PBG-Based All-OLGs

The PBG of the architecture is utilized to detect hidden frequency ranges that cannot flow through the structure in PBG-based all-optical OLGs. By adding various forms of defects, one of the suppressed frequencies can transmit across the structure. Multimode interference (MMI), nonlinear Kerr effect, and interference are commonly used in the design of PBG-based OLGs. In [150], MMI-based AND and XOR OLGs are introduced, as depicted in Figure 8.8(a), with A and B I/P points and X and Y O/P points. It was constructed utilizing triangular lattice geometry, SiO_2 as the foundation material, and Si rods. An I/P signal with an out-phasing of π was transmitted at point A. On point B, the signal was emitted with an out-phasing of $-\pi/2$, which represented logic "1" and creates logic "1" at the O/P point. When there was a phase of π at point A, which indicated logic "0," and another signal with out-phasing $\pi/2$ at point B, which expressed logic "0," the O/P generated logic "1." In another scenario, when a signal with an out-phasing of π expressed logic "0" at I/P point A while a signal with an out-phasing of $\pi/2$ stated logic "0" at point B, logic "0" was identified at the O/P. When both I/Ps at point A and point B displayed logic

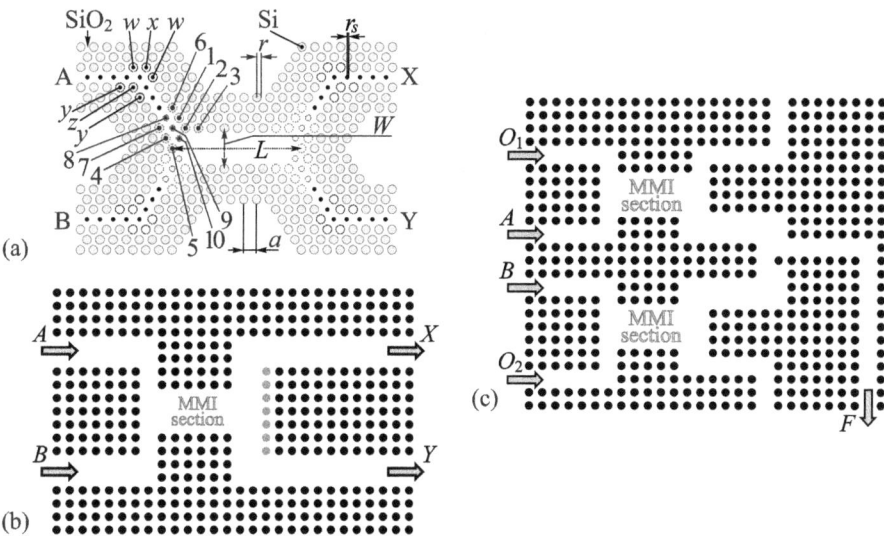

FIGURE 8.8 Schematic of OLGs established on MMI. Inspired by [150], (a) AND/XOR OLG, (b) XOR/XNOR, (c) AND and NOR OLGs. [Inspired by [151].]

"1" with out-phasings of 0 and $-\pi/2$, the O/P point sensed logic "0." The AND OLG was constructed using the same concept by modifying the length of MMI and appropriately choosing I/P signal phases. Both AND and XOR OLGs obtained a CR of roughly 6.79 dB.

The self-imaging mechanism underpins the operation of MMI apparatuses. The tiny field at the I/P activates guided modes in the effective region, which constitutes interference in that region, in this phenomenon. Binary phase-shift keyed (BPSK) signals are utilized as I/P logic values because, in MMI, I/P values are often expressed by the phase of the I/P signals, and O/P logic levels are expressed using amplitude independent of phase. Silicon material was employed for the rods in the suggested device, with air as the background oriented in a square lattice shape [151]. As indicated in Figure 8.8(b), A and B are the I/P points of the XOR/XNOR OLG arrangement, whereas X and Y are the O/P points. When there is an out-phasing of π at point A, logic "0" was stated as signal I/P, while logic "1" is explicit with phase 0. When there was an out-phasing of $3\pi/2$ at point B, logic "0" was expressed as signal I/P, while logic "1" was explicit with phase 0. To realize the AND OLG indicated in Figure 8.8(c), I/Ps A and B with phase π represented logic "0," whereas I/Ps A and B with phase 0 represented logic "1." With an out-phasing of $3\pi/2$, logic "0" was established at O_1 and O_2. The sole change in designing the NOR OLG was that O_1 and O_2 were locked at logic "1," represented by an out-phasing of $\pi/2$. The AND OLG realized a CR of 21 dB, whereas the NOR operation achieved a CR of 19 dB [151]. Table 8.2 demonstrates the effectiveness of PBG-based all-OLGs on performance metrics like the CR. PBG-based OLGs established on the interference phenomenon have a higher CR and are easier to construct than OLGs established on other concepts [152–155].

TABLE 8.2
Device Performance of Different Types of PBG-Based All-OLGs

Logic Operation	Operational Wavelength (nm)	Lattice Type	Polarization	CR (dB)	Ref.
XOR, XNOR, NAND, OR	1530–1565	Triangular (rods in SiO2)	TM	XOR-28.6, XNOR-28.6, NAND-25, OR-26.6	[152]
XNOR, XOR, OR, NAND	1550	Square (rods in air)	TE	37.4–40.41	[19]
NAND	1554	Square (rods in air)	TM	–	[154]
NOT, AND, OR, XOR, XNOR, NAND	1550	Triangular (holes in Si)	TE	NOT-3.74, AND-11.47, OR-12.48, XOR-6.50, XNOR-6.50	[48]
OR, AND	OR-1529, AND-1538	Triangular (rods in air)	TM	6	[156]
NAND, NOR, XNOR	1550	Square (rods in air)	TE	17.59, 14.3, 10.52	[157]
NOR, AND	1550	Triangular (rods in air)	TE	–	[158]
OR	1287.8	Triangular (rods in air)	TE	7.27	[159]

8.4.6 RESONANT NANOPHOTONIC CONSTRUCTIONS

Nanophotonic components are being suggested as a novel foundation for analog optoelectronic computing [160–163]. It was shown that a layer of a well-designed metamaterial can visually execute several essential mathematical operations (differentiation and integration of light waves concerning a spatial coordinate, convolution of light waves with a predefined core) [161]. It generated a lot of interest and aided the creation of novel nanophotonic structures for AOC. Differentiation (integration) of the pulse envelope is commonly understood in the context of temporal differentiation (integration) of a light wave (optical pulse). The most important findings in fiber-optic network design and production for the differentiation (integration) of optical pulses propagating in OFs were acquired [164–170]. Bragg resonant structures and RRs have been proposed to perform time-domain differentiation (integration) procedures. Because the spectra of reflection and transmission in the proximity of resonances are characterized by the Fano profile, and in a certain frequency, intervals may accurately estimate the transfer functions of differentiating and integrating filters; resonant formations can be used to implement differentiation and integration operations. It is worth noting that more complex systems with many resonators may solve ordinary differential equations of various orders as well as solutions of differential equations in the temporal domain [171–174].

The employment of differentiation and integration of light pulses traveling in free space is also of importance, in addition to the temporal changes of light signals traveling through OFs. V.A. Soifer's research team achieved several significant discoveries in this field. The processes of differentiation (integration) of the pulse envelope may be successfully conducted by employing a resonant diffraction grating, as described in studies [175–180]. High-order derivatives may be calculated quickly using a set of multiple stacked diffraction gratings [179]. The creation of the theory of spatiotemporal transformations of light waves was established on the theoretical explanation of diffraction of light pulses on resonant diffractive assemblies.

An optical correlator can conduct a wide range of spatial filtering functions on light waves. The optical correlator, also known as the coherent optical Fourier processor, is made up of two lenses that execute the FT optically and a spatial filter that encapsulates the transmission function characterizing the incoming light beam's needed spatial transformation. As a spatial filter, a differentiating filter with a complicated transmission function corresponding to the spatial frequency should be utilized to accomplish the differentiation operation in such a scheme. Nevertheless, the optical correlator's significantly larger size limits its practical applicability. Using nanophotonic configurations, spatial differentiators, and integrators with a thickness similar to the wavelength of the modulated light wave may be created. In 2014, the idea of implementing spatial changes of light waves utilizing nanophotonic apparatuses was introduced. The fundamental work in [160] was a theoretical investigation of the execution of spatial differentiation, integration, and convolution operations on light waves. There were two ways offered. The first was to employ tiny analogs of optical correlators, with the traditional Fourier lenses being substituted by tiny layers with a gradient refractive index and an MS serving as the spatial filter storing the requisite transmission function. The second method was to utilize a multilayer structure that was particularly built to accomplish the spatial transformation of the I/P signal indicated by the convolution operator with a certain core.

Professor S.I. Bozhevolnyi's team conducted the first study proving the capability of differentiation and integration regarding a spatial factor using an MS in 2015 [163]. The first strategy was utilized for optical differentiation and integration, in which the processes were carried out in an optical correlator comprised of a lens and a reflecting spatial filter. A reflecting MS encrypting transmission function of a differentiating (integrating) filter was employed as the filter. The MS in question was made up of a series of the metal-insulator-metal resonant circuit [163, 181]. Fabry–Perot resonances of plasmonic modes traveling in metal slots are supported by such resonators. The experimental findings provided in [163] show that optical differentiation (integration) may be implemented with the use of an MS. Simultaneously, it should be highlighted that the MSs in [163] only conduct differentiation and integration tasks in conjunction with a lens, preventing the suggested system from being compact. In subsequent work [182–184], dielectric MSs were used to encode the essential transmission functions of the spatial filter, causing a reduction in losses due to absorption in the metal claddings of nano-resonators and enhanced performance in the deployment of differential and integral conversions.

Because the nanophotonic assembly directly executes the needed spatial change of the I/P signal, the second option is more plausible. In this example, the optical correlator is replaced with a single assembly (with no extra lenses). It's worth noting that the structure's reflection and transmission coefficients as functions of spatial frequency (a tangential component of the incoming wave vector) match the transfer functions defining the incident light beam's conversion [163]. As a result, nanophotonic assemblies with reflection (or transmission) coefficients that approximate the transfer functions of the differentiator or integrator would be required to accomplish the elementary actions of spatial differentiation or integration of the light beam. A resonant reflection or transmission band with a Lorentzian line form can be used to mimic the integrator's transfer function. About the zeros of reflection and transmission happening near the resonance, the differentiator's transfer function is very well modeled.

V.A. Soifer's research team achieved the most important results in this sector. The work of this group initially established that phase-shifted Bragg gratings (PSBGs) [185–187], resonant diffraction gratings [188, 189], PC resonators [190–195], and three-layer arrangements with W-shaped refractive index profiles may efficiently execute the functions of differentiation and integration of the light wave profile [196]. It's worth noting that resonant nanostructures can also be used to compute the Laplace operator optically [187]. In image processing, this method is utilized to recognize edges. The first investigation affirming the potential of differentiating a spatial variable using the resonant diffraction grating was carried out in 2018 using the apparatuses of the Collective Use Center "Nano-Photonics and Diffraction Optics" [197], which was established by a cooperative endeavor of Samara University and the Russian Academy of Sciences' Image Processing Systems Institute [196]. It resulted in a high level of distinction that was far superior to the quality of differentiation attained using MSs [163]. It's also important to note that differentiators and integrators founded on resonant diffraction gratings and PSBGs are not only smaller but also easier to fabricate than equivalent apparatuses that rely on correlators with MSs. The emphasis of [198] is a differentiator comprising of a prism with a metal film placed on one of its sides; specifically, the differentiator is composed of a prism with a metal film-coated on one of its sides [199, 200]. The incident light wave's differentiation is carried out in this scenario in reflection due to the incident light wave's activation of a surface plasmon-polariton on the metal film's surface. The effectiveness of employing such a framework for optical edge detection was proved in experiments.

The first-order differentiation of the transverse profile of an incoming light wave regarding a spatial variable is demonstrated experimentally using a subwavelength diffraction grating as shown in Figure 8.9(a) [189]. Figure 8.9(b) shows a typical SEM picture of the manufactured grating. The experimental findings accord well with the provided theoretical model, implying that the differentiation occurs in transmission at oblique incidence and is linked to the grating's guided-mode resonance. As per this concept, the grating's transfer function about the resonance is similar to that of an exact differentiator. Figure 8.9(c and d) illustrates the incident Gaussian light wave and transmitted light wave profiles, respectively. The incident light wave has a Gaussian shape to it. The precise derivative agrees well with the form of the

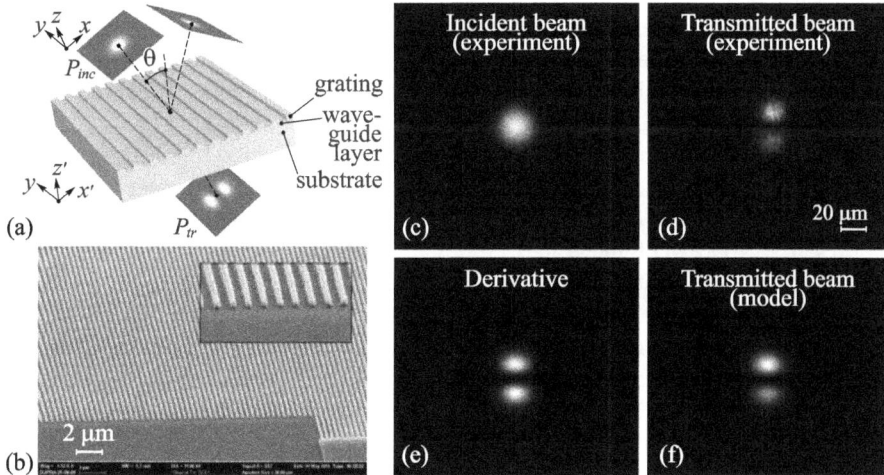

FIGURE 8.9   (a) The diffraction of an optical light wave on a resonant diffraction assembly made up of a grating on top of a slab waveguide layer put on a substrate [189]. (b) The diffraction grating was created using ERP-40 electron resist on top of a TiO_2 layer, as shown by SEM [189]. (c) Measured profile of incident Gaussian light wave [189]. (d) Measured profile of transmitted light wave [189]. (e) Analytically derived derivative of incident light wave [189]. (f) Profile of transmitted light wave estimated taking into account manufactured structural flaws [189].

transmitted light wave (Figure 8.9(e)). The profile of the transmitted light wave estimated considering manufactured structural flaws is also shown in Figure 8.9(f). The
configuration under consideration might be used in the development of novel photonic apparatuses for light wave shaping, optical data processing, and AOC.

8.5 GROWTH IDEAS, CONSTRAINTS, AND MISCONCEPTIONS

For the many methods of OC, there are certain common obstacles. To begin with, large-
scale integration of optical-electrical chips must be manufactured to boost the parallelism of OC systems at the physical level. In addition, optical-electrical co-package
technology is required to decrease the price of data transport between the electrical
and optical domains. Second, current optical transmitters and modulators are primarily intended for optical transmission rather than processing. Because most implementations demand high bit depth I/P data, OC methods allow substantially greater
extinction ratio (ER) and linearity of optical apparatuses than optical communication.

Furthermore, because optical apparatuses with a greater ER and linearity may
allow high-efficiency optical coding for data I/P, the overall performance will be
enhanced. Eventually, a new architectural design is necessary. The optical-electrical
transformation might severely limit the energy efficiency of the hybrid computing
device, making it impossible to take use of the benefits of OC in a traditional computer design. The new architectural design might have a significant speed-up factor while retaining as much customizability as feasible in the meantime. Finally,
there are a few investigations on algorithms that are appropriate for analog OC.
Algorithms are now built using Boole logics, which are ideal for digital computer
systems. They are, nonetheless, impossible to match when it comes to the functionality given by OC. If algorithms for OC are created, their operation overhead and
execution time will be significantly reduced compared to present methods.

Despite the numerous hurdles, OC's potential has been growing. To begin with,
various manufacturing methods have been involved in the development of larger-
scale optical-electrical chip integration. For instance, Light-matter delivered the
world's first 4096 MZI integrated chip, called "Mars," demonstrating the capability of large-scale integration and giving researchers in OC more confidence.
Furthermore, the previously described wavelength division multiplexing (WDM)
and mode division multiplexing (MDM), as well as the spatial optical system, are all
sustainable with increasing parallelism. Moreover, by directly employing a faster-
speed optical device with low-bit depth optical coding, the poor ER and linearity
of optical apparatuses may be adjusted. In terms of data I/P performance, a 2-GHz
optical modulator with OOK and a 1-GHz optical modulator with PAM4 are equal.
This type of compensation, on the other hand, is only possible in computer processes
that can be transformed into a linear combination of low-bit depth operations in the
time domain. Using low-bit depth quantization for application I/P data, on the other
hand, is a common method for making current optical instruments practical in OC.

Light wave looping should be extensively used to maintain data in the optical
domain for as prolonged as feasible in hybrid computing systems to decrease the overhead from optical-electrical conversion. The temporal delay induced by light wave
looping might be minor due to the fast propagation speed of light. New designs can

TABLE 8.3

Some Important Applications of OC

	Applications	References
1	Parallel processing	[202–205]
2	Optical switches	[4, 206, 207]
3	Optical data storage	[201, 208, 209]
4	Data communication	[210, 211]
5	All-optical logic operation	[128, 212, 213]

be inspired by dataflow approaches. Finally, OC techniques might consider the complicated operators available in the optical domain. To minimize implementation risk and processing time, several sets of Boole logic operators in present algorithms can be substituted with a single complicated operator. As a result, integrating complicated operators with Boole logic operators in an algorithm might be a promising technique to construct OC algorithms. OC's possibilities have been expanding. The ever-increasing need for ANNs, as well as their processing requirements, will continue to push research into OC patterns. Optical sensing and communication may provide another opportunity for OC to be used. Furthermore, high-complexity computing algorithms in the optical domain, including FT, convolution, and equation solving, might significantly improve systematic efficiency. Some of the important applications of OC and current state-of-the-art research works are summarized in Table 8.3.

Investigators from the universities of Oxford, Exeter, and Münster have developed a novel technology that allows them to retain more optical data in a smaller space on-chip than formerly conceivable. This method achieves on the phase-change optical memory cell, which utilizes light to record and read data, and might result in a quicker, more energy-efficient computer memory. Unlike today's computers, which utilize electrical impulses to store data in one of two states – zero or one – the optical memory cell stores data via light. The researchers achieved optical memory with over 32 states, or levels, equating to 5 bits. This is a significant step toward the development of an all-optical computer, which is a long-term aim for many researchers in this field [201].

In the coming years, we can expect the creation of all-optical computers in the form of research samples, however industrial prototypes that have a market niche require the implementation of effective technological solutions in electronics. Nevertheless, the advantages of photonic solutions over electronic ones listed in the review paper make it necessary for scientists to continue research in this direction and systematize the accumulated results. From the point of view of the authors, at present, solutions in the optical implementation of certain types of special processors may turn out to be effective, for example, replacing the used electronic analog machines with similar optical ones, providing the solution of certain types of differential equations, contour detection, etc. The proposed solutions can be promising where initially there is an optical signal at the input of a special processor, for example, a video stream. In this case, the optical neural network allows processing and recognizing some images faster than converting them into electronic form and processing it on a digital computer using graphics accelerators.

The fact that processing is a nonlinear process in which several signals must interplay is a key problem for optical computing. Light, which is an EM wave, can only engage with another EM wave in the possession of electrons in a material, and the intensity of this interaction is significantly less for EM waves than for electrical signals in a traditional computer. As a response, processing elements for an optical computer may require more power and have bigger size than processing elements for a traditional electronic computer employing transistors. Another myth is that optical transistors should be susceptible to incredibly high frequencies since light travels far faster than electrons' drift velocity and at frequencies measured in THz. The pace at which an optical transistor may react to a signal is still restricted by its spectral bandwidth, since every EM wave must respect the transform limit. Practical constraints such as dispersion frequently restrict fiber-optic communications channels to bandwidths of tens of GHz, just marginally better than many silicon transistors. To achieve far quicker operation than electronic transistors, effective means of sending ultrashort pulses along extremely dispersive waveguides would be required.

8.6 CONCLUSION

Artificial neural networks (ANNs) have been used to build artificial intelligence in recent times, resulting in unimaginable demands for computing resources. Nevertheless, due to Moore's Law's insufficiency and the failure of Dennard's scaling laws, traditional computer hardware related to electronic transistors and von Neumann architecture would be unable to meet such an incomprehensible demand. Conversely, analog optical computing (AOC) provides an alternate method for unleashing enormous processing power to speed up a wide range of compute-intensive operations. To perform specific computing operations, AOC makes use of physical features of light, including amplitude and phase, as well as the interplay between light and optical devices. Because of the unique mathematical portrayal of computational processes in one specific analog optical computing system, it is a specialized computing system. AOC can achieve higher data processing acceleration in specialized applications, including pattern recognition and numerical calculation, as opposed to traditional digital computing. The popularity of optically implemented neural networks has grown in recent years, attributable to the growing quantity of databases that must be handled, placing a strain on the effectiveness of traditional digital, electronic computers. The key consideration in implementing a viable optical computer (including a neural one) is to integrate the linear part of the system, from which optics derives its competitive advantage, with nonlinear components and I/P–O/P interfaces while preserving the optical interconnections' speed and power efficiency. In this chapter, several optical components such as spatial light modulators, plasmonic switches, neural networks, diffractive neural networks, photonic crystal all-optical logic gates, and resonant nanophotonic structures are used to implement optical computing have been reviewed and their advancements have been discussed. We believe that this chapter will be beneficial to the scientific community working on the topic of optical computing.

REFERENCES

1. de Lima, T. F., Tait, A. N., Mehrabian, A., Nahmias, M. A., Huang, C., Peng, H.-T., Marquez, B. A., Miscuglio, M., El-Ghazawi, T., Sorger, V. J., Shastri, B. J., and Prucnal, P. R. 2020. Primer on silicon neuromorphic photonic processors: Architecture and compiler. *Nanophotonics.* 9(13):4055–73. https://doi.org/10.1515/nanoph-2020-0172.
2. Minzioni, P., Lacava, C., and Tanabe, T. et al. 2019. Roadmap on all-optical processing. *J. Opt.* 21(6):063001. https://doi.org/10.1088/2040-8986/ab0e66.
3. Boyd, G. D., Chirovsky, L. M. F., and Morgan, R. A. 1991. Dynamic optical switching of symmetric self-electro-optic effect devices. *Appl. Phys. Lett.* 59(21):2631. https://doi.org/10.1063/1.105920.
4. Zasedatelev, A. V., and Baranikov, A. V. et al. 2021. Single-photon nonlinearity at room temperature. *Nature.* 597:493–7. https://doi.org/10.1038/s41586-021-03866-9.
5. Brunner, D., Marandi, A., Bogaerts, W., and Ozcan, A. 2020. Photonics for computing and computing for photonics. *Nanophotonics.* 9(13):4053–4. https://doi.org/10.1515/nanoph-2020-0470.
6. Goswami, D. 2003. Optical computing. *Resonance.* 8:8–21. https://doi.org/10.1007/BF02834399.
7. Optalysys. 2022. https://optalysys.com/.
8. Smith, S. 1985. Lasers, nonlinear optics and optical computers. *Nature.* 316:319–24. https://doi.org/10.1038/316319a0.
9. Yeh, P., Chiou, A., Hong, J., Backwith, P., Chang, T., and Khoshnevisan, M. 1989. Photorefractive nonlinear optics and optical computing. *Opt. Eng.* 28(4):328–43. https://doi.org/10.1142/9789812832047_0075.
10. Gillette, J. 1989. CD-ROM data storage technology: Benefits and limitations in document publication. *Book Res. Q* 5:37–43. https://doi.org/10.1007/BF02683798.
11. Goodman, J. 1968. *Introduction to Fourier optics.* San Francisco, CA: McGraw-Hill. ISBN: 978-0-07-023776-6.
12. Maréchal, A., and Croce, P. 1953. Un filtre de frequences spatiales pour l'amelioration du contraste des images optiques. *Comptes Rendus de l'Académie des Sciences* 237(12):607–9.
13. Cutrona, L. J., Leith, E. N., and Palermo, C. J. et al. 1960. Optical data processing and filtering systems. *IEEE Trans. Inf. Theory.* 6(3):386–400. https://doi.org/10.1109/TIT.1960.1057566.
14. Neill, E. 1956. Spatial filtering in optics. *IRE Trans. Inf. Theory.* 2(2):56–65. https://doi.org/10.1109/TIT.1956.1056785.
15. Lugt, A. V. 1964. Signal detection by complex spatial filtering. *IEEE Trans. Inf. Theory.* 10(2):139–45.
16. Weaver, C., and Goodman, J. 1966. A technique for optically convolving two functions. *Appl. Opt.* 5(7):1248–9. https://doi.org/10.1364/AO.5.001248.
17. Rhodes, W., and Sawchuk, A. 1981. Incoherent optical processing. In *Optical information processing: Fundamentals*, ed. S. H. Lee, 69–110. Berlin, Heidelberg: Springer-Verlag. https://doi.org/10.1007/3540105220_10.
18. Leith, E. 1989. Incoherent optical processing and holography. In *Optical processing and computing*, ed. H. Arsenault, 421–440. SanDiego, CA: Academic Press. https://doi.org/10.1016/B978-0-12-064470-4.50017-X.
19. Goodman, J. 1981. Linear space-variant optical data processing. In *Optical information processing: Fundamentals*, ed. S. H. Lee, 235–260. Berlin, Heidelberg: Springer-Verlag. https://doi.org/10.1007/3540105220_13.
20. Ambs, P., Lee, S. H., Tian, Q., and Fainman, Y. 1986. Optical implementation of the Hough transform by a matrix of holograms. *Appl. Opt.* 25(22):4039–45. https://doi.org/10.1364/AO.25.004039.

21. Lee, S. 1981. Nonlinear optical processing. In *Optical information processing: Fundamentals*, ed. S. H. Lee, 261–303. Berlin, Heidelberg: Springer-Verlag. https://doi.org/10.1007/3540105220_14.

22. Ambs, P. 2010. Optical computing: A 60-year adventure. *Adv. Opt. Technol.* 2010:372652. https://doi.org/10.1155/2010/372652.

23. Schwabe, R., Zelinger, S., Key, T., and Phipps, K. 1998. Electronic lighting interference. *IEEE Ind. Appl. Mag.* 4:46–8. https://doi.org/10.1109/2943.692532.

24. Hu, W., Li, X., Yang, J., and Kong, D. 2010. Crosstalk analysis of aligned and misaligned free-space optical interconnect systems. *J. Opt. Soc. Am. A.* 27(2):200–5. https://doi.org/10.1364/JOSAA.27.000200.

25. Carpenter, J. 2012. *Holographic mode division multiplexing in optical fibres.* PhD diss. Cambridge, UK: University of Cambridge.

26. O'Neill, E. L. 1956. Spatial filtering in optics. *IRE Trans. Inform. Theory.* IT-2:56–65. https://doi.org/10.1109/TIT.1956.1056785.

27. Stroke, G. W. 1966. *An introduction to coherent optics and holography.* New York: Academic Press.

28. Preston, K. 1972. *Coherent optical computers.* New York: McGraw-Hill. ISBN: 978-0-07-050785-2.

29. Lugt, A. V. 1974. Coherent optical processing. *Proc. IEEE.* 162(10):1300–19. https://doi.org/10.1109/PROC.1974.9624.

30. Perrin, M., and Metzger, G. 1975. Principles and feasibility of an optical preprocessor in high energy physics. *Nucl. Instrum. Methods.* 126:509–18. https://doi.org/10.1016/0029-554X(75)90801-0.

31. Kotlyar, V. V., Khonina, S. N., and Soifer, V. A. 1998. Light field decomposition in angular harmonics by means of diffractive optics. *J. Mod. Opt.* 45(7):1495–506. https://doi.org/10.1080/09500349808230644.

32. Khonina, S. N., Kotlyar, V. V., Soifer, V. A., Paakkonen, P., and Turunen, J. 2001. Measuring the light field orbital angular momentum using DOE. *Opt. Mem. Neural Netw.* 10(4):241–55.

33. Kotlyar, V. V., Khonina, S. N., Soifer, V. A., and Wang, Y. 2002. Light field orbital angular moment measurement with the help of diffractive optical element. *Avtometriya.* 38(3):33–44.

34. Kotlyar, V. V., Kovalev, A. A., and Volyar, A. V. 2020. Topological charge of optical vortices and their superpositions. *Comput. Opt.* 44(2):145–54. https://doi.org/10.18287/2412-6179-CO-685.

35. Reddy, A. N. K., Anand, V., Khonina, S. N., Podlipnov, V. V., and Juodkazisk, S. 2021. Robust demultiplexing of distinct orbital angular momentum infrared vortex beams into different spatial geometry over a broad spectral range. *IEEE Access.* 9:143341–8. https://doi.org/10.1109/ACCESS.2021.3120836.

36. Porfirev, A. P., and Khonina, S. N. 2016. Experimental investigation of multi-order diffractive optical elements matched with two types of Zernike functions. *Proc. SPIE.* 9807:98070E. https://doi.org/10.1117/12.2231378.

37. Degtyarev, S. A., Porfirev, A. P., and Khonina, S. N. 2017. Zernike basis-matched multi-order diffractive optical elements for wavefront weak aberrations analysis. *Proc. SPIE.* 10337:103370Q. https://doi.org/10.1117/12.2269218

38. Khorin, P. A., Volotovskiy, S. G., and Khonina, S. N. 2021. Optical detection of values of separate aberrations using a multi-channel filter matched with phase Zernike functions. *Comput. Opt.* 45(4):525–33. https://doi.org/10.18287/2412-6179-CO-906.

39. Khonina, S. N., Karpeev, S. V., and Porfirev, A. P. 2020. Wavefront aberration sensor based on a multichannel diffractive optical element. *Sensors.* 20(14):3850. https://doi.org/10.3390/s20143850.

40. Khorin, P. A., Porfirev, A. P., and Khonina, S. N. 2022. Adaptive detection of wave aberrations based on the multichannel filter. *Photonics.* 9(3):204. https://doi.org/10.3390/photonics9030204.

41. Soifer, V. A., Golub, M. A., and Khonina, S. N. 1993. Decorrelated features of images extracted with the aid of optical Karhunen-Loeve expansion. *Pattern Recognit. Image Anal.* 3(3):289–95.

42. Soifer, V. A., and Khonina, S. N. 1994. Stability of the Karhunen-Loeve expansion in the problem of pattern recognition. *Pattern Recognit. Image Anal.* 4(2):137–48.

43. Soifer, V. A., Kotlyar, V. V., and Khonina, S. N. 1996. An optical method of directions field construction. *Avtometriya.* 1:31–6.

44. Soifer, V. A., Kotlyar, V. V., Khonina, S. N., and anad Skidanov, R. V. 1996. Optical methods of fingerprints identification. *Comput. Opt.* 16:78–89.

45. Soifer, V., Kotlyar, V., Khonina, S., and Skidanov, R. 1998. Optical-digital methods of fingerprint identification. *Opt. Lasers Eng.* 29(4–5):351–9. https://doi.org/10.1016/S0143-8166(97)00122-X.

46. Khonina, S. N., Kotlyar, V. V., Skidanov, R. V., and Soifer, V. A. 2003. Optodigital system for identifying fingerprints in real time. *J. Opt. Technol.* 70(8):586–9. https://doi.org/10.1364/JOT.70.000586.

47. Khonina, S. N., and Demidov, A. S. 2014. Extended depth of focus through imaging system's phase apodization in coherent and incoherent cases. *Opt. Mem. Neural Netw.* 23(3):130–9. https://doi.org/10.3103/S1060992X14030035.

48. Khonina, S. N., and Ustinov, A. V. 2015. Generalized apodization of an incoherent imaging system aimed for extending the depth of focus. *Pattern Recognit. Image Anal.* 25(4):626–31. https://doi.org/10.1134/S1054661815040100.

49. Khonina, S. N., Ustinov, A. V., and Porfirev, A. P. 2019. Dynamic focal shift and extending depth of focus based on the masking of the illuminating beam and using an adjustable axicon. *J. Opt. Soc. Am. A.* 36(6):1039–47. https://doi.org/10.1364/JOSAA.36.001039.

50. Khonina, S. N., Volotovskiy, S. G., Dzyuba, A. P., Serafimovich, P. G., Popov, S. B., and Butt, M. A. 2021. Power phase apodization study on compensation defocusing and chromatic aberration in the imaging system. *Electronics.* 10(11):1327. https://doi.org/10.3390/electronics10111327.

51. Khonina, S. N., Kotlyar, V. V., Skinkaryev, M. V., Soifer, V. A., and Uspleniev, G. V. 1992. The phase rotor filter. *J. Mod. Opt.* 39(5):1147–54. https://doi.org/10.1080/09500349214551151.

52. Ananin, M. A., and Khonina, S. N. 2009. Modelling of optical processing of images with use of the vortical spatial filter. *Comput. Opt.* 33(4):466–72.

53. Davis, J. A., McNamara, D. E., Cottrell, D. M., and Campos, J. 2000. Image processing with the radial Hilbert transform: Theory and experiments. *Opt. Lett.* 25(2):99–101. https://doi.org/10.1364/OL.25.000099.

54. Guo, C.-S., Han, Y.-J., Xu, J.-B., and Ding, J. 2006. Radial Hilbert transform with Laguerre-Gaussian spatial filters. *Opt. Lett.* 31(10):1394–6. https://doi.org/10.1364/OL.31.001394.

55. Lazarev, G., and Hermerschmidt, A. 2012. LCOS spatial light modulators: Trends and applications. In *Optical imaging and metrology: Advanced technologies*, eds. W. Osten, and N. Reingand, 1–29. Weinheim: Wiley-VCH.

56. Pivnenko, M., Li, K., and Chu, D. Sub-millisecond switching of multi-level liquid crystal on silicon spatial light modulators for increased information bandwidth. *Opt. Express.* 2021;29(16):24614–28. https://doi.org/10.1364/OE.429992.

57. Jin, D., Zhou, R., Yaqoob, Z., and So, P. T. C. 2018. Dynamic spatial filtering using a digital micromirror device for high-speed optical diffraction tomography. *Opt. Express.* 26(1):428–37. https://doi.org/10.1364/OE.26.000428.

58. Turtaev, S., Leite, I. T., Mitchell, K. J., Padgett, M. J., Phillips, D. B., and Cizmar, T. 2017. Comparison of nematic liquid-crystal and DMD based spatial light modulation in complex photonics. *Opt. Express.* 25(24):29874–84. https://doi.org/10.1364/OE.25.029874.

59. Phase spatial light modulator LCOS-SLM. 2012. *Opto-semiconductor handbook*, 1–14. Tokyo: Hamamstsu Photonics.

60. Kovachev, M. et al. 2006. Reconstruction of computer generated holograms by spatial light modulators. In *Multimedia content representation, classification and security*, ed. B. Gunsel, A. K. Jain, A. M. Tekalp, B. Sankur, 706–713. Berlin, Heidelberg: Springer. https://doi.org/10.1007/11848035_93.

61. Younus, S., Hussein, A., Alresheedi, M., and Elmirghani, J. 2017. CGH for indoor visible light communication system. *IEEE Access.* 5:24988–5004. https://doi.org/10.1109/ACCESS.2017.2765378.

62. Torii, Y., Balladares-Ocana, L., and Martinez-Castro, J. 2013. An iterative Fourier transform algorithm for digital hologram generation using phase-only information and its implementation in a fixed-point digital signal processor. *Optik.* 124(22):5416–21. https://doi.org/10.1016/j.ijleo.2013.03.112.

63. Ripoll, O., Kettunen, V., and Herzig, H. P. 2004. Review of iterative Fourier-transform algorithms for beam shaping applications. *Opt. Eng.* 43(11):2549–56. https://doi.org/10.1117/1.1804543.

64. Roelens, M. A. F. et al. 2008. Dispersion trimming in a reconfigurable wavelength selective switch. *J. Lightw. Technol.* 26(1):73–8. https://doi.org/10.1109/JLT.2007.912148.

65. Wang, M. et al. 2017. LCoS SLM study and its application in wavelength selective switch. *Photonics.* 4(2):22. https://doi.org/10.3390/photonics4020022.

66. Turunen, J., and Wyrowski, F. 1998. *Diffractive optics for industrial and commercial applications*. Berlin: Wiley-VCH Verlag GmbH. ISBN: 978-3-05-501733-9.

67. Osten, W., Kohler, C., and Liesener, J. 2005. Evaluation and application of spatial light modulators for optical metrology. *Optica Pura y Aplicada.* 38(3):71–81.

68. Varga, J. J. M. et al. 2015. Preparing arbitrary pure states of spatial qudits with a single phase-only spatial light modulator. *J. Phys. Conf. Ser.* 605:012035. 10.1088/1742-6596/605/1/012035.

69. Schröder, J., Roelens, M., Du, L., Lowery, A., and Eggleton, B. 2012. LCOS based waveshaper technology for optical signal processing and performance monitoring. *17th Opto-Electronics and Communications Conf:* 859–60. https://doi.org/10.1109/OECC.2012.6276666.

70. Barbier, P., and Moddel, G. 1997. Spatial light modulators: Processing light in real time. *Opt. Photon. News.* 8(3):16–21. https://doi.org/10.1364/OPN.8.3.000016.

71. Shrestha, P., Chun, Y., and Chu, D. 2015. A high-resolution optically addressed spatial light modulator based on ZnO nanoparticles. *Light. Sci. Appl.* 4:e259. https://doi.org/10.1038/lsa.2015.32.

72. Chen, P., Ma, L.-L., Hu, W., Shen, Z.-X., Bisoyi, H., Wu, S.-B., Ge, S.-J., Li, Q., and Lu, Y.-Q. 2019. Chirality invertible superstructure mediated active planar optics. *Nat. Commun.* 10(1):2518. https://doi.org/10.1038/s41467-019-10538-w.

73. Shih, M., Shishido, A., and Khoo, I. 2001. All-optical image processing by means of a photosensitive nonlinear liquid-crystal film: Edge enhancement and image addition–subtraction. *Opt. Lett.* 26(15):1140–2. https://doi.org/10.1364/OL.26.001140.

74. Zhang, J., Wang, H., Yoshikado, S., and Aruga, T. 1997. Incoherent-to-coherent conversion by use of the photorefractive fanning effect. *Opt. Lett.* 22(21):1612–4. https://doi.org/10.1364/OL.22.001612.

75. Woods, D., and Naughton, T. 2012. Photonic neural networks. *Nat. Phys.* 8:257–9. https://doi.org/10.1038/nphys2283.

76. Solodar, A., Kumar, T., Sarusi, G., and Abdulhalim, I. 2016. Infrared to visible image up-conversion using optically addressed spatial light modulator utilizing liquid crystal and InGaAs photodiodes. *Appl. Phys. Lett.* 108(2):021103. https://doi.org/10.1063/1.4939903.

77. Gelbaor, K., Matvey, K., Victor, L., Neil, C., and Abdulhalim, I. 2014. Liquid crystal high-resolution optically addressed spatial light modulator using a nanodimensional chalcogenide photosensor. *Opt. Lett.* 39(7):2048–51. https://doi.org/10.1364/OL.39.002048.

78. Li, S.-Q., Xu, X., Veetil, R., Valuckas, V., Paniagua-Domnguez, R., and Kuznetsov, A. 2019. Phase-only transmissive spatial light modulator based on tunable dielectric metasurface. *Science* 364(6445):1087–90. https://doi.org/10.1126/science.aaw6747.

79. Li, J., Yu, P., Zhang, S., and Liu, N. 2020; Electrically-controlled digital metasurface device for light projection displays. *Nat. Commun.* 11(1):3574. https://doi.org/10.1038/s41467-020-17390-3.

80. Park, J., and Jeong, B. G. et al. 2020. All-solid-state spatial light modulator with independent phase and amplitude control for three-dimensional lidar applications. *Nat. Nanotechnol.* 16:69–76. https://doi.org/10.1038/s41565-020-00787-y.

81. Gong, S., Ren, M., Wu, W., Cai, W., and Xu, J. 2021. Optically addressed spatial light modulator based on nonlinear metasurface. *Photonics Res.* 9(4):610–4. https://doi.org/10.1364/PRJ.416189.

82. Heck, M. 2018. Optical computers light up the horizon. https://phys.org/news/2018-03-optical-horizon.html.

83. Waldrop, M. 2016. The chips are down for Moore's law. *Nat. News.* 530:144. https://doi.org/10.1038/530144a.

84. Atabaki, A. H. et al. 2018. Integrating photonics with silicon nanoelectronics for the next generation of systems on a chip. *Nature.* 556:349–54. https://doi.org/10.1038/s41586-018-0028-z.

85. Paniccia, M. 2011. A perfect marriage: Optics and silicon. Integrated silicon-based photonics now running at 50 Ggps, with Terabit speeds on the horizon. *Optik Photon.* 6(2):34–8. https://doi.org/10.1002/opph.201190327.

86. Liu, K., Ye, C., Khan, S., and Sorger, V. 2015. Review and perspective on ultrafast wavelength-size electro-optic modulators. *Laser Photon. Rev.* 9(2):172–94. https://doi.org/10.1002/lpor.201400219.

87. Chung, S., Nakai, M., and Hashemi, H. 2019. Low-power thermo-optic silicon modulator for large-scale photonic integrated systems. *Opt. Express.* 27(9):13430–59. https://doi.org/10.1364/OE.27.013430.

88. Lee, B., Biberman, A., Chan, J., and Bergman, K. 2009. High-performance modulators and switches for silicon photonic networks-on-chip. *IEEE J. Sel. Top. Quantum Electron.* 16(1):6–22. https://doi.org/10.1109/JSTQE.2009.2028437.

89. Ying, Z. et al. 2019. Integrated multi-operand electro-optic logic gates for optical computing. *Appl. Phys. Lett.* 115(17):171104. https://doi.org/10.1063/1.5126517.

90. Xiao, X. et al. 2013. High-speed, low-loss silicon Mach-Zehnder modulators with doping optimization. *Opt. Express.* 21(4):4116–25. https://doi.org/10.1364/OE.21.004116.

91. Shu, H., Jin, M., Tao, Y., and Wang, X. 2019. Graphene-based silicon modulators. *Front. Inf. Technol. Electron. Eng.* 20:458–71. https://doi.org/10.1631/FITEE.1800407.

92. Zhang, Q. et al. 2018. Broadband non-volatile photonic switching based on optical phase change materials: Beyond the classical figure-of-merit. *Opt. Lett.* 43(1):94–7. https://doi.org/10.1364/OL.43.000094.

93. Yang, Z., and Ramanathan, S. 2015. Breakthroughs in photonics 2014: Phase change materials for photonics. *IEEE Photon. J.* 7(3):0700305. https://doi.org/10.1109/JPHOT.2015.2413594.

94. Badri, S., and Farkoush, S. 2021. Subwavelength grating waveguide filter based on cladding modulation with a phase-change material grating. *Appl. Opt.* 60(10):2803–10. https://doi.org/10.1364/AO.419587.

95. Wang, J., Wang, L., and Liu, J. 2020. Overview of phase-change materials based photonic devices. *IEEE Access.* 8:121211–45. https://doi.org/10.1109/ACCESS.2020. 3006899.

96. Badri, S., Gilarlue, M., Farkoush, S., and Rhee, S. 2021. Reconfigurable bandpass optical filters based on subwavelength grating waveguides with a Ge2Sb2Te5 cavity. *J. Opt. Soc. Am. B.* 38(4):1283–9. https://doi.org/10.1364/JOSAB.419475.

97. Ghosh, R., and Dhawan, A. 2021. Integrated non-volatile plasmonic switches based on phase-change-materials and their application to plasmonic logic circuits. *Sci. Rep.* 11:18811. https://doi.org/10.1038/s41598-021-98418-6.

98. Prieto, A. et al. 2016. Neural networks: An overview of early research, current frameworks and new challenges. *Neurocomputing.* 214:242–68. https://doi.org/10.1016/j. neucom.2016.06.014.

99. LeCun, Y., Bengio, Y., and Hinton, G. 2015. Deep learning. *Nature.* 521(7553):436–44. https://doi.org/10.1038/nature14539.

100. Krizhevsky, A., Sutskever, I., and Hinton, G. 2017. Imagenet classification with deep convolutional neural networks. *Commun. ACM.* 60(6):84–90. https://doi. org/10.1145/3065386.

101. Nitta, T. 2004. Orthogonality of decision boundaries in complex-valued neural networks. *Neural Comput.* 16(1):73–97. https://doi.org/10.1162/08997660460734001.

102. Aizenberg, I. 2011. *Complex-valued neural networks with multi-valued neurons.* Berlin, Heidelberg: Springer-Verlag. ISBN: 978-3-642-20352-7.

103. Peng, H.-T., Nahmias, M., De Lima, T., Tait, A., and Shastri, B. 2018. Neuromorphic photonic integrated circuits. *IEEE J. Sel. Top. Quantum Electron.* 24(6):6101715. https://doi.org/10.1109/JSTQE.2018.2840448.

104. Sze, V., Chen, Y.-H., Yang, T.-J., and Emer, J. 2017. Efficient processing of deep neural networks: A tutorial and survey. *Proc. IEEE.* 105:2295–329. https://doi.org/10.1109/ JPROC.2017.2761740.

105. de Lima, T. F. et al. 2019. Machine learning with neuromorphic photonics. *J. Lightw. Technol.* 37(5):1515–34. https://doi.org/10.1109/JLT.2019.2903474.

106. Lin, X. et al. 2018. All-optical machine learning using diffractive deep neural networks. *Science.* 361(6406):1004–8. https://doi.org/10.1126/science.aat8084.

107. Vandoorne, K. et al. 2014. Experimental demonstration of reservoir computing on a silicon photonics chip. *Nat. Commun.* 5:3541. https://doi.org/10.1038/ncomms4541.

108. Harris, N. C. et al. 2018. Linear programmable nanophotonic processors. *Optica.* 5(12):1623–31. https://doi.org/10.1364/OPTICA.5.001623.

109. Williamson, I. A. D. et al. 2019. Reprogrammable electro-optic nonlinear activation functions for optical neural networks. *IEEE J. Sel. Top. Quantum Electron.* 26(1):7700412. https://doi.org/10.1109/JSTQE.2019.2930455.

110. Bueno, J. et al. 2018. Reinforcement learning in a large-scale photonic recurrent neural network. *Optica.* 5(6):756–60. https://doi.org/10.1364/OPTICA.5.000756.

111. Ambrogio, S. et al. 2018. Equivalent-accuracy accelerated neural-network training using analogue memory. *Nature.* 558:60–7. https://doi.org/10.1038/s41586-018-0180-5.

112. Li, C. et al. 2018. Efficient and self-adaptive in-situ learning in multilayer memristor neural networks. *Nat. Commun.* 9:2385. https://doi.org/10.1038/s41467-018-04484-2.

113. Yao, P. et al. 2020. Fully hardware-implemented memristor convolutional neural network. *Nature* 577:641–6. https://doi.org/10.1038/s41586-020-1942-4.

114. Rakkiyappan, R., Velmurugan, G., and Li, X. 2015. Complete stability analysis of complex-valued neural networks with time delays and impulses. *Neural Process. Lett.* 41:435–68. https://doi.org/10.1007/s11063-014-9349-6

115. Wang, H., Duan, S., Huang, T., Wang, L., and Li, C. 2016. Exponential stability of complex-valued memristive recurrent neural networks. *IEEE Trans. Neural Netw. Learn. Syst.* 28(3):766–71. https://doi.org/10.1109/TNNLS.2015.2513001.
116. Velmurugan, G., Rakkiyappan, R., Vembarasan, V., Cao, J., and Alsaedi, A. 2017. Dissipativity and stability analysis of fractional-order complex-valued neural networks with time delay. *Neural Netw.* 86:42–53. https://doi.org/10.1016/j.neunet.2016.10.010.
117. Hirose, A. 1994. Applications of complex-valued neural networks to coherent optical computing using phase-sensitive detection scheme. *Inf. Sci. Appl.* 2(2):103–17. https://doi.org/10.1016/1069-0115(94)90014-0.
118. Zhang, H., Gu, M., and Jiang, X. D. et al. 2021. An optical neural chip for implementing complex-valued neural network. *Nat. Commun.* 12:457. https://doi.org/10.1038/s41467-020-20719-7.
119. Wei, H. et al. 2011. Cascaded logic gates in nanophotonic plasmon networks. *Nat. Commun.* 2:387. https://doi.org/10.1038/ncomms1388.
120. Wei, H. et al. 2011. Quantum dot-based local field imaging reveals plasmon-based interferometric logic in silver nanowire networks. *Nano Lett.* 11(2):471–5. https://doi.org/10.1021/nl103228b.
121. Fu, Y. L. et al. 2012. All-optical logic gates based on nanoscale plasmonic slot waveguides. *Nano Lett.* 12(11):5784–90. https://doi.org/10.1021/nl303095s.
122. Sang, Y. G. et al. 2018. Broadband multifunctional plasmonic logic gates. *Adv. Opt. Mater.* 6(13):1701368. https://doi.org/10.1002/adom.201701368.
123. Xu, Q., and Lipson, M. 2007. All-optical logic based on silicon micro-ring resonators. *Opt. Express.* 15(3):924–9. https://doi.org/10.1364/OE.15.000924.
124. Raeker, B., and Grbic, A. 2019. Compound metaoptics for amplitude and phase. *Phys. Rev. Lett.* 122(11):113901. https://doi.org/10.1103/PhysRevLett.122.113901.
125. Khonina, S. N., Kazanskiy, N. L., Karpeev, S. V., and Butt, M. A. 2020. Bessel beam: Significance and applications–A progressive review. *Micromachines.* 11(11):997. https://doi.org/10.3390/mi11110997.
126. Qian, C., Lin, X., Lin, X., Xu, J., Sun, Y., Li, E., Zhang, B., and Chen, H. 2020. Performing optical logic operations by a diffractive neural network. *Light Sci. Appl.* 9:59. https://doi.org/10.1038/s41377-020-0303-2.
127. Onifade, A. 2004. History of the computer. *IEEE Conf. on the History of Electronics*: 21.
128. Cotter, D. et al. 1999. Nonlinear optics for high-speed digital information processing. *Science.* 286(5444):1523–8. https://doi.org/10.1126/science.286.5444.1523.
129. Suzuki, Y., Shimada, J., and Yamashita, H. 1985. High-speed optical-optical logic gate for optical computers. *Electron. Lett.* 21(4):161–2. https://doi.org/10.1049/el:19850114.
130. Meindl, J. 1995. Low power microelectronics: Retrospect and prospect. *Proc. IEEE.* 83(4):619–35. https://doi.org/10.1109/5.371970.
131. Yavuz, D. 2006. All-optical femtosecond switch using two-photon absorption. *Phys. Rev. A.* 74(5):053804. https://doi.org/10.1103/PhysRevA.74.053804.
132. Kim, J. H. et al. 2002. All-optical XOR gate using semiconductor optical amplifiers without additional input beam. *IEEE Photonics Technol. Lett.* 14(10):1436–8. https://doi.org/10.1109/LPT.2002.801841.
133. Sharaiha, A., Topomondzo, J., and Morel, P. 2006. All-optical logic AND-NOR gate with three inputs based on cross-gain modulation in a semiconductor optical amplifier. *Opt. Commun.* 265(1):322–5. https://doi.org/10.1016/j.optcom.2006.03.036.
134. Nesset, D., Tatham, M., and Cotter, D. 1995. All-optical AND gate operating on 10 Gbit/s signal-sat the same wavelength using four-wave mixing in a semiconductor laser amplifier. *Electron. Lett.* 31(11):896–7. https://doi.org/10.1049/el:19950600.
135. Webb, R. P. et al. 2003. 40 Gbit/s all-optical XOR gate based on hybrid-integrated Mach-Zehnder interferometer. *Electron. Lett.* 39(1):79–81. https://doi.org/10.1049/el:20030010.

136. Wu, Y.-D., Shih, T.-T., and Chen, M.-H. 2008. New all-optical logic gates based on the local non-linear Mach-Zehnder interferometer. *Opt. Express.* 16(1):248–57. https://doi.org/10.1364/OE.16.000248.

137. McGeehan, J., Giltrelli, M., and Willner, A. 2007. All-optical digital 3-input AND gate using sum- and difference-frequency generation in PPLN waveguide. *Electron. Lett.* 43(7):409–10. https://doi.org/10.1049/el:20073430.

138. Wang, J., Sun, J., and Sun, Q. 2007. Single-PPLN-based simultaneous half-adder, half-subtracter, and OR logic gate: Proposal and simulation. *Opt. Express.* 15(4):1690–9. https://doi.org/10.1364/OE.15.001690.

139. Butt, M. A., and Kazanskiy, N. L. 2021. Two-dimensional photonic crystal hetero-structure for light steering and TM-polarization maintaining applications. *Laser Phys.* 31(3):036201. https://doi.org/10.1088/1555-6611/abd8ca.

140. Butt, M. A., Khonina, S. N., and Kazanskiy, N. L. 2021. 2D-photonic crystal het-erostructures for the realization of compact photonic devices. *Photonics Nanostruct.* 44:100903. https://doi.org/10.1016/j.photonics.2021.100903.

141. Kazanskiy, N. L., Butt, M. A., and Khonina, S. N. 2021. 2D-heterostructure photonic crystal formation for on-chip polarization division multiplexing. *Photonics.* 8(8):313. https://doi.org/10.3390/photonics8080313.

142. Kazanskiy, N. L., and Butt, M. A. 2020. One-dimensional photonic crystal waveguide based on the SOI platform for transverse magnetic polarization-maintaining devices. *Photonics Lett. Poland.* 12(3):85–7. https://doi.org/10.4302/plp.v12i3.1044.

143. John, S. 1987. Strong localization of photons in certain disordered dielectric superlat-tices. *Phys. Rev. Lett.* 58(23):2486. https://doi.org/10.1103/PhysRevLett.58.2486.

144. Zhang, Y., Zhang, Y., and Li, B. 2007. Optical switches and logic gates based on self-collimated beams in two-dimensional photonic crystals. *Opt. Express.* 15(15):9287–92. https://doi.org/10.1364/OE.15.009287.

145. Fan, R. et al. 2016. 2D photonic crystal logic gates based on self-collimated effect. *J. Phys. D.* 49(32):325104. https://doi.org/10.1088/0022-3727/49/32/325104.

146. Christina, X. S., and Kabilan, A. P. 2012. Design of optical logic gates using self-colli-mated beams in 2D photonic crystal. *Photonic Sens.* 2(2):173–9. https://doi.org/10.1007/s13320-012-0054-7.

147. Hou, J. et al. 2009. Polarization insensitive self-collimation waveguide in square lat-tice annular photonic crystals. *Opt. Commun.* 282(15):3172–6. https://doi.org/10.1016/j.optcom.2009.04.051.

148. Kabilan, A. P., Christina, X. S., and Caroline, P. E. 2010. Photonic crystal based all optical OR and XO logic gates. *2010 Second Int. Conf. on Computing, Communication and Networking Technologies*: 1–4. https://doi.org/10.1109/ICCCNT.2010.5591766.

149. Xavier, S. C. et al. 2016. Compact photonic crystal integrated circuit for all-optical logic operation. *IET Optoelectron.* 10(4):142–7. https://doi.org/10.1049/iet-opt.2015.0072.

150. Ishizaka, Y. et al. 2011. Design of ultra compact all-optical XOR and AND logic gates with low power consumption. *Opt. Commun.* 284(14):3528–33. https://doi.org/10.1016/j.optcom.2011.03.069.

151. Lin, Y. et al. 2014. Design and optimization of all-optical AND and NOR logic gates in a two-dimensional photonic crystal for binary-phase-shift-keyed signals. *7th Int. Conf. on Biomedical Engineering and Informatics*: 965–969. https://doi.org/10.1109/BMEI.2014.7002912.

152. Liu, W. et al. 2013. Design of ultra compact all-optical XOR, XNOR, NAND and OR gates using photonic crystal multi-mode interference waveguides. *Opt. Laser Technol.* 50:55–64. https://doi.org/10.1016/j.optlastec.2012.12.030.

153. Shaik, E., and Rangaswamy, N. 2017. Multi-mode interference-based photonic crystal logic gates with simple structure and improved contrast ratio. *Photonic Netw. Commun.* 34(1):140–8. https://doi.org/10.1007/s11107-016-0683-7.

154. Serajmohammadi, S., and Absalan, H. 2016. All optical NAND gate based on non-linear photonic crystal ring resonator. *Inf. Process. Agric.* 3(2):119–23. https://doi.org/10.1016/j.inpa.2016.04.002.

155. Rani, P. et al. 2017. Realization of all optical logic gates using universal NAND gates on photonic crystal platform. *Superlattices Microstruct.* 109:619–25. https://doi.org/10.1016/j.spmi.2017.05.046.

156. Younis, R. M., Areed, N. F. F., and Obayya, S. S. A. 2014. Fully integrated AND and OR optical logic gates. *IEEE Photonics Technol. Lett.* 26(19):1900–3. https://doi.org/10.1109/LPT.2014.2340435.

157. Rao, D. G. S., Swarnakar, S., and Kumar, S. 2020. Performance analysis of all-optical NAND, NOR, and XNOR logic gates using photonic crystal waveguide for optical computing applications. *Opt. Eng.* 59(5):057101. https://doi.org/10.1117/1.OE.59.5.057101.

158. Parandin, F., and Karhanehchi, M. 2017. Terahertz all-optical NOR and AND logic gates based on 2D photonic crystals. *Superlattices Microstruct.* 101:253–60. https://doi.org/10.1016/j.spmi.2016.11.038.

159. Pirzadi, M., Mir, A., and Bodaghi, D. 2016. Realization of ultra-accurate and compact all-optical photonic crystal or logic gate. *IEEE Photonics Technol. Lett.* 28(21):2387–90. https://doi.org/10.1109/LPT.2016.2596580.

160. Sun, C., Wade, M., and Lee, Y. et al. 2015. Single-chip microprocessor that communicates directly using light. *Nature.* 528:534–8. https://doi.org/10.1038/nature16454.

161. Silva, A., Monticone, F., Castaldi, G., Galdi, V., Alu, A., and Engheta, N. 2014. Performing mathematical operations with metamaterials. *Science.* 343(6167):161–3. https://doi.org/10.1126/science.1242818.

162. Solli, D., and Jalali, B. 2015. Analog optical computing. *Nat. Photonics.* 9:704–6. https://doi.org/10.1038/nphoton.2015.208.

163. Pors, A., Nielsen, M., and Bozhevolnyi, S. 2015. Analog computing using reflective plasmonic metasurfaces. *Nano Lett.* 15(1):791–7. https://doi.org/10.1021/nl5047297.

164. Ferrera, M., Park, Y., and Razzari, L. et al. 2010. On-chip CMOS-compatible all-optical integrator. *Nat. Commun.* 1(3):29. https://doi.org/10.1038/ncomms1028.

165. Pasquazi, A., and Peccianti, M. et al. 2011. Sub-picosecond phase-sensitive optical pulse characterization on a chip. *Nat. Photonics.* 5(10):618–23. https://doi.org/10.1038/nphoton.2011.199.

166. Fernández-Ruiz, M., Carballar, A., and Azaña, J. 2013. Design of ultrafast all-optical signal processing devices based on fiber Bragg gratings in transmission. *J. Lightw. Technol.* 31(10):1593–600. https://doi.org/10.1109/JLT.2013.2254467.

167. Asghari, M., and Azana, J. 2011. Photonic integrator-based optical memory unit. *IEEE Photon. Technol. Lett.* 23(4):209–11. https://doi.org/10.1109/LPT.2010.2096806.

168. Ashrafi, R., and Azaña, J. 2012. Figure of merit for photonic differentiators. *Opt. Express.* 20(3):2626–39. https://doi.org/10.1364/OE.20.002626.

169. Rutkowska, K. A., Duchesne, D., Strain, M. J., Morandotti, R., Sorel, M., and Azana, J. 2011. Ultrafast all-optical temporal differentiators based on CMOS-compatible integrated-waveguide Bragg gratings. *Opt. Express* 19(20):19514–22. https://doi.org/10.1364/OE.19.019514.

170. Li, M., Janner, D., Yao, J., and Pruneri, V. 2009. Arbitrary-order all-fiber temporal differentiator based on a fiber Bragg grating: Design and experimental demonstration. *Opt. Express* 17(22):19798–807. https://doi.org/10.1364/OE.17.019798.

171. Tan, S., Wu, Z., Lei, L., Hu, S., Dong, J., and Zhang, X. 2013. All-optical computation system for solving differential equations based on optical intensity differentiator. *Opt. Express.* 21(6):7008–13. https://doi.org/10.1364/OE.21.007008.

172. Tan, S., Xiang, L., Zou, J., Zhang, Q., Wu, Z., Yu, Y., Dong, J., and Zhang, X. 2013. High-order all-optical differential equation solver based on microring resonators. *Opt. Lett.* 38(19):3735–8. https://doi.org/10.1364/OL.38.003735.

173. Lu, L., Wu, J., Wang, T., and Su, Y. 2012. Compact all-optical differential-equation solver based on silicon microring resonator. *Front. Optoelectron.* 5:99–106. https://doi.org/10.1007/s12200-012-0186-9.

174. Yang, T., Dong, J., Lu, L., Zhou, L., Zheng, A., Zhang, X., and Chen, J. 2014. All-optical differential equation solver with constant-coefficient tunable based on a single microring resonator. *Sci. Rep.* 4:5581. https://doi.org/10.1038/srep05581.

175. Bykov, D. A., Doskolovich, L. L., and Soifer, V. A. 2012. Integration of optical pulses by resonant diffraction gratings. *JETP Lett.* 95(1):6–9. https://doi.org/10.1134/S0021364012010031.

176. Soifer, V. A., ed. 2017. *Diffractive optics and nanophotonics.* Boca Raton, FL: CRC Press. ISBN: 978-1-4978-5447-7.

177. Bykov, D. A., Doskolovich, L. L., Golovastikov, N. V., and Soifer, V. A. 2013. Time-domain differentiation of optical pulses in reflection and in transmission using the same resonant grating. *J. Opt.* 15(10):105703. https://doi.org/10.1088/2040-8978/15/10/105703.

178. Bykov, D. A., Doskolovich, L. L., and Soifer, V. A. 2011. Temporal differentiation of optical signals using resonant gratings. *Opt. Lett.* 36(11):3509–11. https://doi.org/10.1364/OL.36.003509.

179. Bykov, D. A., Doskolovich, L. L., and Soifer, V. A. 2012. On the ability of resonant diffraction gratings to differentiate a pulsed optical signal. *J. Exp. Theor. Phys.* 114(5):724–30. https://doi.org/10.1134/S1063776112030028.

180. Bugaev, A. 2020. Resonant nanophotonic structures for analog optical computing. *2020 Int. Conf. on Information Technology and Nanotechnology (ITNT):* 1–5. https://doi.org/10.1109/ITNT49337.2020.9253358.

181. Pors, A., and Bozhevolnyi, S. I. 2016. Plasmonic metasurfaces for efficient phase control in reflection. *Opt. Express.* 21(22):27438–51. https://doi.org/10.1364/OE.21.027438

182. Chizari, A., Abdollahramezani, S., Jamali, M. V., and Salehi, J. A. 2016. Analog optical computing based on a dielectric meta-reflect array. *Opt. Lett.* 41(15):3451–4. https://doi.org/10.1364/OL.41.003451.

183. Abdollahramezani, S., Chizari, A., Dorche, A. E., Jamali, M. V., and Salehi, J. A. 2017. Dielectric metasurfaces solve differential and integro-differential equations. *Opt. Lett.* 42:1197–200. https://doi.org/10.1364/OL.42.001197.

184. Babashah, H., Kavehvash, Z., Koohi, S., and Khavasi, A. 2017. Integration in analog optical computing using metasurfaces revisited: Toward ideal optical integration. *J. Opt. Soc. Am. B.* 34(6):1270–9. https://doi.org/10.1364/JOSAB.34.001270.

185. Doskolovich, L. L., Bykov, D. A., Bezus, E. A., and Soifer, V. A. 2014. Spatial differentiation of optical beams using phase-shifted Bragg grating. *Opt. Lett.* 39(5):1278–81. https://doi.org/10.1364/OL.39.001278.

186. Golovastikov, N., Bykov, D., Doskolovich, L., and Bezus, E. 2015. Spatial optical integrator based on phase-shifted Bragg gratings. *Opt. Commun.* 338:457–60. https://doi.org/10.1016/j.optcom.2014.11.007.

187. Bykov, D. A., Doskolovich, L. L., Bezus, E. A., and Soifer, V. A. 2014. Optical computation of the Laplace operator using phase-shifted Bragg grating. *Opt. Express.* 22(21):25084–92. https://doi.org/10.1364/OE.22.025084.

188. Golovastikov, N. V., Bykov, D. A., and Doskolovich, L. L. 2014. Resonant diffraction gratings for spatial differentiation of optical beams. *Quantum Electron.* 44(10):984–8. https://doi.org/10.1070/QE2014v044n10ABEH015477.

189. Bykov, D. A., Doskolovich, L. L., Morozov, A. A., Podlipnov, V. V., Bezus, E. A., Verma, P., and Soifer, V. A. 2018. First-order optical spatial differentiator based on a guided-mode resonant grating. *Opt. Express.* 26(8):10997–1006. https://doi.org/10.1364/OE.26.010997.

190. Kazanskiy, N. L., Serafimovich, P. G., and Khonina, S. N. 2013. Use of photonic crystal cavities for temporal differentiation of optical signals. *Opt. Lett.* 38(7):1149–51. https://doi.org/10.1364/OL.38.001149.

191. Kazanskiy, N. L., Serafimovich, P. G., and Khonina, S. N. 2012. Use of photonic crystal resonators for differentiation of optical impulses in time. *Comput. Opt.* 36(4):474–8.

192. Kazanskiy, N. L., and Serafimovich, P. G. 2014. Coupled-resonator optical waveguides for temporal integration of optical signals. *Opt. Express.* 22(11):14004–13. https://doi.org/10.1364/OE.22.014004.

193. Serafimovich, P. G., and Kazanskiy, N. L. 2015. Active photonic crystal cavities for optical signal integration. *Opt. Mem. Neural Networks.* 24(4):260–71. https://doi.org/10.3103/S1060992X15040050.

194. Serafimovich, P. G., and Kazanskiy, N. L. 2016. Optical modulator based on coupled photonic crystal cavities. *J. Mod. Opt.* 63(13):1233–8. https://doi.org/10.1080/09500340.2015.1135258.

195. Serafimovich, P. G., Stepikhova, M. V., Kazanskiy, N. L., Gusev, S. A., Egorov, A. V., Skorokhodov, E. V., and Krasilnik, Z. F. 2016. On a silicon-based photonic-crystal cavity for the near-IR region: Numerical simulation and formation technology. *Semiconductors.* 50(8):1112–6. https://doi.org/10.1134/S1063782616080212.

196. Golovastikov, N. V., Doskolovich, L. L., Bezus, E. A., Bykov, D. A., and Soifer, V. A. 2018. An optical differentiator based on a three-layer structure with a W-shaped refractive index profile. *J. Exp. Theor. Phys.* 127(2):202–9. https://doi.org/10.1134/S1063776118080174.

197. Kazanskiy, N. L., and Skidanov, R. V. 2019. Technological line for creation and research of diffractive optical elements. *Proc. SPIE.* 11146:111460W. https://doi.org/10.1117/12.2527274.

198. Kazanskiy, N. L. 2017. Efficiency of deep integration between a research university and an academic institute. *Procedia Eng.* 201:817–31. https://doi.org/10.1016/j.proeng.2017.09.604.

199. Zhu, T., Zhou, Y., Lou, Y., Ye, H., Qiu, M., Ruan, Z., and Fan, S. 2017. Plasmonic computing of spatial differentiation. *Nat. Commun.* 8:15391. https://doi.org/10.1038/ncomms15391.

200. Wesemann, L., Panchenko, E., Singh, K., Gaspera, E. D., Gomez, D. E., Davis, T. J., and Roberts, A. 2019. Selective near-perfect absorbing mirror as a spatial frequency filter for optical image processing. *APL Photon.* 4:100801. https://doi.org/10.1063/1.5113650.

201. Li, X., Youngblood, N., Rios, C., Cheng, Z., Wright, C. D., Pernice, W. H. P., and Bhaskaran, H. 2019. Fast and reliable storage using a 5 bit, nonvolatile photonic memory cell. *Optica.* 6(1):1–6. https://doi.org/10.1364/OPTICA.6.000001.

202. Stucke, G. 1989. Parallel architecture for a digital optical computer. *Appl. Opt.* 28(2):363–70. https://doi.org/10.1364/AO.28.000363.

203. Jenkins, B. K., and Giles, C. L. 1986. Parallel processing paradigms and optical computing. *Proc. SPIE.* 625:22–9. https://doi.org/10.1117/12.963476.

204. Feldmann, J. et al. 2021. Parallel convolutional processing using an integrated photonic tensor core. *Nature.* 589:52–8. https://doi.org/10.1038/s41586-020-03070-1.

205. Stark, H. 1975. An optical-digital computer for parallel processing of images. *IEEE Trans. Comput.* C-24(4):340–7. https://doi.org/10.1109/T-C.1975.224227.

206. White, I., Aw, E. T., Williams, K., Wang, H., Wonfor, A., and Penty, R. 2009. Scalable optical switches for computing applications. *J. Opt. Netw.* 8(2):215–24. https://doi.org/10.1364/JON.8.000215.

207. Rashed, A. N. Z., Mohammed, A. E. N., Zaky, W. F., Amiri, I. S., and Yupapin, P. 2019. The switching of optoelectronics to full optical computing operations based on nonlinear metamaterials. *Results Phys.* 13:102152. https://doi.org/10.1016/j.rinp.2019.02.088.

208. Gu, M., Li, X., and Cao, Y. 2014. Optical storage arrays: A perspective for future big data storage. *Light Sci. Appl.* 3:e177. https://doi.org/10.1038/lsa.2014.58.
209. Alexoudi, T., Kanellos, G. T., and Pleros, N. 2020. Optical RAM and integrated optical memories: A survey. *Light Sci. Appl.* 9:91. https://doi.org/10.1038/s41377-020-0325-9.
210. Zhu, H. H., and Zou, J. et al. 2022. Space-efficient optical computing with an integrated chip diffractive neural network. *Nat. Commun.* 13:1044. https://doi.org/10.1038/s41467-022-28702-0.
211. Salmani, M., Eshaghi, A., Luan, E., and Saha, S. 2021. Photonic computing to accelerate data processing in wireless communications. *Opt. Express.* 29(14):22299–314. https://doi.org/10.1364/OE.423747.
212. Qiu, C., Xiao, H., Wang, L., and Tian, Y. 2022. Recent advances in integrated optical directed logic operations for high performance optical computing: A review. *Front. Optoelectron.* 15:1. https://doi.org/10.1007/s12200-022-00001-y.
213. Lee, K.-Y. et al. 2011. Optical logic operation for AND Gate based on planar photonic crystal circuit. *Proc. SPIE.* 8308:83081S. https://doi.org/10.1117/12.905637.

General Conclusion

Nikolay Lvovich Kazanskiy[1,2]
[1]Samara National Research University, Samara, Russia
[2]Image Processing Systems Institute – Branch of the Federal
 Scientific Research Centre "Crystallography and Photonics"
 of Russian Academy of Sciences, Samara, Russia

Notwithstanding considerable advances made in designing nanophotonic structures, devices, and components, their practical uses have been limited so far. It is the shared opinion of this book's authors that before long, the situation will start changing rapidly. In support of this view, one may mention numerous experiments that have corroborated theoretical and numerical estimations and first successful attempts of replicating devices comprising nanophotonic components. Because of this, the contributors to the monograph have chosen not to discuss methods for solving Maxwell's equations, which are described in a number of well-known books [1–3], and other general approaches employed in diffractive nanophotonics [4–5], focusing instead on a number of promising research directions relating to the development of sensorics and optical transformations.

Silicon photonic waveguides discussed in Chapter 1 are attractive thanks to the mature and affordable manufacturing technology. The authors have analyzed several types of silicon-based optical fibers (OF), including conventional ridge, comb, slit, hybrid, and suspended waveguides. Conventional OF designs, such as comb and ridge, can offer better confinement of propagating modes, making them convenient for using as optical interconnects in a crystal chip. Directly structuring the silicon material enables optical interconnects to be implemented in laser signal transmission systems that can carry enormous data volumes, while featuring low energy consumption and not suffering from heating and signal deterioration. Silicon-based slit and hybrid plasmonic OFs are extremely sensitive to the environment, which makes them ideal candidates for sensor applications. Advantages of silicon OFs include a low cost of the material, mature and well-established processing and manufacturing techniques in the microelectronics industry, as well as the possibility of easy integration into electric devices and components in a single substrate.

In Chapter 2, we numerically and experimentally studied potentialities that comb photonic-crystal waveguides have for solving a wide range of problems. Among such problems, special mention should be made of temporal differentiation and integration (including the multiple one) of optical signals, development of an active cavity enabling vertical electric or plane optical pumping, enhancing the sensitivity of an optical sensor system, designing an optical modulator, and so forth. A technique for manufacturing a photonic crystal resonator by ion etching of a hole array in a silicon strip waveguide was also described.

In Chapter 3, we focused on the analysis of plasmonic refractive index sensors. Cutting-edge developments in special-type plasmonic structures based on metal-insulator-metal

DOI: 10.1201/9781003439165-9

waveguides and their application in refractive index sensors were discussed. Various types of plasmonic waveguides, their geometry, materials, and manufacturing procedures, as well as potential energy losses were analyzed. Design and numerical simulation of two sensor configurations based on a metal-insulator-metal waveguide that can be used both for refractive index measurements and as a plasmonic filter were discussed. An important part of the chapter was concerned with the discussion of sensorics potentialities based on the combined use of surface plasmon resonances and nanoparticles. In the concluding part of the chapter, we made an attempt to predict the future of plasmonic sensors, coming to a conclusion that the future looks bright.

In Chapter 4, the authors conducted an analysis of plasmonic nanolasers, covering various modern aspects. In the recent decade, great advances were made both in the theory and experimental demonstrations of nanolasers, attracting significant interest from researchers and industry. The authors described surface plasmons, typical amplifying materials, and an operation principle of plasmonic nanolasers. They analyzed experimental demonstrations of the devices, classified according to a confinement technique used for light enhancement, including a solitary particle, a Fabry–Perot resonator, a whispering gallery mode resonator, and a metal particle/hole array, before giving a concise description of major parameters of nanolasers. Next, the authors discussed the use of plasmonic nanolasers in integrated photonic circuits, sensors, and biology. In conclusion, the authors outlined directions of further development and assessed prospects of plasmonic nanolasers.

In Chapter 5, we presented several important practical uses of metasurfaces (MS). An MS is a two-dimensional or planar version of a metamaterial of subwavelength thickness. Thanks to their small size and light weight, MSs have been extensively studied and utilized in electromagnetic devices. MSs possess a remarkable ability to block, absorb, concentrate, scatter, or direct waves ranging from a millimeter to the visible spectrum of wavelengths at grazing, normal, and tilted incidence. We set forth methods for MS design, analyzed currently used types of metasurfaces (antennae, metalenses, absorbers, sensors, camouflage MS, modulators, and polarizers), their limitations, and their advantages.

In Chapter 6, the authors analyzed several types of fiber-optic sensors. Major benefits of these sensors include the compatibility with the existing fiber-optic data transmission networks, high resistance to the external chemical and radiation factors, the immunity to electromagnetic interference, high sensitivity to the design parameters, and easy integration into measurement systems. Depending on a particular type of fiber-optic sensor, the output signal can be in the form of either an output amplitude or a phase/frequency shift. For the amplitude modulation of light, multimode optical fibers can be utilized, thus making it possible to enhance the sensitivity and adjust measurement ranges by using diffractive optical elements incorporated in the measurement system. Under illumination by coherent light, the amplitude sensors are sensitive to noise and perform well only in short-length fiber-optic lines, especially when filtering individual modes. Fiber-optic sensors that rely on a frequency shift of transmitted radiation show essentially higher immunity to disturbances thanks to operating on a principal wavelength. They also can easily be combined into measurement circuits in which each individual sensor is identifiable, enabling measurements at a considerable distance from the object of interest. In these sensors, fibers are microstructured to create fiber Bragg gratings. These days, microstructuring techniques have been rapidly developing and were also dealt with in this chapter.

In Chapter 7, building on the expertise accumulated studying wavefront aberrations in the human eye, the authors demonstrated prospects of developing diffractive aberrometers and wavefront sensors for professional ophthalmology. The approaches and devices proposed enable the optical system of the human eye to be better characterized, leading to a more accurate computer-aided diagnosis and thus making possible a transition to the personified high-tech medicine.

Chapter 8 is based on our recent review [6]. I felt it relevant to include the review material in this monograph for a possibility of real-time processing of data coming from optical sensors, while omitting a stage of optical-digital transformation followed by computer-aided numerical analysis, by directly using a specialized optical chip seems enticing. To date, an optical computer has not yet been developed and mass-produced, with good solutions still waiting to be offered for optical transistors, optical memory, and many more, which would allow overcoming the enormous inertia of the many well-established manufacture procedures in electronics. Nonetheless, the urgent demand for optical technology relies on the fact that the time response of electrical circuits limits today's computers [5, 7]. A solid transmission medium restricts signal speed and volume while also generating heat that destroys components. The extreme downsizing of microscopic electronic apparatuses also causes "cross talk" or signal mistakes that compromise a system's dependability. These and other challenges have prompted researchers to look for photonic solutions. The advantages of the photonic solutions over the electronic ones set forth in this chapter give researchers an impetus to pursue their efforts in this direction, systematizing the accumulated bulk of knowledge. In this chapter, we reviewed several optical components, such as spatial light modulators, plasmonic switches, neural networks, diffractive neural networks, photonic crystal all-optical logic gates, and resonant nanophotonic structures that are used to implement optical computing, and discussed their advantages. From our viewpoint, at the time being, efficient solutions can be found through developing separate types of specialized optical processors capable of replacing currently used electronic-analogous computers for solving some types of differential equations, contour detection and extraction, and others. The proposed solutions may show promise in situations where an optical signal (e.g. a video stream) is directly fed to the input of a specialized processor. In this case, an optical neural network enables some images to be processed and recognized in a faster and more economic way than by previously digitizing and digitally processing them using graphics processing units.

It gave me great pleasure to collaborate on the book project with a wonderful team of contributing authors. I hope that reading this book will also be a pleasant process for the readers.

REFERENCES

1. Soifer, V. A., ed. 2014. *Diffractive nanophotonics*. Boca Raton, FL: CRC Press. ISBN: 978-1-4665-9070-0.
2. Soifer, V. A., ed. 2017. *Diffractive optics and nanophotonics*. Boca Raton, FL: CRC Press. ISBN: 978-1-4987-5447-7.

3. Soifer, V. A., ed. 2013. *Computer design of diffractive optics.* Cambridge, UK: Woodhead Publishing. ISBN: 978-1-84569-635-1.

4. Soifer, V. A., Kotlyar, V. V., and Doskolovich, L. L. 2009. Diffractive optical elements in nanophotonics devices. *Comput. Opt.* 33(4):352–68.

5. Soifer, V. A. 2014. Diffractive nanophotonics and advanced information technologies. *Her. Russ. Acad. Sci.* 84(1):9–20. https://doi.org/10.1134/S1019331614010067.

6. Kazanskiy, N. L., Butt, M. A., and Khonina, S. N. 2022. Optical computing: Status and perspectives. *Nanomaterials.* 12(13):2171. https://doi.org/10.3390/nano12132171.

7. Brunner, D., Marandi, A., Bogaerts, W., and Ozcan, A. 2020. Photonics for computing and computing for photonics. *Nanophotonics.* 9(13):4053–4. https://doi.org/10.1515/nanoph-2020-0470.

Appendix A: Listing for Calculating the Sensor Matched with the Zernike Basis Functions

```
// radial part of the Zernike basis functions
double Z (int n, int m, double r, double r0){
    double Zer = 0;
    m = abs(m);
    for(int p=0;p<=(n-m)/2;p++){
        Zer += (pow(-1,p) * fact(n-p) * pow((fact(p)*fact((n+m)/
        2-p)*fact((n-m)/2-p)),-1) * pow(r/r0,n-2*p));
    }
    return Zer;
}

// the Zernike basis functions in trigonometric
representation
complex <double> Z_trigon_cortes (int n, int m, int N, int j,
int k, double a){
    double hx = 2*a/N;
    double xj=-a+j*hx;
    double hy = 2*a/N;
    double yk=-a+k*hy;
    complex <double> coef (m*atan2(yk,xj),0);
    if (m>=0){
        complex <double> expp = cos(coef);
        double r = sqrt(xj*xj+yk*yk);
        if (r>a) return 0;
        complex <double> Zer = Z(n,m,r,a)*expp;
        return Zer;
    }
    if (m<0){
        complex <double> expp = sin(coef);
        double r = sqrt(xj*xj+yk*yk);
        if (r>a) return 0;
        complex <double> Zer = Z(n,m,r,a)*expp;
        return Zer;
    }
}
```

```cpp
// array of values of Zernike basis functions in Cartesian
representation
void Z_exp_cortes_Arr (complex <double> **arr, int n, int m,
int N, double a){
   for(int j=0;j<=N*2;j++){
      for(int k=0;k<=N*2;k++){
         arr[j][k] = Z_exp_cortes(n,m,N,j,k,1);
      }
   }
}


// array of values of Zernike basis functions in trigonometric
representation
void Z_trigon_cortes_Arr (complex <double> **arr, int n, int
m, int N, double a){
   for(int j=0;j<=N*2;j++){
      for(int k=0;k<=N*2;k++){
         arr[j][k] = Z_trigon_cortes(n,m,N,j,k,1);
      }
   }
}


// optical Zernike analyzer (multichannel)
void Zernik_Optic_Analizer_Multy_Channel(complex <double> **E,
complex <double> **Filtr, int N, int n, int m, double a,
Bitmap *img1, Bitmap *img2){

   complex <double> **Z1 = new complex <double> *[2*N];
   for (int i=0;i<=2*N;i++)
      Z1[i] = new complex <double>[2*N];

   complex <double> **Z16 = new complex <double> *[2*N];
   for (int i=0;i<=2*N;i++)
      Z16[i] = new complex <double>[2*N];

   complex <double> Psi1;
   complex <double> Psi2;

   double hx = 2*a/N;
   double hy = 2*a/N;

   for(int j=0;j<=N;j++){
      for(int k=0;k<=N;k++){
         double xj=-a+j*hx;
         double yk=-a+k*hy;

         Psi1 = Z_exp_cortes(2,0,N,j,k,1);
         Psi2 = Z_exp_cortes(2,2,N,j,k,1);

         complex <double> coef (0,3*atan2(yk,xj));
      }
   }
```

```
//Z_exp_cortes_Ampl_Phase(E,img1,img2,2,2,N,a);
Z_exp_cortes_Ampl_Phase(Z1,img1,img2,1,1,N,a);
Z_exp_cortes_Ampl_Phase(Z16,img1,img2,7,7,N,a);

for(int j=0;j<2*N;j++){
   for(int k=0;k<2*N;k++){

      double xj=-a+j*hx;

      double yk=-a+k*hy;

      complex <double> e0 (0,0);

      complex <double> e1 (0,2*100*xj);
      complex <double> e2 (0,2*70*yk);
      complex <double> e3 (0,2*70*xj+2*70*yk);

      complex <double> e4 (0,-2*70*xj);
      complex <double> e5 (0,-2*70*yk);
      complex <double> e6 (0,-2*70*xj+2*70*yk);

      complex <double> e7 (0,2*70*xj-2*70*yk);
      complex <double> e8 (0,-2*70*xj-2*70*yk);


      complex <double> e9 (0,4*70*xj);
      complex <double> e10 (0,4*70*yk);
      complex <double> e11 (0,4*70*xj+4*70*yk);

      complex <double> e12 (0,-4*70*xj);
      complex <double> e13 (0,-4*70*yk);
      complex <double> e14 (0,-4*70*xj+4*70*yk);

      complex <double> e15 (0,4*70*xj-4*70*yk);
      complex <double> e16 (0,-4*70*xj-4*70*yk);

      Filtr[j][k]=E[j][k]*(conj(Z1[j][k])*exp(e1)+
conj(Z2[j][k])*exp(e2)+conj(Z3[j][k])*exp(e3)+conj(Z4[j][k])*
exp(e4)+conj(Z5[j][k])*exp(e5)+conj(Z6[j][k])*exp(e6)+
conj(Z7[j][k])*exp(e7)+conj(Z8[j][k])*exp(e8)+conj(Z9[j][k])*
exp(e9)+conj(Z10[j][k])*exp(e10)+conj(Z11[j][k])*exp(e11)+
conj(Z12[j][k])*exp(e12)+conj(Z13[j][k])*exp(e13)+conj(Z14[j]
[k])*exp(e14)+conj(Z15[j][k])*exp(e15)+conj(Z16[j][k])*
exp(e16));
      }
   }

   double max = 0;
   for(int i=0;i<N;i++){
      for(int j=0;j<N;j++){
         if (max < abs(Filtr[i][j])) max = abs(Filtr[i][j]);
      }
   }
```

```
int a;
for(int i=0;i<=N;i++){
    for(int j=0;j<=N;j++){
        a = (int) (abs(Filtr[i][j])*255/max);
        img1->SetPixel(i, j, Color(a,a,a));
    }
    printf("%d \n", a);
}

fft_draw(Filtr,N,N,img1,1,2);


}


// Zernike math analyzer
complex <double> *Zernik_Math_Analize___txt_to_excel(FILE *m,
FILE *m1, complex <double> **E, int N, double a){


    complex <double> **Z1 = new complex <double> *[2*N];
    for (int i=0;i<=2*N;i++)
        Z1[i] = new complex <double>[2*N];
    complex <double> **Z16 = new complex <double> *[2*N];
    for (int i=0;i<=2*N;i++)
        Z16[i] = new complex <double>[2*N];

    complex <double> *_C = new complex <double> [2*N];

    complex <double> *norm = new complex <double> [2*N];

    Z_trigon_cortes_Arr(Z1,0,0,N,a);
    Z_trigon_cortes_Arr(Z15,4,4,N,a);

    for(int j=0;j<=N;j++){
        for(int k=0;k<=N;k++){
            norm[0] += abs(Z1[j][k]*Z1[j][k]);
            norm[14] += abs(Z15[j][k]*Z15[j][k]);

        }
    }

    for(int j=0;j<=N;j++){
        for(int k=0;k<=N;k++){
            _C[0] += (E[j][k]*conj(Z1[j][k]) / norm[0]);
            _C[14] += (E[j][k]*conj(Z15[j][k]) / norm[14]);
        }
    }

    printf("Math_Analizer_Zernike \n");
```

```
fprintf (m, "C[%d] [%d] = %f", 0, 0,abs(_C[0]),"w","w");

fprintf (m, "C[%d] [%d] = %f", 4, 4,abs(_C[14]),"w","w");

fprintf (m, "\n", "w", "w");

fprintf (m1,"%f",abs(_C[0]), "w", "w");
fprintf (m1,"%f",abs(_C[14]), "w", "w");

fprintf (m1, "\n", "w", "w");


return _C;
}
```

Appendix B: Listing for Calculating a Distorted Image

```cpp
// simulation of imaging in the focal plane of an optical
system with introduced aberrations
complex <double>** Wavefront_FRONT_BACK__PSF(complex <double>
C00, complex <double> C11,complex <double> C1_1,
          complex <double> C22, complex <double> C20,
complex <double> C2_2,
          complex <double> C33, complex <double> C31,
complex <double> C3_1, complex <double> C3_3,
          complex <double> C44, complex <double> C42,
complex <double> C40, complex <double> C4_2, complex <double>
C4_4,
          complex <double> _C00, complex <double> _
C11,complex <double> _C1_1,
          complex <double> _C22, complex <double> _C20,
complex <double> _C2_2,
          complex <double> _C33, complex <double> _C31,
complex <double> _C3_1, complex <double> _C3_3,
          complex <double> _C44, complex <double> _C42,
complex <double> _C40, complex <double> _C4_2, complex
<double> _C4_4,
          int N, Bitmap *ampl1, Bitmap *phase1, Bitmap
*_ampl1, Bitmap *_phase1, Bitmap *ampl2, Bitmap *phase2,
Bitmap *_ampl2, Bitmap *_phase2, Bitmap *fft0, Bitmap *fft1,
Bitmap *_fft0, Bitmap *_fft1, Bitmap *ampl3, Bitmap *phase3,
Bitmap *fft2, Bitmap *fft3, double a, float zeros ){

    complex <double> **arr1 = new complex <double> *[2*N];
    for (int i=0;i<=2*N;i++)
       arr1[i] = new complex <double>[2*N];

    complex <double> **_arr1 = new complex <double> *[2*N];
    for (int i=0;i<=2*N;i++)
       _arr1[i] = new complex <double>[2*N];

    complex <double> **arr2 = new complex <double> *[2*N];
    for (int i=0;i<=2*N;i++)
       arr2[i] = new complex <double>[2*N];

    complex <double> **_arr2 = new complex <double> *[2*N];
    for (int i=0;i<=2*N;i++)
       _arr2[i] = new complex <double>[2*N];
```

```
complex <double> **arr3 = new complex <double> *[2*N];
for (int i=0;i<=2*N;i++)
   arr3[i] = new complex <double>[2*N];


         C00 *= 1E3;
         C4_4 *= 1E3;

         _C00 *= 1E3;
         _C4_4 *= 1E3;

//sum(Cij*Zij)
      for(int i = 0; i<N; i++){
         for(int j = 0; j<N; j++){
            arr1[i][j] = C00*Z_trigon_cortes(0,0,N,i,j,1)+
            C11*Z_trigon_cortes(1,1,N,i,j,1)+
            C4_4*Z_trigon_cortes(4,-4,N,i,j,1);
            arr1[i][j] *= 2.*M_PI;


         }
      }

      for(int i = 0; i<N; i++){
         for(int j = 0; j<N; j++){
            _arr1[i][j] = _C00*Z_trigon_cortes(0,0,N,i,j,1)+
            _C11*Z_trigon_cortes(1,1,N,i,j,1)+
            _C4_4*Z_trigon_cortes(4,-4,N,i,j,1);

            _arr1[i][j] *= 2.*M_PI;


         }
      }
//amplitude and phase sum(Cij*Zij)
      Ampl_Phase(arr1,N,ampl1,phase1,0);
      Ampl_Phase(_arr1,N,_ampl1,_phase1,0);

//reduce circle
//exp(i*sum(Cij*Zij))
      complex <double> pokasatel = 0;
      complex <double> _pokasatel = 0;
      complex <double> I (0,1);
      double a = 1;
      double hx = 2*a/N;
      double hy = 2*a/N;
      for(int i = 0; i<N; i++){
         for(int j = 0; j<N; j++){
            double xi =-a+i*hx;
            double yj =-a+j*hy;
            double r = sqrt(xi*xi+yj*yj);
               if (r>a) arr2[i][j] = 0;
```

```
                else {
                    pokasatel = I*arr1[i][j];
                    _pokasatel = I*_arr1[i][j];
                    arr2[i][j] = exp(pokasatel);
                    _arr2[i][j] = exp(_pokasatel);

                    arr3[i][j] = exp(pokasatel + _pokasatel);
                }
            }
        }

    //amplitude and phase exp(i*sum(Cij*Zij))
        Ampl_Phase(arr2,N,ampl2,phase2,0);
        Ampl_Phase(_arr2,N,_ampl2,_phase2,0);
        Ampl_Phase(arr3,N,ampl3,phase3,0);

    fft(arr2,N,N,1,zeros);
    fft(_arr2,N,N,1,zeros);
    fft(arr3,N,N,1,zeros);
    Ampl_Phase(arr2,N,fft0,fft1,0);
    Ampl_Phase(_arr2,N,_fft0,_fft1,0);
    Ampl_Phase(arr3,N,fft2,fft3,0);


return arr2;
}


// modeling an image distorted by aberrations
void Aber_Image (complex <double> **arr1, complex <double>
**arr, Bitmap *bm, Bitmap *bm1, Bitmap *bm2, Bitmap *bm3,
Bitmap *bm4, Bitmap *bm5, Bitmap *bm6, int n, int m, double
Cnm, int N,  double a){

    complex <double> I (0,1);
    int jj=30;
    int kk=30;

    /*n=4;
    m=-4;
    Cnm=1;*/

    int background = 0;
    int figure = 255;

    for(int j=0;j<=N*2;j++){
        for(int k=0;k<=N*2;k++){
            arr1[j][k] = background;
        }
    }
```

```
   for(int j=-jj;j<=jj;j++){
      for(int k=-kk;k<=kk;k++){
         arr1[(int)N/2][(int)N/2+k] = figure;
         arr1[(int)N/2+j][(int)N/2] = figure;

         arr1[(int)N/2-1][(int)N/2+k] = figure;
         arr1[(int)N/2+j][(int)N/2+1] = figure;

         arr1[(int)N/2+1][(int)N/2+k] = figure;
         arr1[(int)N/2+j][(int)N/2-1] = figure;
      }
   }

   Ampl_Phase(arr1,N,bm,bm1,1);

   fft_draw(arr1,N,N,bm1,-1,1);

   for(int j=0;j<=N*2;j++){
      for(int k=0;k<=N*2;k++){
         arr[j][k] = Z_trigon_cortes(n,m,N,j,k,1);
      }
   }

   Ampl_Phase(arr,N,bm2,bm3,1);

   for(int j=0;j<=N*2;j++){
      for(int k=0;k<=N*2;k++){
         arr[j][k] = exp(Cnm*2.*M_PI*I*Z_trigon_cortes(n,m,N,
j,k,1));
      }
   }

   Ampl_Phase(arr,N,bm4,bm5,1);

   for(int j=0;j<=N*2;j++){
      for(int k=0;k<=N*2;k++){
         arr[j][k] *= arr1[j][k];
      }
   }

   fft_draw(arr,N,N,bm6,-1,1);

}
```

Index

Note: Locators in *italics* represent figures and **bold** indicate tables in the text.

For Product Safety Concerns and Information please contact our EU
representative GPSR@taylorandfrancis.com
Taylor & Francis Verlag GmbH, Kaufingerstraße 24, 80331 München, Germany

www.ingramcontent.com/pod-product-compliance
Lightning Source LLC
Chambersburg PA
CBHW060331220326
41598CB00023B/2679